Lecture Notes in Computer Science 7922

Commenced Publication in 1973
Founding and Former Series Editors:
Gerhard Goos, Juris Hartmanis, and Jan van Leeuwen

Editorial Board

T0218348

Johannes Fischer   Peter Sanders (Eds.)

Combinatorial Pattern Matching

24th Annual Symposium, CPM 2013
Bad Herrenalb, Germany, June 17-19, 2013
Proceedings

 Springer

Volume Editors

Johannes Fischer
Peter Sanders
Karlsruhe Institute of Technology
Fakultät für Informatik
Institut für Theoretische Informatik
Am Fasanengarten 5, 76131 Karlsruhe, Germany
E-mail: {johannes.fischer, peter.sanders}@kit.edu

ISSN 0302-9743   e-ISSN 1611-3349
ISBN 978-3-642-38904-7   e-ISBN 978-3-642-38905-4
DOI 10.1007/978-3-642-38905-4
Springer Heidelberg Dordrecht London New York

Library of Congress Control Number: 2013939776

CR Subject Classification (1998): I.5, F.2, G.1, G.2.1-2, G.2, E.1, E.4, H.3.3, I.2.7, H.2.8, J.3

LNCS Sublibrary: SL 1 – Theoretical Computer Science and General Issues

Typesetting: Camera-ready by author, data conversion by Scientific Publishing Services, Chennai, India

Printed on acid-free paper

Springer is part of Springer Science+Business Media (www.springer.com)

Preface

This volume contains the papers presented at the 24th Annual Symposium on Combinatorial Pattern Matching (CPM 2013) held in Bad Herrenalb near Karlsruhe, Germany, during June 17–19, 2013. The hosting university was the Karlsruhe Institute of Technology.

There were 51 submissions from 22 countries. Each submission was reviewed by at least three Program Committee members. The committee decided to accept 21 papers, corresponding to an acceptance rate of 41%. We thank the members of the Program Committee and all additional external reviewers for their hard work that resulted in this excellent program. Their names are listed on the following pages. The whole submission and review process was carried out with the invaluable help of the EasyChair conference system.

The program also included two invited talks by Moshe Lewenstein from Bar Ilan University, Israel, and by Gene Myers from the MPI for Molecular Cell Biology and Genetics, Dresden, Germany. We thank the invited speakers for their contributions.

2013 marks the 40th anniversary of Peter Weiner's foundational paper "Linear pattern matching algorithms" on suffix trees, in those days called "bi-trees" [14th Annual Symposium on Switching and Automata Theory (SWAT; nowadays FOCS), pp. 1–11, 1973]. CPM 2013 celebrated this event with a special session, organized by Martin Farach-Colton and S. Muthukrishnan, both from Rutgers University, USA. It included talks by Edward M. McCreight, Vaughan R. Pratt, Peter Weiner, and Jacob Ziv. This special session was accompanied by an invited contribution "Forty Years of Text Indexing" by Alberto Apostolico, Maxime Crochemore, Martin Farach-Colton, Zvi Galil, and S. Muthukrishnan.

The objective of the annual CPM meetings is to provide an international forum for research in combinatorial pattern matching and related applications. It addresses issues of searching and matching strings and more complicated patterns such as trees, regular expressions, graphs, point sets, and arrays. The goal is to derive non-trivial combinatorial properties of such structures and to exploit these properties in order to either achieve superior performance for the corresponding computational problems or pinpoint conditions under which searches cannot be performed efficiently. The meeting also deals with problems in computational biology, data compression and data mining, coding, information retrieval, natural language processing, and pattern recognition.

The Annual Symposium on Combinatorial Pattern Matching started in 1990, and has since taken place every year. Previous CPM meetings were held in Paris, London (UK), Tucson, Padova, Asilomar, Helsinki, Laguna Beach, Aarhus, Piscataway, Warwick, Montreal, Jerusalem, Fukuoka, Morelia, Istanbul, Jeju Island, Barcelona, London (Ontario, Canada), Pisa, Lille, New York, Palermo, and Helsinki again. This year's meeting was the first in Germany. Starting from the third meeting, proceedings of all meetings have been published in the LNCS series, as volumes 644, 684, 807, 937, 1075, 1264, 1448, 1645, 1848, 2089, 2373, 2676, 3109, 3537, 4009, 4580, 5029, 5577, 6129, 6661, and 7354.

We thank SAP (Walldorf, Germany) and the German Research Foundation (DFG) for their financial support.

April 2013   Johannes Fischer
                                                             Peter Sanders

Organization

Program Committee

Rolf Backofen	University of Freiburg, Germany
Philip Bille	Technical University of Denmark
Christina Boucher	Colorado State University, USA
Francisco Claude	University of Waterloo, Canada
Johannes Fischer	Karlsruhe Institute of Technology, Germany (Co-chair)
Travis Gagie	University of Helsinki, Finland
Paweł Gawrychowski	MPI Saarbrücken, Germany
Meng He	Dalhousie University, Canada
Jan Holub	Czech Technical University in Prague, Czech Republic
Jesper Jansson	Kyoto University, Japan
Juha Kärkkäinen	University of Helsinki, Finland
Tsvi Kopelowitz	Weizmann Institute of Science, Israel
Alejandro López-Ortiz	University of Waterloo, Canada
Gonzalo Navarro	University of Chile
Kunsoo Park	Seoul National University, South Korea
Mike Paterson	University of Warwick, UK
Rajeev Raman	University of Leicester, UK
Benjamin Sach	University of Warwick, UK
Kunihiko Sadakane	National Institute of Informatics, Japan
Peter Sanders	Karlsruhe Institute of Technology, Germany (Co-chair)
Jens Stoye	University of Bielefeld, Germany
Rossano Venturini	University of Pisa, Italy
Oren Weimann	University of Haifa, Israel

Steering Committee

Alberto Apostolico	Georgia Institute of Technology, USA
Maxime Crochemore	Université Paris-Est, France, and King's College London, UK
Zvi Galil	Georgia Institute of Technology, USA

Organizing Committee

Timo Bingmann	Karlsruhe Institute of Technology, Germany
Johannes Fischer	Karlsruhe Institute of Technology, Germany

Peter Sanders   Karlsruhe Institute of Technology, Germany
Nodari Sitchinava   Karlsruhe Institute of Technology, Germany

Additional Reviewers

Amir, Amihood
Amit, Mika
Arimura, Hiroki
Arroyuelo, Diego
Baier, Jan
Bannai, Hideo
Bingmann, Timo
Braga, Marilia
Canovas, Rodrigo
Chitsaz, Hamidreza
Clifford, Raphael
Cording, Patrick Hagge
Cormode, Graham
Crochemore, Maxime
Cunial, Fabio
Cygan, Marek
Doerr, Daniel
Dorrigiv, Reza
Fernau, Henning
Goto, Keisuke
Guillemot, Sylvain
Inenaga, Shunsuke
Jahn, Katharina
Jalsenius, Markus
Jeż, Artur
Jurdzinski, Tomasz
Kempa, Dominik
Kennedy, Sean
Krusche, Peter
Kucherov, Gregory
Ladra, Susana
Lokshtanov, Daniel

Manea, Florin
Martinez-Prieto, Miguel A.
Mäkinen, Veli
Nicholson, Patrick K.
Osipov, Vitaly
Pedersen, Christian Nørgaard Storm
Philip, Geevarghese
Polach, Radomir
Ponty, Yann
Porat, Ely
Pritchard, David
Puglisi, Simon
Sakamoto, Hiroshi
Salinger, Alejandro
Salmela, Leena
Savari, Serap
Schmiedl, Christina
Seco, Diego
Shibuya, Tetsuo
Sirén, Jouni
Thankachan, Sharma
Tiskin, Alexander
Transier, Frederik
Travnicek, Jan
Truszkowski, Jakub
Vildhøj, Hjalte Wedel
Walen, Tomasz
Will, Sebastian
Wittler, Roland
Zhou, Gelin
Zhu, Binhai
Ziv-Ukelson, Michal

Table of Contents

Forty Years of Text Indexing

Alberto Apostolico[1], Maxime Crochemore[2,3], Martin Farach-Colton[4],
Zvi Galil[1], and S. Muthukrishnan[4]

[1] College of Computing, Georgia Institute of Technology, 801 Atlantic Drive,
Atlanta, GA 30318, USA
{axa,galil}@cc.gatech.edu
[2] King's College London, Strand, London WC2R 2LS, UK
maxime.crochemore@kcl.ac.uk
[3] Université Paris-Est, Institut Gaspard-Monge,
77454 Marne-la-Vallée Cedex 2, France
[4] Department of Computer Science, Rutgers University, Piscataway, NJ 08854, USA
{farach,muthu}@cs.rutgers.edu

Abstract. This paper reviews the first 40 years in the life of textual
inverted indexes, their many incarnations, and their applications. The
paper is non-technical and assumes some familiarity with the structures
and constructions discussed. It is not meant to be exhaustive. It is meant
to be a tribute to a ubiquitous tool of string matching — the suffix tree
and its variants — and one of the most persistent subjects of study in
the theory of algorithms.

Keywords: pattern matching, string searching, bi-tree, suffix tree, dawg,
suffix automaton, factor automaton, suffix array, FM-index, wavelet tree.

1 Prolog

When William Legrand finally decrypted the string:

53‡‡†305))6*,48264‡.)4‡);806",48†8P60))85;1‡(;:‡*8†83(88)5*†,46(;88*96
?;8)‡(;485);5*†2:*‡(;4956*2(5*4)8P8*;4069285);)6‡8)4‡‡;1(‡9;48081;8:8
‡1;48 85;4)485†528806*81(ddag9;48;(88;4(‡?34;48)4‡;161;:188;‡?;

it did not seem to make much more sense than it did before. The decoded message
read: "A good glass in the bishop's hostel in the devil's seat forty-one degrees
and thirteen minutes northeast and by north main branch seventh limb east side
shoot from the left eye of the death's-head a bee line from the tree through the
shot fifty feet out." But at least it did sound more like natural language, and
eventually guided the main character of Edgar Allan Poe's "The Gold Bug" [56]
to discover the treasure he had been after. Legrand solved a substitution ci-
pher using symbol frequencies. He first looked for the most frequent symbol and
changed it into the most frequent letter of English, then similarly treated the
second most frequent symbol, and so on.

J. Fischer and P. Sanders (Eds.): CPM 2013, LNCS 7922, pp. 1–10, 2013.

Both before and after 1843, the natural impulse when faced with some mysterious message has been to count frequencies of individual tokens or subassemblies in search of a clue. Perhaps one of the most intense and fascinating subjects for this kind of scrutiny has been bio-sequences. As soon as some such sequences became available, statistical analyses tried to link characters or blocks of characters to relevant biological function. With the early examples of whole genomes emerging in the mid 90's, it seemed natural to count the occurrences of all blocks of size 1, 2, etc. up to any desired length, looking for statistical characterizations of coding regions, promoter regions, etc. [67].

This review is not about cryptography. It is about a data structure and its variants, and the many surprising and useful features it carries. Among these is the fact that, to set up a statistical table of occurrences for *all* substrings, of *any* length, of a text string of n characters, it only takes time and space linear in the length of the text string. While nobody would be so foolish to solve the problem by generating all exponentially many possible substrings and then counting their occurrences one by one, a text string may still contain $O(n^2)$ distinct substrings, so that tabulating all of them in linear space, never mind linear time, seems already puzzling.

Over the years, such structures have held center stage in text searching, indexing, statistics, and compression as well as in the assembly, alignment and comparison of biosequences. Their range of scope extends to areas as diverse as detecting plagiarism, finding surprising substrings in a text, testing the unique decipherability of a code, and more. Their impact on Computer Science and IT at large cannot be overstated. Text and Web searching and Bioinformatics would not be the same without them. In 2013, the Combinatorial Pattern Matching symposium celebrates the 40th anniversary of the appearance of Wiener's paper with a special session entirely dedicated to that event.

2   History Bits and Pieces

At the dawn of "stringology", Don Knuth conjectured that the problem of finding the longest substring common to two long text sequences of total length n required $\Omega(n \log n)$ time. An $O(n \log n)$-time had been provided by Karp, Miller and Rosenberg [40]. That construction was destined to play a role in parallel pattern matching [6, 24, 31, 32], but Knuth's conjecture was short lived: in 1973, Peter Weiner showed that the problem admitted an elegant linear-time solution [68], as long as the alphabet of the string was fixed. Such a solution was actually a byproduct of a construction he had originally set up for a different purpose, i.e., identifying any substring of a textfile without specifying all of them. In doing so, Weiner introduced a notion of a textual inverted index that would elicit refinements, analyses and applications for forty years and counting, a feature hardly shared by any other data structure.

Weiner's original construction processed the textfile from right to left. As each new character was read in, the structure, which he called a "bi-tree", would be updated to accommodate longer and longer suffixes of the textfile. Thus this was an inherently offline construction, since the text had to be known in its entirety before the construction could begin. Alternatively, one could say that the algorithm would build online the structure for the reverse of the text. About three years later, Ed McCreight provided a left-to-right algorithm and changed the name of the structure to "suffix tree", a name that would stick [52]. In unpublished lecture notes of 1975, Vaughan Pratt displayed the duality of this structure and Weiner's "repetition finder" [57]. McCreight's algorithm was still inherently offline, and it immediately triggered a craving for an online version. Some partial attempts at an on-line algorithm were performed [42, 49], but such a variant had to wait almost two decades for Esko Ukkonen's paper in 1995 [65]. In all linear constructions, linearity was based on the assumption of a finite alphabet and took $O(n \log n)$ time in general. In 1997, Martin Farach introduced an algorithm that abandoned the one-suffix-at-time approach prevalent until then; this algorithm gives a linear-time reduction from suffix-tree construction to character sorting, and thus runs in linear time, for example, even when the alphabet is of size polynomial in the input size [26].

Around 1984, Anselm Blumer, et al. [12–14] and Maxime Crochemore [19] exposed the surprising result that the smallest finite automaton recognizing all and only the suffixes of a string of n characters has only $O(n)$ states and edges. Initially coined as a directed acyclic word graph (DAWG), it can even be reduced if all states are terminal states. It then accepts all subsstrings of the string and is called the factor/substring automaton. Although it has never been fully elucidated, it seems that Anatoli Slissenko [59, 60] ended up with a similar structure for his work on the detection of repetitions in strings. These automata provided another more efficient counterexample to Knuth's conjecture when they are used, against the grain, as pattern matching machines (see [20]).

The appearance of suffix trees dovetailed with some interesting and independent developments in information theory. In his famous approach to the notion of information, Kolmogorov [45] equated the information or structure in a string to the length of the shortest program that would be needed to produce that string by a Universal Turing Machine. The unfortunate thing is that this measure is not computable and even if it were, most long strings would be incompressible (would lack a short program producing them), since there are increasingly many long strings and comparatively much fewer short programs (themselves strings).

The regularities exploited by Kolmogorov's universal and omniscient machine could be of any conceivable kind, but what if one limited them to the syntactic redundancies affecting a text in form of repeated substrings? If a string is repeated many times one could profitably encode all occurrences by a pointer to a common copy. This copy could be internal or external to the text. In the latter case one could have pointers going in both directions or only in one, allow or forbid nesting of pointers, etc. In his doctoral thesis, Jim Storer [61–63] showed that virtually all such "macro schemes" are intractable, *except one*. Not long before that, in a landmark paper entitled "On the Complexity of Finite Sequences" [48], Abraham Lempel and Jacob Ziv had proposed a variable-to-block encoding based on a simple parsing of the text and with the feature that the compression achieved would match in the limit that produced by a compressor tailored to the source probabilities. Thus, by a remarkable alignment of stars, the compression method brought about by Lempel and Ziv was not only optimal in the information theoretic sense; it found an optimal, linear-time implementation by the suffix tree, as was detailed immediately by Michael Rodeh, Vaugham Pratt and Simon Even [58].

In his original paper, Weiner listed a few applications of his "bi-tree", including most notably off-line string searching: preprocessing a text file to support queries that return the occurrences of a given pattern in time linear in the length of the pattern. And of course, the "bi-tree" addressed Knuth's conjecture, by showing how to find a longest substring common to two files in linear time for finite alphabet. There followed unpublished notes by Pratt entitled "Improvements and Applications for the Weiner Repetition Finder" [57]. A decade later, Alberto Apostolico would list more applications in a paper entitled "The Myriad Virtues of Suffix Trees" [3], and two decades later suffix trees and companion structures elicited with their applications several chapters in reference books by Crochemore and Rytter [25], Dan Gusfield [35], and Crochemore, Hancart, Lecroq [21].

The space required by suffix trees has been a nuisance in applications where they were needed the most. With genomes in the order of gigabytes, for instance, it makes a big difference to need space 20 times bigger than the source versus, say, only 11 times that big. For a few lusters, Stefan Kurtz and his co-workers devoted their effort to cleverly allocating the tree and some of its companion structures [46]. In 2001 David R. Clark, J. Ian Munro proposed one of the best space-saving methods on secondary storage [18]. Clark and Munro's "succinct suffix tree" sought to preserve as much of the structure of the suffix tree as possible. Udi Manber and Eugene W. Myers took a different approach, however. In 1990 [50, 51], they introduced the "suffix array," which eliminated most of the structure of the suffix tree, but was still able to implement many of the same operations, at a cost of only twice the input size. Although the suffix array seemed at first to be a different data structure than the suffix tree, over time they have come to be more and more similar. For example, Manber and Myers's original construction of the suffix array took $O(n \log n)$ time for any alphabet, but the suffix array could be constructed in linear time from the suffix tree for any alphabet. In 2001, Toru Kasai et al. [41] showed that the suffix tree could be constructed in linear time from the suffix array. The suffix array was shown to be a succinct representation of the suffix tree. In 2003, three groups [39, 43, 44] modified in three different ways Farach's algorithm for suffix tree construction to give the first linear-time algorithms for directly constructing the suffix array, that is, the first linear-time algorithms for computing suffix arrays that did not first compute the full suffix tree. With fast construction algorithms and small space, the suffix array is the suffix-tree variant that has gained the most widespread adoption in software systems.

Actually, the history of inverted indexes and compression is tightly intertwined. This should not come as a surprise, since the redundancies that pattern discovery tries to unearth are ideal candidates to be removed for purposes of compression. In 1994, M. Burrows and D. J. Wheeler proposed a puzzling data compression method in a report that "as it happens sometimes to results that are just too simple and elegant" ended it up never finding an archival venue [16]. Circa 1995, Amihood Amir, Gary Benson and Martin Farach posed the problem of searching in compressed texts [1, 2]. In 2000, Paolo Ferragina and Giovanni Manzini introduced a compressed full text index based on the Burrows-Wheeler transform [29, 30]. In the same year, Roberto Grossi and Jeffrey Scott Vitter presented compressed versions of suffix trees and suffix arrays [34]. These structures supported searching without decompression while being possibly smaller than the source file. This was extended to compressed tree indexing problems in [28] using a modification of the Burrows-Wheeler transform.

3   Fallout, Extensions and Challenges

As highlighted in our prolog, there has been hardly any application of text processing that did not need these indexes at one point or another. A prominent case has been searching with errors, a problem first efficiently tackled in 1985 by Gad Landau in his PhD thesis [47]. In this kind of searches, one looks for substrings of the text that differ from the pattern in a limited number of errors such as a single character deletion, insertion or substitution. To efficiently solve this problem, Landau combined Suffix Trees with a clever solution to the so-called lowest common ancestor (LCA) problem. The LCA problem assumes that a rooted tree is given and then for any pair of nodes, it seeks the lowest node in the tree that is an ancestor of both [37] (see [11] and references therein for subsequent, simpler constructions). It is seen that following a linear time preprocessing of the tree any LCA query can be answered in constant time. Landau used LCA queries on Suffix Trees to perform contant-time jumps over segments of the text that would be guaranteed to match the pattern. When k errors are allowed, the search for an occurrence at any given position can be abandoned after k such jumps. This leads to an algorithm that searches for a pattern with k errors in a text of n characters in $O(nk)$ steps.

Among the basic primitives supported by suffix trees and arrays one finds of course searching for a pattern in a text in time proportional to the length of the pattern rather than the text. In fact, it is even possible to enumerate occurrences in time proportional to their number and, with trivial preprocessing of the tree, tell the total number of occurrences for any query pattern in time proportional to the pattern size. The problem of finding the longest substring appearing twice in a text or shared between two files has been already mentioned: this is probably where it all started. A germane problem is that of detecting squares, repetitions and maximal periodicities in a text, a problem rooted in work by Axel Thue dated more than a century ago [64], a problem with multiple contemporary applications in compression and DNA analysis. A square is a pattern consisting of two consecutive occurrences of the same string. Suffix trees have been used to detect in optimal $O(n \log n)$ time all squares (or repetitions) in a text, each with its set of starting positions [7], and later to find and store all distinct square substrings in a text in linear time [36]. Squares play a role in an augmentation of the suffix tree suitable to report, for any query pattern, the number of its non-overlapping occurrences [8, 15].

There are multiple uses of suffix trees in setting up some kind of signature for texstrings, as well as measures of similarity or difference. Among the latter, there is the problem of computing the forbidden or absent words of a text, which are strings that do not appear in the text while all of their substrings do [10, 22]. Such words subtend, among other things, to an original approach to text compression [23]. Once regarded as the succinct representation of the "bag-of-words" of a text, suffix trees can be used to assess the similarity of two textfiles, thereby supporting clustering, document classification and even phylogeny [5, 17, 66]. Intuitively, this is done by assessing how much the trees relative the two input sequences have in common. Suitably enriched with the probability affecting the substring ending at each node, a tree can be used to detect surprisingly over-represented substrings of any length [4], e.g., in the quest of promoter regions in biosequences [67].

The suffix tree of the concatenation of say, $k \geq 2$ texfiles, supports efficient solutions to problems arising in domains ranging from plagiarism detection to motif discovery in biosequences. The need for k distinct end-markers poses some subtleties in maintaining linear time, for which the reader is referred to [35]. In its original form, the problem was called "color problem" and seeks to report, for any given query string and in time linear in the query, how many documents out of the total of k contain each at least one occurrence of the query. A simple and elegant solution was given in 1992 by Lucas C. K. Hui [38]. More recently, it was extended to a variety of document listing problems, where once a set of text documents are preprocessed, one can return the list of all documents that contain a query pattern in time proportional to the number of such documents, not the total number of occurrences [53].

One surprising variant of the suffix tree was introduced by Brenda Baker for purposes of detection of plagiarism in student reports as well as optimization in software development [9] . This variant of pattern matching, called "parameterized matching", enables to find program segments that are identical up to a systematic change of parameters, or substrings that are identical up to a systematic relabeling or permutation of the characters in the alphabet.

One obvious extension of the notion of a suffix tree is to more than one dimension, albeit the mechanics of the extension itself is far from obvious [54,55] Among more distant relatives, one finds "wavelet trees". Originally proposed as a representation of compressed suffix arrays [33], wavelet trees enable to perform on general alphabets the ranking and selection primitives previously limited to bit vectors, and more [27].

The list could go on and on, but the scope of this paper was not meant be exhaustive. Actually, after forty years of unrelenting developments, it is fair to assume that the list will continue to grow. On the other hand, many of the observed sequences are expressed in numbers rather than characters, and in both cases are affected by various types of errors. While the outcome of a two character comparison is just one bit, two numbers can be more or less close, depending on their difference or some other metric. Likewise, two textstrings can be more or less similar, depending on the number of elementary steps necessary to change one in the other. The most disruptive aspect of this framework is the loss of the transitivity property that subtends to the most efficient exact string matching solutions. And yet indexes capable of supporting fast and elegant approximate pattern queries of the kind just highlighted would be immensely useful. Hopefully, they will come up soon and, in time, get their own 40th anniversary celebration.

References

1. Amir, A., Benson, G., Farach, M.: Let sleeping files lie: Pattern matching in Z-compressed files. In: Proceedings of the 5th ACM-SIAM Annual Symposium on Discrete Algorithms, Arlington, VA, pp. 705–714 (1994)
2. Amir, A., Benson, G., Farach, M.: Let sleeping files lie: Pattern matching in Z-compressed files. J. Comput. Syst. Sci. 52(2), 299–307 (1996)
3. Apostolico, A.: The myriad virtues of suffix trees. In: Apostolico, A., Galil, Z. (eds.) Combinatorial Algorithms on Words. NATO Advanced Science Institutes, Series F, vol. 12, pp. 85–96. Springer, Berlin (1985)
4. Apostolico, A., Bock, M.E., Lonardi, S.: Monotony of surprise and large-scale quest for unusual words. Journal of Computational Biology 10(3/4), 283–311 (2003)
5. Apostolico, A., Denas, O., Dress, A.: Efficient tools for comparative substring analysis. Journal of Biotechnology 149(3), 120–126 (2010)
6. Apostolico, A., Iliopoulos, C., Landau, G.M., Schieber, B., Vishkin, U.: Parallel construction of a suffix tree with applications. Algorithmica 3, 347–365 (1988)
7. Apostolico, A., Preparata, F.P.: Optimal off-line detection of repetitions in a string. Theor. Comput. Sci. 22(3), 297–315 (1983)
8. Apostolico, A., Preparata, F.P.: Data structures and algorithms for the strings statistics problem. Algorithmica 15(5), 481–494 (1996)
9. Baker, B.S.: Parameterized duplication in strings: Algorithms and an application to software maintenance. SIAM J. Comput. 26(5), 1343–1362 (1997)
10. Béal, M.-P., Mignosi, F., Restivo, A.: Minimal forbidden words and symbolic dynamics. In: Puech, C., Reischuk, R. (eds.) STACS 1996. LNCS, vol. 1046, pp. 555–566. Springer, Heidelberg (1996)
11. Bender, M.A., Farach-Colton, M.: The ICA problem revisited. In: Gonnet, G.H., Panario, D., Viola, A. (eds.) LATIN 2000. LNCS, vol. 1776, pp. 88–94. Springer, Heidelberg (2000)
12. Blumer, A., Blumer, J., Ehrenfeucht, A., Haussler, D., Chen, M.T., Seiferas, J.: The smallest automaton recognizing the subwords of a text. Theor. Comput. Sci. 40(1), 31–55 (1985)
13. Blumer, A., Blumer, J., Ehrenfeucht, A., Haussler, D., McConnell, R.: Building a complete inverted file for a set of text files in linear time. In: Proceedings of the 16th ACM Symposium on the Theory of Computing, pp. 349–351. ACM Press, Washington, D.C. (1984)
14. Blumer, A., Blumer, J., Ehrenfeucht, A., Haussler, D., McConnell, R.: Complete inverted files for efficient text retrieval and analysis. J. Assoc. Comput. Mach. 34(3), 578–595 (1987)

15. Brodal, G.S., Lyngsø, R.B., Östlin, A., Pedersen, C.N.S.: Solving the string statistics problem in time $\mathcal{O}(n \log n)$. In: Widmayer, P., Triguero, F., Morales, R., Hennessy, M., Eidenbenz, S., Conejo, R. (eds.) ICALP 2002. LNCS, vol. 2380, pp. 728–739. Springer, Heidelberg (2002)

16. Burrows, M., Wheeler, D.J.: A block-sorting lossless data compression algorithm. Technical Report 124, Digital Equipments Corporation (May 1994)

17. Chairungsee, S., Crochemore, M.: Using minimal absent words to build phylogeny. Theoretical Computer Science 450(1), 109–116 (2012)

18. Clark, D.R., Munro, J.I.: Efficient suffix trees on secondary storage. In: Proceedings of the 7th ACM-SIAM Annual Symposium on Discrete Algorithms, Atlanta, Georgia, pp. 383–391 (1996)

19. Crochemore, M.: Transducers and repetitions. Theor. Comput. Sci. 45(1), 63–86 (1986)

20. Crochemore, M.: Longest common factor of two words. In: Ehrig, H., Kowalski, R., Levi, G., Montanari, U. (eds.) CAAP 1987 and TAPSOFT 1987. LNCS, vol. 249, pp. 26–36. Springer, Heidelberg (1987)

21. Crochemore, M., Hancart, C., Lecroq, T.: Algorithms on Strings. Cambridge University Press, Cambridge (2007)

22. Crochemore, M., Mignosi, F., Restivo, A.: Automata and forbidden words. Information Processing Letters 67(3), 111–117 (1998)

23. Crochemore, M., Mignosi, F., Restivo, A., Salemi, S.: Data compression using antidictonaries. Proceedings of the I.E.E.E. 88(11), 1756–1768 (2000); Special issue Lossless data compression, Storer, J. (ed.)

24. Crochemore, M., Rytter, W.: Usefulness of the Karp-Miller-Rosenberg algorithm in parallel computations on strings and arrays. Theor. Comput. Sci. 88(1), 59–82 (1991)

25. Crochemore, M., Rytter, W.: Text algorithms. Oxford University Press (1994)

26. Farach, M.: Optimal suffix tree construction with large alphabets. In: Proceedings of the 38th IEEE Annual Symposium on Foundations of Computer Science, Miami Beach, FL, pp. 137–143 (1997)

27. Ferragina, P., Giancarlo, R., Manzini, G.: The myriad virtues of wavelet trees. Inf. Comput. 207(8), 849–866 (2009)

28. Ferragina, P., Luccio, F., Manzini, G., Muthukrishnan, S.: Compressing and indexing labeled trees, with applications. J. ACM 57(1) (2009)

29. Ferragina, P., Manzini, G.: Opportunistic data structures with applications. In: FOCS, pp. 390–398 (2000)

30. Ferragina, P., Manzini, G.: Indexing compressed text. J. ACM 52(4), 552–581 (2005)

31. Galil, Z.: Optimal parallel algorithms for string matching. In: Proceedings of the 16th ACM Symposium on the Theory of Computing, pp. 240–248. ACM Press, Washington, D.C. (1984)

32. Galil, Z.: Optimal parallel algorithms for string matching. Inf. Control 67(1-3), 144–157 (1985)

33. Grossi, R., Gupta, A., Vitter, J.S.: High-order entropy-compressed text indexes. In: SODA, pp. 841–850 (2003)

34. Grossi, R., Vitter, J.S.: Compressed suffix arrays and suffix trees with applications to text indexing and string matching. In: Proceedings of the ACM Symposium on the Theory of Computing, Portland, Oregon, pp. 397–406. ACM Press (2000)

35. Gusfield, D.: Algorithms on strings, trees and sequences: computer science and computational biology. Cambridge University Press, Cambridge (1997)

36. Gusfield, D., Stoye, J.: Linear time algorithms for finding and representing all the tandem repeats in a string. J. Comput. Syst. Sci. 69(4), 525–546 (2004)
37. Harel, D., Tarjan, R.E.: Fast algorithms for finding nearest common ancestors. SIAM J. Comput. 13(2), 338–355 (1984)
38. Hui, L.C.K.: Color set size problem with applications to string matching. In: Apostolico, A., Crochemore, M., Galil, Z., Manber, U. (eds.) CPM 1992. LNCS, vol. 644, pp. 230–243. Springer, Heidelberg (1992)
39. Kärkkäinen, J., Sanders, P.: Simple linear work suffix array construction. In: Baeten, J.C.M., Lenstra, J.K., Parrow, J., Woeginger, G.J. (eds.) ICALP 2003. LNCS, vol. 2719, pp. 943–955. Springer, Heidelberg (2003)
40. Karp, R.M., Miller, R.E., Rosenberg, A.L.: Rapid identification of repeated patterns in strings, trees and arrays. In: Proceedings of the 4th ACM Symposium on the Theory of Computing, pp. 125–136. ACM Press, Denver, CO (1972)
41. Kasai, T., Lee, G., Arimura, H., Arikawa, S., Park, K.: Linear-time longest-common-prefix computation in suffix arrays and its applications. In: Amir, A., Landau, G.M. (eds.) CPM 2001. LNCS, vol. 2089, pp. 181–192. Springer, Heidelberg (2001)
42. Kempf, M., Bayer, R., Güntzer, U.: Time optimal left to right construction of position trees. Acta. Inform. 24(4), 461–474 (1987)
43. Kim, D.K., Sim, J.S., Park, H., Park, K.: Constructing suffix arrays in linear time. J. Discrete Algorithms 3(2-4), 126–142 (2005)
44. Ko, P., Aluru, S.: Space efficient linear time construction of suffix arrays. J. Discrete Algorithms 3(2-4), 143–156 (2005)
45. Kolmogorov, A.N.: Three approaches to the quantitative definition of information. Problems of Information Transmission 1(1), 1–7 (1965)
46. Kurtz, S.: Reducing the space requirements of suffix trees. Softw. Pract. Exp. 29(13), 1149–1171 (1999)
47. Landau, G.M.: String matching in erroneus input. Ph. D. Thesis, Department of Computer Science, Tel-Aviv University (1986)
48. Lempel, A., Ziv, J.: On the complexity of finite sequences. IEEE Trans. Inf. Theory 22, 75–81 (1976)
49. Majster, M.E., Ryser, A.: Efficient on-line construction and correction of position trees. SIAM J. Comput. 9(4), 785–807 (1980)
50. Manber, U., Myers, G.: Suffix arrays: a new method for on-line string searches. In: Proceedings of the 1st ACM-SIAM Annual Symposium on Discrete Algorithms, San Francisco, CA, pp. 319–327 (1990)
51. Manber, U., Myers, G.: Suffix arrays: a new method for on-line string searches. SIAM J. Comput. 22(5), 935–948 (1993)
52. McCreight, E.M.: A space-economical suffix tree construction algorithm. J. Algorithms 23(2), 262–272 (1976)
53. Muthukrishnan, S.: Efficient algorithms for document listing problems. In: Proceedings of the 13th ACM-SIAM Annual Symposium on Discrete Algorithms, pp. 657–666 (2002)
54. Na, J.C., Ferragina, P., Giancarlo, R., Park, K.: Two-dimensional pattern indexing. In: Encyclopedia of Algorithms (2008)
55. Na, J.C., Giancarlo, R., Park, K.: On-line construction of two-dimensional suffix trees in o(n^2 log n) time. Algorithmica 48(2), 173–186 (2007)
56. Poe, E.A.: The Gold-Bug and Other Tales. Dover Thrift Editions Series. Dover (1991)
57. Pratt, V.: Improvements and applications for the Weiner repetition finder, Manuscript (1975)

58. Rodeh, M., Pratt, V., Even, S.: Linear algorithm for data compression via string matching. J. Assoc. Comput. Mach. 28(1), 16–24 (1991)
59. Slisenko, A.O.: Determination in real time of all the periodicities in a word. Sov. Math. Dokl. 21, 392–395 (1980)
60. Slisenko, A.O.: Detection of periodicities and string matching in real time. J. Sov. Math. 22, 1316–1386 (1983)
61. Storer, J.A.: NP-completeness results concerning data compression. Report 234, Princeton University (1977)
62. Storer, J.A., Szymanski, T.G.: The macro model for data compression. In: Proceedings of the 10th ACM Symposium on the Theory of Computing, San Diego, CA, pp. 30–39. ACM Press (1978)
63. Storer, J.A., Szymanski, T.G.: Data compression via textual substitution. J. Assoc. Comput. Mach. 29(4), 928–951 (1982)
64. Thue, A.: Über die gegenseitige lage gleicher teile gewisser zeichenreichen. Nor. Vidensk. Selsk. Skr. Mat. Nat. Kl. 1, 1–67 (1912)
65. Ukkonen, E.: On-line construction of suffix trees. Algorithmica 14(3), 249–260 (1995)
66. Ulitsky, I., Burstein, D., Tuller, T., Chor, B.: The average common substring approach to phylogenomic reconstruction. Journal of Computational Biology 13(2), 336–350 (2006)
67. van Helden, J., André, B., Collado-Vides, J.: Extracting regulatory sites from the upstream region of the yeast genes by computational analysis of oligonucleotides. J. Mol. Biol. 281, 827–842 (1998)
68. Weiner, P.: Linear pattern matching algorithm. In: Proceedings of the 14th Annual IEEE Symposium on Switching and Automata Theory, Washington, DC, pp. 1–11 (1973)

LCP Magic

Moshe Lewenstein

Bar Ilan University, Ramat Gan, Israel

Abstract. LCP, the Longest Common Prefix, is ubiquitous in the realm of data structures for indexing. This surprisingly simple minute definition – magically – plays a big role in Pattern Matching. We will try to understand this "magic" by examining a couple of interesting applications.

J. Fischer and P. Sanders (Eds.): CPM 2013, LNCS 7922, p. 11, 2013.
© Springer-Verlag Berlin Heidelberg 2013

Discrete Methods for Image Analysis Applied to Molecular Biology

Gene Myers

Max Planck Institute for Molecular Cell Biology and Genetics, Dresden, Germany

Abstract. The field of image analysis and signal processing originally developed in the engineering community and is thus dominated by methods appealing to continuous mathematics. As a discrete mathematician recently entering this domain in the context of analyzing biological images, primarily from various forms of microscopy, I have found that discrete techniques involving trees and graphs better solve some segmentation and tracking problems than their continuous competitors. We illustrate this with three examples: component trees for adaptively segmenting nuclei in *C. elegans* 3D stacks, progress graph merging for segmenting cells in a 2D image of a fly wing, and shortest paths for segmenting and modeling individual neurons in a fly brain.

J. Fischer and P. Sanders (Eds.): CPM 2013, LNCS 7922, p. 12, 2013.

Locating All Maximal Approximate Runs in a String

Mika Amit[1,*], Maxime Crochemore[3,4], and Gad M. Landau[1,2,**]

[1] Department of Computer Science, University of Haifa, Mount Carmel, Haifa, Israel
mika.amit2@gmail.com, landau@cs.haifa.ac.il
[2] Department of Computer Science and Engineering, NYU-Poly, Brooklyn NY, USA
[3] King's College London, Strand, London WC2R 2LS, UK
maxime.crochemore@kcl.ac.uk
[4] Université Paris-Est, Institut Gaspard-Monge,
77454 Marne-la-Vallée Cedex 2, France

Abstract. An exact run in a string, T, is a non-empty substring of T that can be divided into adjacent non-overlapping identical substrings. Finding exact runs in strings is an important problem and therefore a well studied one in the strings community. For a given string T of length n, finding all maximal exact runs in the string can be done in $O(n \log n)$ time or $O(n)$ time on integer alphabets. In this paper, we investigate the maximal approximate runs problem: for a given string T and a number k, find every non-empty substring T' of T such that changing at most k letters in T' transforms it into a maximal exact run in T. We present an $O(nk^2 \log k \log \frac{n}{k})$ algorithm.

Keywords: algorithms on strings, pattern matching, repetitions, tandem repeats, runs.

1 Introduction

The domain of Algorithms on strings is fond of combinatorial properties on words. They are used to analyze the behavior of algorithms in conjunction with statistical results, and often lead to improving their design up to optimal characteristics. Conversely, some combinatorial properties on words, obtained without any algorithmic objectives, yield algorithms that are surprisingly efficient according to various aspects (time, space, design, etc.).

The most central properties relate to periodicities in words and pop up in many examples. The notion is doubtlessly at the core of many stringology questions. It constitutes a fundamental area of string combinatorics due to important applications to text algorithms, data compression, biological sequences analysis, or music analysis. Indeed, periods are ubiquitous in string algorithms because

* Partially supported by the Israel Science Foundation grant 347/09.
** Partially supported by the National Science Foundation Award 0904246, Israel Science Foundation grant 347/09, Yahoo and Grant No. 2008217 from the United States-Israel Binational Science Foundation (BSF).

J. Fischer and P. Sanders (Eds.): CPM 2013, LNCS 7922, pp. 13–27, 2013.
© Springer-Verlag Berlin Heidelberg 2013

stuttering is likely to slow down any of the algorithms for pattern matching, text compression, and genome assembly, for example.

A maximal exact run in a string is informally an occurrence of a non-extensible segment having a small period. The concept captures all the power of local periodicities and repetitions and therefore has attracted a lot of studies.

Several methods are available to detect all the occurrences of exact repetitions in strings with some small variations on the elements they target (see [1–3]). For a given string T, of length n, these algorithms run in $O(n \log n)$ time, which is optimal because some strings contain this number of elements. Selecting some of their occurrences or just distinct repetitions regardless of their number of occurrences (see [4, 5]) paved the path to faster algorithms.

Runs have been introduced by Iliopoulos, Moore, and Smyth [6] who showed that Fibonacci words contain only a linear number of them according to their length. Kolpakov and Kucherov [7] (see also [8], Chapter 8) proved that the property holds for any string. Meanwhile they designed an algorithm to compute all runs in a string of length n over an alphabet Σ in $\mathcal{O}(n \log(|\Sigma|))$ time. Their algorithm extends Main's algorithm [9], which itself extends the method in [10] (see also [8]).

The design of a linear-time algorithm for building the Suffix Array of a string on an integer alphabet (see [8]) and the introduction of another related data structure (the Longest Previous Factor table in [11]) have eventually led to a linear-time solution (independent of the alphabet size) for computing all runs in a string.

Finding approximate runs is more sensible than finding exact runs in some applications. A typical example is genetic sequence analysis. This problem was widely researched and many different measurements have been used in order to find such runs ([12–15]). The k-approximate run problem can be defined as follows: given a string x and a number k, divide x into non-empty substrings, $x = u_1 u_2 \cdots u_t$, such that the distance between every two adjacent substrings, u_i and u_{i+1}, is not greater than k; or such that the distance between *every* two substrings u_i and u_j in x cannot exceed k; or find a consensus substring u such that the distance between u and every u_i is not greater than k.

Another definition of approximate run is as follows: a string x is a k-approximate run if $x = u_1 u_2 \cdots u_t$ and the removal of the same k positions from all u_i generates an exact run.

In this work, we solve the following definition of the approximate run problem: given a string T, find all non-empty substrings that by the modification of at most k letters form an exact run. In other words, for each such substring there exists a consensus string, u, such that the sum over all Hamming distances between u and u_i is not greater than k. We present an $O(nk^2 \log k \log \frac{n}{k})$-time algorithm that solves this problem.

Roadmap: we start in section 2 with definitions and notations that will be used throughout the paper. In section 3, we present a simple $O(n^2)$ algorithm for finding all approximate runs in the string. Then, in section 4 we continue to an improved $O(nk^3 \log \frac{n}{k})$ algorithm. Finally, on section 5, we present the $O(nk^2 \log k \log \frac{n}{k})$ algorithm.

2   Definitions and Notations

Let $T = T[1]T[2]\cdots T[n]$ be a string of size n defined over the constant alphabet Σ. In this abstract the size of the alphabet is constant. We denote the i'th letter in T as $T[i]$, and the substring of T that starts at position i and ends at position j as $T[i..j] = T[i]T[i+1]\cdots T[j]$. Let T^R be the reversed substring of T.

An *exact run* is a non empty string, T, that can be divided into a number of identical adjacent non overlapping substrings $T = u_1 u^t u_2$, where the first substring u_1 can be a suffix of u, and the last substring u_2 can be a prefix of u. u is called a *period* and its length is denoted as the *period length*, $p = |u|$. The exponent of the run is of size $\frac{|T|}{p}$ and it is greater or equal to 2. For instance, *'ababababa'*, has exact runs with period length 2 and 4. Their exponent are 4.5 and 2.25, respectively.

A *maximal exact run* is an exact run that cannot be extended to either right or left. For instance, *'dabababac'*, has a maximal exact run starting at position 2 with period length 2 and exponent 3.5. If a substring T contains a maximal exact run starting at index i, it means that either $i = 1$ or $T[i-1] \neq T[i+p-1]$, for otherwise, the exact run is not maximally extended. Similarily, if the maximal exact run ends at position j, then either $j = n$ or $T[j+1] \neq T[j+1-p]$.

A *k-maximal approximate run (k-MAR)* is a non empty string, T, such that the modification of at most k letters in T generates a maximal exact run. For instance, *'abaabcaba'*, is a 1-maximal approximate run with period length 3 and exponent 3. In this example, the letter in position 6 is a *modified* letter, since modifying it from *'c'* to *'a'* generates an exact run.

For the rest of this paper we distinguish between two notations: a *mismatch* and a *modified letter*, that have a very strong relation between each other, but, as we will explain shortly, are not always identical in their meanings. If two letters $T[i]$ and $T[i+p]$ are not identical, we say that there is a *mismatch* between positions i and $i+p$ in T. We mark position $i+p$ as the position in which the mismatch occurs. A *modified letter* corresponds to a modification of a letter in an approximate run in order to convert it to an exact run. For instance, for period length $p = 3$ and the substring *'abcabcabd'*, there is a *mismatch* at position 9 (between *'c'* and *'d'*) and there is a modified letter in the same position, 9.

Observe that while a mismatch can imply that a modified letter needs to be used in the k-MAR and vice versa, it is not always straightforward:

A1. Two mismatches can imply only one modified letter. For instance, for a substring *'abaabcaba'* with period length 3, there are two mismatches in positions (6) and (9), but only one modified letter is needed in position 6 in order to convert the original substring into an exact run (by modifying the letter 'c' to 'a').

A2. One mismatch can imply at most $\frac{n}{2p}$ modified letters. For instance, in the 2-MAR with period length 3 of the string *'abcabcabcdabdabd'*, there is only one mismatch in position 9 and two modified letters on positions 3 and 6.

We continue with a definition of Parikh matrix (see also [16]) that will be used throughout the paper: A *Parikh matrix*, $P[1..|\Sigma|, 1..p]$, is a two-dimensional array defined over a substring $T[i..j]$ and a period length p. An entry $P[a,c]$ contains the number of occurrences of $a \in \Sigma$ in the column c of the period. In addition, for each column c, we keep an additional variable, $win(c)$, that contains the *winner* letter - the letter that occurs more than any other letter in this column (breaking ties arbitrarily).

We use the Parikh matrix in order to count the number of modified letters used in a k-MAR of period length p that starts at position i and ends at position j. For simplicity, we set the first column of the period to be position 1 in the text. This means that an approximate run can start at any column of the period, and for period length p, the column of index i is $Column(i) = i \bmod p$ (with the exception of the case where $i \bmod p = 0$, in which $Column(i) = p$). The number of modified letters associated with a column is the sum over all letters that are not the winner letter. i.e. $\Sigma_{a \neq win(c)} P[c,a]$. If the sum of modified letters in a column is greater than 0, the column is denoted as a *problematic column*. The number of modified letters associated with the k-MAR is the sum of modified letters over all its problematic columns.

Figure 1 shows examples of Parikh matrices for three period lengths computed for the same prefix of a string.

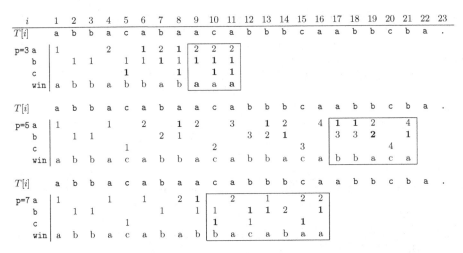

Fig. 1. Examples of maximal approximate runs with the maximum of 5 modified letters. The longest run has length 11 for period length 3, 21 for period length 5, and 16 for period length 7. The three Parikh matrices are computed from left to right.

2.1 Problem Definition

In this paper we present three algorithms that solve the following problem:

Definition 1 (The k-Maximal Approximate Runs Problem). *Given a string T of size n defined over the alphabet Σ, and a number k, find all k-MAR in the string T of all period lengths p, $1 \leq p \leq \frac{n}{2}$.*

3 A Simple $O(n^2)$ Algorithm Using Parikh Matrices

We start with a simple algorithm for computing all k-MAR in a string T. The algorithm iterates over all period lengths p, $1 \le p \le \frac{n}{2}$, and for each period length computes all k-MAR that exist in the string. We describe the procedure for finding all k-MAR with period length p:

FINDKMAR(T, p)

```
 1   Initialize Parikh matrix P
 2   T[n + 1] ← $
 3   ℓ ← 1, r ← 1, count ← 0, newKmar ← true
 4   Move r :
 5   while r ≤ n + 1 do
 6       if Winner(r) = true or count < k then
 7           if Winner(r) = false then
 8               count ← count + 1
 9           update Parikh matrix according to T[r]
10           r ← r + 1
11           newKmar ← true
12       else \ * r cannot be increased *\
13           if newKmar then
14               announce k-MAR T[ℓ .. r − 1]
15               newKmar ← false
16           goto Move ℓ
17
18   Move ℓ :
19   if r = n + 1 then
20       return
21   update Parikh matrix according to T[ℓ] delete
22   if Winner(ℓ) = false then
23       count ← count − 1
24   ℓ ← ℓ + 1
25   goto Move r
26
27   return
```

In the $FindKmar$ procedure described above we keep two pointers, ℓ and r, on the string T, such that they define possible leftmost and rightmost positions of a k-MAR, respectively. The computation uses Parikh matrix, P, of size $|\Sigma| \times p$. The initialization of the Parikh matrix (line 1) consists of setting the winners of all columns of p according to the letters in $T[1 .. p]$.

The procedure uses an auxiliary function, $Winner()$, that gets a position i in the text as a parameter and returns true if the number of occurrences of the letter $T[i]$ in its column is equal to the number of occurrences of the winner letter in it (lines 6, 7, 22). Recall that the first column of the period is set to position 1 in the text.

Each iteration call checks whether r position can be increased: either r is one of the winners in its column (thus no modified letters are used when r is increased) or the number of used modified letters is less than the maximum allowed. When r position cannot be increased, ℓ position is increased. Note that as long as $r \le n$ each update of ℓ position calls *Move r* sub procedure. The reason for that is that the deletion of $T[\ell]$ can either release a modified letter (line 22) or change the current winners list of ℓ's column. In the case where these two cases have not occurred, *Move r* procedure does not increase r and the iteration returns to *Move ℓ* sub procedure (line 16).

A new k-MAR is announced whenever r position cannot be increased and its position has moved since the last k-MAR announcement. In this case $T[\ell \mathinner{..} r - 1]$ is announced as a k-MAR (line 14). Observe that in every string T there is always a k-MAR that consists of n position. This case is handled as follows: an extra letter, \$, (that is not included in the alphabet) is added to the text in position $n + 1$ (line 2). This way, all k-MARS, including the rightmost one, will be announced by the procedure.

An example of the procedure run is demonstrated in figure 2.

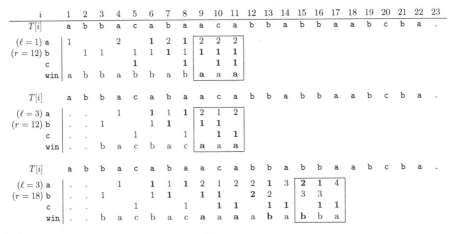

Fig. 2. One iteration of finding 5-MAR with period length 3. The first Parikh matrix is computed until 5 modified letters are in use. Then, ℓ is moved to the right until one modified letter is released. Then, r is moved to the right until one modified letter is used. At the end of this iteration the 5-MAR $T[3 \mathinner{..} 17]$ is announced. In the next iteration, when ℓ is moved to the right ($\ell = 4$), there is a tie between the letters a and b in column 3, therefore, r can be moved to the right without using modified letters until position 20.

Time Complexity: each Parikh matrix entry computation takes $O(1)$, either when adding a letter (moving r index to the right) or when subtracting a letter (moving ℓ to the right). Therefore, the time complexity depends on the number of such updates. For a single period length, p, each position i, $1 \le i \le n$, is updated in the Parikh matrix at most twice: when $\ell = i$ and when $r = i$. This gives a total of $O(n)$ time per period. p goes from 1 to $\frac{n}{2}$, which gives a total time complexity of $O(n^2)$ for the entire algorithm.

4   An $O(nk^3 \log \frac{n}{k})$ Algorithm

The general strategy of the improved algorithm is a divide-and-conquer recursive scheme similar to the one used by Main and Lorentz [3] for computing squares and to that of [12] for computing approximate repetitions. It works as follows: during the first step, we locate all k-MAR that contain the middle position of T, i.e. $\frac{n}{2}$. This is computed for all possible period lengths p such that $1 \le p \le \frac{n}{2}$. Then, the same procedure for finding all k-MAR with period lengths $1 \le p \le \frac{n}{4}$ is applied independently on the first half ($T[1 .. \frac{n}{2} - 1]$) and on the second half of T ($T[\frac{n}{2} + 1 .. n]$).

High level description of the algorithm is described in algorithm ApproxRep.

APPROXREP($T, start, end$)
  1   $n \leftarrow (end - start)$
  2   **if** $n > 2k$ **then**
  3       $mid \leftarrow start + \lfloor n/2 \rfloor$
  4   **for** $p \leftarrow 1$ **to** $\lfloor n/2 \rfloor$ **do**
  5   Find all k-MAR with period length p that contain mid
  6   APPROXREP($T, start, mid - 1$)
  7   APPROXREP($T, mid + 1, end$)

In the rest of this section we describe the algorithm for finding all k-MAR with period length p that exist on a text T of size n and contains position $\frac{n}{2}$ (line 5 in the above algorithm).

4.1   Initialization Step: Defining the Substring Boundaries

We start with a simple observation regarding the boundaries of the substring that should be handled on each iteration. Let $T[i .. j]$ be a k-MAR that contains position $\frac{n}{2}$. According to observation A1 in section 2, since one modified letter can imply on at most 2 mismatches, there are no more than $2k$ mismatches such that $T[i] \ne T[i + p]$ in the substring $T[i .. j]$. Thus, all k-MAR substrings that contain position $\frac{n}{2}$ must start at a position that is right to the position of the $2k + 2$ mismatch going from $\frac{n}{2}$ to the left on the reversed substring of T. More precisely, if x is the position of the $2k + 1$ mismatch going from $\frac{n}{2}$ to the left, then i must be right to position $x - p$ (this is due to the fact that by definition, a mismatch between two positions, i and $i + p$, is marked in $i + p$ position and not in i). Let ℓ_1 be the position $x - p$. In addition, j must be left to the position of the $2k + 1$ mismatch going from $\frac{n}{2}$ to the right. Let r_1 be the position of this mismatch. Any extension to these positions cannot contain the position $\frac{n}{2}$ in the k-MAR.

As an initialization step, the algorithm uses the technique described in [12], using [17] and [18], in order to find $2k + 1$ mismatches between two copies of T

shifted by p positions starting at index $\frac{n}{2}$ going right, and $2k + 1$ mismatches starting at index $\frac{n}{2}$ going left. This is done by constructing two suffix trees (or suffix arrays), one for the string T and one for its reversed string T^R, and using the "kangaroo" jumps of [18] (i.e., using suffix trees and LCA algorithm for a constant time "jump" over equal substrings of the aligned copies of T) in order to find the positions of $2k+1$ mismatches between the suffixes $T^R[\frac{n}{2} \mathinner{.\,.} n - p]$ and $T^R[\frac{n}{2} + p \mathinner{.\,.} n]$, and the positions of $2k + 1$ mismatches between the substrings $T[\frac{n}{2} \mathinner{.\,.} n - p]$ and $T[\frac{n}{2} + p \mathinner{.\,.} n]$.

Observe that the mismatches define the *problematic columns* that should be handled in the current iteration over p. A modified letter can only be used in one of these columns. Hence, there are at most $4k + 2 = O(k)$ such columns.

We use two sorted lists: *ColumnList* that contains the problematic columns in the substring $T[\ell_1 \mathinner{.\,.} r_1]$, and *MismatchList* that contains the mismatch positions in the substring. The mismatch positions are inserted in a sorted manner to the *MismatchList*, such that m_1 is the leftmost mismatch. In addition, *ColumnList* is updated with all problematic columns as follows: for each mismatch in position i, we add its column to *ColumnList*.

Time Complexity: in order to find all $4k + 2$ mismatches, two suffix trees are constructed. This is done once for the entire algorithm in $O(n)$ time. The algorithm for finding $O(k)$ mismatch positions between the substrings is done in $O(k)$ time, including the update of both *MismatchList* and *ColumnList*.

4.2   Main Procedure: Finding All k-MAR in the Substring

The main procedure of the algorithm uses both *MismatchList* and *ColumnList* in order to find all k-MAR in between the boundaries ℓ_1 and r_1. It is similar to the simple algorithm described in section 3 which uses the Parikh matrix in order to find k-MAR, and that it keeps two pointers on the string, ℓ and r, that are increased in turns. The difference between the two algorithms is that in this algorithm we take advantage of the fact that not all positions in the substring need to be visited, the only relevant positions are the ones of the *problematic columns* in all periods.

Observe that the mismatch positions in *MismatchList* divide the text $T[\ell_1 \mathinner{.\,.} r_1]$ into $4k + 1$ adjacent non overlapping substrings. We denote each such substring as a *zone*. The leftmost zone is the substring $T[\ell_1 \mathinner{.\,.} m_1 - 1]$, the second zone is the substring $T[m_1 \mathinner{.\,.} m_2 - 1]$, and so on. Each *zone* presents either an exact run (since there are no mismatches in it), or a prefix of an exact run (if the zone's length is smaller than $2p$). Note that because the mismatch position of $(i, i + p)$ is marked in position $i + p$ by definition, these exact runs are not maximally extended - they can be extended to the left by $p - 1$ letters.

Consider a possible k-MAR $T[\ell + 1 \mathinner{.\,.} r - 1]$, such that $\ell_1 < \ell < \frac{n}{2}$ and $\frac{n}{2} < r < r_1$: from the definition of k-MAR we get that both substrings $T[\ell \mathinner{.\,.} r - 1]$ and $T[\ell + 1 \mathinner{.\,.} r]$ contain $k + 1$ modified letters (for otherwise, the approximate run is not maximally extended). We denote the *zone* in which ℓ is contained as L and the *zone* in which r is contained as R. Let m_i be the mismatch position in L zone, and let m_j be the mismatch position in R zone (see figure 3).

We continue with an observation regarding the possible positions of ℓ and r:

Observation 1 *ℓ position can only be on the $k+1$ rightmost periods in L zone. r position can only be on the $k+1$ leftmost periods in R zone. Denote these periods as the $(k+1)$-periods.*

Proof. We prove this property by contradiction: assume that L zone contains $z > k+1$ periods and that ℓ position is left to the rightmost $k+1$ periods in L. We consider two cases regarding the problematic columns in L:

Case 1: there exists at least one letter in L that is a non-winner letter of its column - this means that the number of modified letters in $T[\ell+1..r-1]$ exceeds $k+1$, a contradiction.

Case 2: all letters in L are winners of their columns - note that the substring $T[m_i - p+1..m_{i+1}-1]$ is an exact run, therefore, in order for $T[\ell+1..r-1]$ to be maximally extended, ℓ has to be equal to $m_i - p$. This position is not in L zone (and it was visited on an earlier iteration), a contradiction. In a similar way, it is easy to show that r position can only be on the leftmost $k+1$ periods in R.

Recall that a *problematic column* is a column that contains at least two different letters in the substring $T[\ell_1..r_1]$. We continue with a definition of a *problematic position*: a problematic position is a position in the substring T such that its column is a problematic column and it is in the $(k+1)$-periods of L or R zones. There are a total of $O(k)$ zones and each one contains $O(k)$ periods that needs to be visited. Each period contains $O(k)$ problematic columns. Hence, the total number of problematic positions in the substring is $O(k^3)$. Note that a problematic position is a position in which a modified letter might be used. These positions are the positions that ℓ and r are assigned to in the improved algorithm.

The algorithm works in a similar way to the simple $O(n^2)$ algorithm with the two following differences: first, the Parikh matrix contains only the problematic columns of *ColumnList*. The second difference is that on each iteration over ℓ and r, their position is increased to the next *problematic position* (and not necessarily increased by 1).

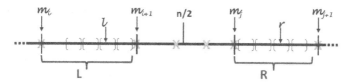

Fig. 3. L, R zones, and the positions ℓ and r in them. The mismatch positions are marked with 'X' and the boundaries of the substrings are marked with vertical lines. For $k = 3$, the relevant periods in each of the zones are marked with brackets.

This complicates the Parikh matrix update procedure, and we distinguish between two cases of it:

Case 1: moving a position inside a zone: moving ℓ in L (or r in R) updates only one column in the Parikh matrix, which is decreased (or increased) by at most 1.

Case 2: moving a position in between zones: assume that ℓ is moved from L zone to L' zone on its right. Suppose that L' contains $q > k + 1$ periods, then ℓ position is increased to the position of the first problematic column in L' such that ℓ is in the rightmost $k + 1$ periods of L'. All problematic columns in the Parikh matrix need to be updated according to the number of times the column occurs in the leftmost $q - (k + 1)$ periods of L'. The situation is similar when moving r from R zone to R' zone: if R contains more than $k + 1$ periods, then the increase of r position should update the entire Parikh matrix and compute the total number of modified letters used in the new substring.

The procedure stops when either $\ell > \frac{n}{2}$ or $r = r_1$.

Time Complexity: each position move of either ℓ or r updates the Parikh matrix. If the position move is in the same zone, the update takes $O(1)$ time. If a position is increased to a different zone, the entire Parikh matrix is computed, which takes $O(k)$ time. On each zone there are at most $O(k^2)$ positions, and there are a total of $O(k)$ zones. This gives a total of $O(k^3)$ updates in the same zone and a total of $O(k)$ updates of zone changing. Therefore, the total time complexity of the main procedure is $O(k^3)$.

4.3   Total Time Complexity Analysis

Finding the initial boundaries of the substring takes $O(k)$ time. The time complexity of the main procedure is $O(k^3)$. The total time complexity for finding all k-MAR with period length p that exist in a substring T is $O(k^3)$. There are at most $O(n \log \frac{n}{k})$ iterations in the ApproxRep algorithm, and each iteration is done in $O(k^3)$ time, which gives a total time complexity of $O(nk^3 \log \frac{n}{k})$ for the entire algorithm.

In the next section, we describe an algorithm that improves the time complexity of the main procedure: instead of going over all $O(k^2)$ possible problematic positions for ℓ and r in each zone, the algorithm only visits $O(k \log k)$ such positions.

5   An improved $O(nk^2 \log k \log \frac{n}{k})$ Algorithm

In this algorithm we improve the main procedure of the previous algorithm. The main idea behind the improvement is that there are still redundant positions that were visited in the previous algorithm: in this algorithm we only visit positions in L and in R that are the positions of modified letters. In addition, we visit each one of the $(k + 1)$-periods in L. The iteration over ℓ and r positions and the update of the Parikh matrix according to them is done in a similar way to the previous algorithm. The only difference is in the initialization and the handling of the relevant positions that need to be visited.

Let $T[\ell+1..r-1]$ be a k-MAR, and let m_i, m_j be the mismatch positions in L and R, respectively. We denote the substring $T[m_{i+1}..m_j-1]$ as M (see figure 4).

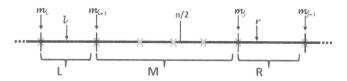

Fig. 4. L, M and R zones, and the positions ℓ and r in them. The mismatch positions are marked with 'X' and the boundaries of the substrings are marked with vertical lines.

Note that the k modified letters in the k-MAR are spread among L, M and R.

In order to handle only the relevant problematic positions in the substring, the following sub procedures are added to the previous algorithm: at first, when either ℓ or r reaches the leftmost position of its zone, an initialization is done, in which initial positions that need to be visited on L and R are found. Second, when iterating over ℓ position - on each one of the $(k+1)$-periods in L, additional columns are added to the list of problematic columns that are relevant for the rest of the iteration. The third sub procedure is added to the iteration over r position - problematic columns are added or removed from the lists of problematic columns in L and R.

5.1   Sub Procedure 1: Updating the Problematic Positions

This sub procedure is called whenever ℓ or r moves between zones. Then, the substrings L, M and R are redefined according to the substring $T[\ell..r]$, and the procedure begins.

Note that there are two different options for positions ℓ and r: either ℓ is on the $(k+1)$'th period of L and r is a position in between the leftmost $k+1$ periods in R, or ℓ is a position in between the $k+1$ rightmost periods in L and r is the mismatch position of R zone. In both options, the Parikh matrix is updated according to the substring $T[\ell..r]$.

The algorithm uses two lists, *LeftList* and *RightList*, that keep the problematic columns that need to be visited in L and in R, respectively. In addition, for each period in the $(k+1)$-periods of L, we keep a list, *newColumnList*, that contains added problematic columns that need to be visited in L starting this period.

Let c be a problematic column, and let x be the letter of column c in L, having $|x|$ occurrences in $T[\ell..r]$. Let y be the letter of c in R having $|y|$ occurrences in $T[\ell..r]$, and z be the letter with the maximum number of occurrences, $|z|$, in c column in M.

We first define the positions in L that need to be visited by the algorithm. The following cases regarding the majority of x in the initial substring $T[\ell..r]$ are considered:

- Case 1: x is the loser of its column - in this case, the column c is added to *LeftList*, since it implies on modified letters. Note that in the case where $x = y$, a losing letter might gain majority and become a winner of its column.
- Case 2: x is the winner of its column, but might lose its majority. This case can be either one of the following two sub cases (or both):
  - x might lose its majority to y - this case can happen when $|y| < |x| \leq |y| + 2k$. The column c is added to *newColumnList* of the $|y|$'th period of L.
  - x might lose its majority to z - this case can happen when $|z| < |x| \leq |z| + k$. The column c is added to *newColumnList* of the $|z|$'th period of L.

  Note that in the case that x can lose to both y and z, the column c is added to *newColumnList* of the $max\{|y|, |z|\}$ period of L only.
- Case 3: x is the winner of its column and will win throughout the entire iteration - in this case, the column is not added to any list.

The number of columns in both *LeftList* and *newColumnList* cannot exceed $O(k)$, since there are at most $O(k)$ problematic columns in $T[\ell .. r]$. Furthermore, in the initialization step, each column in *LeftList* implies on a column that will be visited throughout the entire iteration over ℓ. The number of such visited positions cannot exceed $O(k)$, since they are the positions of the modified letters in L. Also note that each column in *newColumnList* of a period i of L implies on i visited positions in L. These are the positions that x might lose its majority to either z or y. The number of such visited positions cannot exceed $O(k)$ either, since these positions imply on modified letters in either M or R. Thus, the total number of the initial visited positions in L is bounded by $O(k)$.

We continue with the visited positions in R. We again consider the following cases regarding the majority of y in the initial substring $T[\ell .. r]$:

- Case 1: y is the loser of its column - in this case, the column c is added to *RightList*, since it may imply on modified letters.
- Case 2: y is the winner of its column, but might lose its majority to z - this case can only happen when $y = x$ and $|z| < |y| \leq |z| + k$. The column c is added to *RightList*, with a special flag that says that this column needs to be visited until r reaches the $|z|$'th period of R.
- Case 3: y is the winner of its column and will win throughout the entire iteration - in this case, the column is not added to any list.

The number of columns in *RightList* is bounded by $O(k)$.

Time Complexity: each column is added to a list in $O(1)$ time and there are $O(k)$ such columns, which gives a total time complexity of $O(k)$ for the update phase.

5.2   Sub Procedure 2: Period Handling in L Zone

This step is called whenever ℓ position moves from one period to the one on its right (in the same zone). The procedure is very simple: it goes over all columns in *newColumnList*, and adds them to *LeftList*. The total number of columns that can be added throughout the entire iteration over ℓ in L is $O(k)$.

Time Complexity: each column is added to *LeftList* in a sorted manner in $O(\log k)$ time and there are $O(k)$ such columns, which gives a total time complexity of $O(k \log k)$ for all period handling steps.

5.3 Sub Procedure 3: Iterating over r in R and ℓ in L

In this algorithm, in a similar way to the algorithm described in section 4, positions ℓ and r are increased to the next problematic positions. The difference is that in this algorithm, the problematic positions are taken from *LeftList* and *RightList*. On each position visit, the Parikh matrix is updated in the same manner as before.

This procedure is called whenever r position is increased. Let c be the problematic column of position r, and let y be the letter in this column. Since the number of occurrences of y was increased by one, the two following cases need to be checked:

Adding a column to LeftList: if y is not the winner of its column and it loses to the letter x in L, then a new wakeup call needs to be added to the respected period in L: column c is added to *newColumnList* of the $|y|$'th period of L, and is removed from the previous *newColumnList* (on period $|y| - 1$ in L), if such exists. Each insertion is done in $O(1)$ time. Note the special case in which the column is added to a period that ℓ is currently in. In this case, the column is added straight to *LeftList* in $O(\log k)$ time.

Removing a column from RightList: if y is the winner of its column, we consider the two cases: $x = y$ and $x \neq y$. If $x \neq y$, it means that y will continue to win throughout the entire iteration, and therefore its column is removed from *RightList*. If $x = y$, the situation is a bit more complicated since y can again lose its majority as ℓ position is increased. Therefore, before the column is removed from *RightList*, a check to see whether r position is on a period greater than $|z|$. If yes, the column is removed from the list. Otherwise, it is not removed, and will be visited on the next period, as well. Each deletion from *RightList* is done in a sorted manner in $O(\log k)$ time.

Time Complexity: adding a column to *newColumnList* is done in $O(1)$ time, since the list does not need to be sorted. adding or removing a column from *LeftList* and *RightList* in a sorted manner is done in $O(\log k)$ time. But, each problematic column is added or removed from the lists at most once, and there are $O(k)$ such columns. This gives a total time complexity of $O(k \log k)$ for the entire iteration over r in R and ℓ in L. In addition, each Parikh matrix update is done in $O(1)$ time (when ℓ and r stay in the same zone), and as proved in Lemma 1 below, there are at most $O(k \log k)$ such updates for both ℓ and r.

Recall that the iteration over ℓ and r visits the following positions: for a letter x in L, as long as x is the winner of its column, the algorithm visits only the periods in L, in which it might lose its majority to z (or y). In a similar way, for a letter y in R, as long as y is the winner of its column, the algorithm visits only the periods in R, in which it might lose its majority to z (this can only happen when $x = y$). If x (or y) is the loser of its column, then all the positions in which it loses (there are at most $k + 1$ such positions) are visited.

Note that although new columns are never added to *RightList*, additional columns are added to *LeftList* from *newColumnList*. This means that the number of positions visited in L (and as a result in R) can be greater than $O(k)$. Lemma 1 proves that this number is bounded by $O(k \log k)$:

Lemma 1. *The total number of visited positions in L and in R is bounded by $O(k \log k)$.*

Proof. We start by counting the number of visited positions in R, and distinguish between the two cases when $x \neq y$ and when $x = y$:

- Case 1: $x \neq y$
  Assume that on the initial substring $T[\ell \mathbin{..} r]$, there are no modified letters in R, and as the iteration over ℓ and r continues, all k modified letters are moved into R. In this situation, the number of visited positions in R is k. Additional visited positions are added to R when a loser letter y becomes a winner, this way r can be extended to the right without using more than k modified letters.
  Suppose that when r is increased to the next problematic position, y becomes a winner of its column. Let a be the number of problematic columns in R ($a \leq k$), then k/a additional modified letters can be visited in R. Now, there are at most $a - 1$ problematic columns in R. The next time it happens, at most $k/(a - 1)$ modified letters that can be visited are added to R, and so on. Thus, the maximal number of problematic positions in R is equal to $\Sigma_{i=1}^{a} k/i$, which gives a total of $O(k \log k)$.
- Case 2: $x = y$
  If y becomes a winner in its column, it does not necessarily mean that it will continue to win throughout the iteration, as ℓ position is increased. This situation is handled in the initialization step (see case 2), and the column is checked until the number of y occurrences is greater than $|z|$. As mentioned above, the total number of such visited positions cannot exceed $O(k)$, as it is bounded by the number of modified letters used in M.

Thus, the total number of visited positions in R cannot exceed $O(k \log k)$. In a similar way, it is easy to show that the total number of visited positions in L cannot exceed $O(k \log k)$.

5.4   Total Time Complexity Analysis

The difference between this algorithm and the previous $O(nk^3 \log \frac{n}{k})$ algorithm is in the number of visited positions of the main procedure. For each L and R possible substring definition, the algorithm visits at most $O(k \log k)$ positions, and updates the column lists of L and R in $O(k \log k)$ time. There are at most $O(k)$ such combinations of L and R, which gives a $O(k^2 \log k)$ time complexity for this step. In addition, moving between zones takes $O(k)$ time, which does not add to the total time. Thus, the total time complexity of the main procedure of the algorithm is $O(k^2 \log k)$. This gives a total time complexity $O(nk^2 \log k \log \frac{n}{k})$ for the entire algorithm.

References

1. Crochemore, M.: An optimal algorithm for computing the repetitions in a word. Inf. Process. Lett. 12, 244–250 (1981)
2. Apostolico, A., Preparata, F.P.: Optimal off-line detection of repetitions in a string. Theor. Comput. Sci. 22, 297–315 (1983)
3. Main, M.G., Lorentz, R.J.: An o(n log n) algorithm for finding all repetitions in a string. J. Algorithms 5, 422–432 (1984)
4. Kosaraju, S.R.: Computation of squares in a string (preliminary version). In: Crochemore, M., Gusfield, D. (eds.) CPM 1994. LNCS, vol. 807, pp. 146–150. Springer, Heidelberg (1994)
5. Gusfield, D., Stoye, J.: Linear time algorithms for finding and representing all the tandem repeats in a string. J. Comput. Syst. Sci. 69, 525–546 (2004)
6. Iliopoulos, C.S., Moore, D., Smyth, W.F.: A characterization of the squares in a fibonacci string. Theor. Comput. Sci. 172, 281–291 (1997)
7. Kolpakov, R.M., Kucherov, G.: Finding maximal repetitions in a word in linear time. In: Foundations of Computer Science, pp. 596–604 (1999)
8. Crochemore, M., Hancart, C., Lecroq, T.: Algorithms on Strings, 392 pages. Cambridge University Press (2007)
9. Main, M.G.: Detecting leftmost maximal periodicities. Discrete Applied Mathematics 25, 145–153 (1989)
10. Crochemore, M.: Recherche linéaire d'un carré dans un mot. C. R. Acad. Sc. Paris Sér. I Math. 296, 781–784 (1983)
11. Crochemore, M., Ilie, L.: Computing longest previous factors in linear time and applications. Information Processing Letters 106, 75–80 (2008), doi:10.1016/j.ipl.2007.10.006
12. Landau, G.M., Schmidt, J.P., Sokol, D.: An algorithm for approximate tandem repeats. Journal of Computational Biology 8, 1–18 (2001)
13. Sim, J.S., Iliopoulos, C.S., Park, K., Smyth, W.F.: Approximate periods of strings. In: Crochemore, M., Paterson, M. (eds.) CPM 1999. LNCS, vol. 1645, pp. 123–133. Springer, Heidelberg (1999)
14. Kolpakov, R.M., Kucherov, G.: Finding approximate repetitions under Hamming distance. Theor. Comput. Sci. 1, 135–156 (2003)
15. Amir, A., Eisenberg, E., Levy, A.: Approximate periodicity. In: Cheong, O., Chwa, K.-Y., Park, K. (eds.) ISAAC 2010, Part I. LNCS, vol. 6506, pp. 25–36. Springer, Heidelberg (2010)
16. Parikh, R.J.: On context-free languages. Journal of the ACM 13, 570–581 (1966)
17. Landau, G.M., Vishkin, U.: Fast parallel and serial approximate string matching. Journal of Algorithms 10, 157–169 (1989)
18. Galil, Z., Giancarlo, R.: Improved string matching with k mismatches. SIGACT News 17, 52–54 (1986)

On Minimal and Maximal Suffixes of a Substring

Maxim Babenko[1,2,3], Ignat Kolesnichenko[2,3], and Tatiana Starikovskaya[3]

[1] Higher School of Economics, Moscow, Russia
[2] Yandex LCC, Moscow, Russia
[3] Moscow State University, Moscow, Russia

Abstract. Lexicographically minimal and lexicographically maximal suffixes of a string are fundamental notions of stringology. It is well known that the lexicographically minimal and maximal suffixes of a given string S can be computed in linear time and space by constructing a suffix tree or a suffix array of S. Here we consider the case when S is a substring of another string T of length n. We propose two linear-space data structures for T which allow to compute the minimal suffix of S in $O(\log^{1+\varepsilon} n)$ time (for any fixed $\varepsilon > 0$) and the maximal suffix of S in $O(\log n)$ time. Both data structures take $O(n)$ time to construct.

1 Introduction

Non-empty lexicographically minimal and lexicographically maximal suffixes of a string are fundamental notions of stringology. Given a string S, a straightforward way to compute its minimal non-empty and maximal suffixes involves constructing the suffix tree or the suffix array for S. (Both of the latter capture the lexicographic order of all suffixes of S, see [7].) This way, the problem can be solved in an optimal linear time.

Now suppose that S is a substring extracted from a longer text T of length n. Then the information about T can be used to speed up the computation of the desired suffixes of S. For example, Duval [6] showed that the minimal non-empty and the maximal suffixes of *all prefixes* of T can be found in $O(n)$ time and space. (Note that computing the answers for each prefix of T separately would take $O(n^2)$ time.) Duval's result was later generalized to parallel machines by Apostolico and Crochemore [2]. It was shown that the minimal non-empty and the maximal suffixes of all prefixes of T can be computed on a CRCW PRAM with n processors in $O(\log n)$ time and linear space. In both cases the minimal non-empty and the maximal suffixes are found as a by-product of Lyndon factorization.

This paper focuses on the RAM model and concerns the problems of computing the minimal non-empty and the maximal suffixes of an arbitrary substring of a given string. Namely, let T be a string of length n. For any substring $T[i..j]$ of T starting in position i and ending in position j, we consider the following problems:

1. MINSUF: find the minimal non-empty suffix of $T[i..j]$;
2. MAXSUF: find the maximal suffix of $T[i..j]$.

J. Fischer and P. Sanders (Eds.): CPM 2013, LNCS 7922, pp. 28–37, 2013.
© Springer-Verlag Berlin Heidelberg 2013

For any given $\varepsilon > 0$ we propose a data structure that solves MinSuf within $O(\log^{1+\varepsilon} n)$ time per query. We also present a data structure to solve MaxSuf in $O(\log(j - i + 1)) = O(\log n)$ time per query. Both data structures involve linear preprocessing and occupy linear space.

Our results align nicely with a number of related substring problems that were earlier studied in the literature. In particular, Crochemore et al [4] and Karhumäki et al [8] concern the problem of computing primitive periods of substrings of a given string. Kociumaka et al [10] focus on computing all periods of substrings of a given string.

The paper is organized as follows. Section 2 gives a formal background and introduces some basic notation and definitions. Section 3 presents the data structure for solving MinSuf, which is conceptually simpler. Section 4 addresses MaxSuf. Finally Section 5 discusses the relation between MinSuf and MaxSuf.

2 Preliminaries

We start by introducing some standard notation and definitions. Let Σ be a finite ordered non-empty set (called an *alphabet*). The elements of Σ are *letters*.

A finite ordered sequence of letters (possibly empty) is called a *string*. Letters in a string are numbered starting from 1, that is, a string T of *length* k consists of letters $T[1], T[2], \ldots, T[k]$. The length k of T is denoted by $|T|$. For $i \leq j$, $T[i..j]$ denotes the *substring* of T from position i to position j (inclusively). If $i > j$, $T[i..j]$ is defined to be the empty string. Also, if $i = 1$ or $j = |T|$ then we omit these indices and we write just $T[..j]$ and $T[i..]$. Substring $T[..j]$ is called a *prefix* of T, and $T[i..]$ is called a *suffix* of T.

We assume the standard RAM model of computation [1]. Letters are treated as integers in range $\{1, \ldots, |\Sigma|\}$; a pair of letters can be compared in $O(1)$ time. This *lexicographic* order on Σ is linear and can be extended in a standard way to the set of strings in Σ. Namely, $T_1 \prec T_2$ if either (i) T_1 is a prefix of T_2; or (ii) there exists $0 \leq i < \min(|T_1|, |T_2|)$ such that $T_1[..i] = T_2[..i]$, and $T_1[i + 1] < T_2[i + 1]$.

Let Suf be the set of all suffixes of a string T. The *suffix array* SA of a string T is a permutation on $\{1, \ldots, |T|\}$ defining the lexicographic order on Suf. More precisely, $SA[r] = i$ iff the rank of $T[i..]$ in the lexicographic order on Suf is r. The inverse permutation is denoted by ISA; it reduces lexicographic comparison of suffixes $T[i..]$ and $T[j..]$ to integer comparison of their ranks $ISA[i]$ and $ISA[j]$. For a string T, both SA and ISA occupy linear space and can be constructed in linear time (see [11] for a survey).

A string T is called *periodic with period* β if $T = \beta^s \beta'$ for an integer $s \geq 1$ and a (possibly empty) prefix β' of β. When this leads to no confusion the length of β will also be called a period of T.

A *border* of a string T is a string that is both a prefix and a suffix of T and differs from T. A string T that has no non-empty border is called *border-free*. Borders and periods are dual notions; namely, if T has period β then it has a border of length $|T| - |\beta|$, and vice versa (see, e.g., [5]).

3   Computing Minimal Suffix

Consider a string T of length n. As a warm-up, in this section we show how to preprocess T so that given its substring $T[i..j]$ we can compute the lexicographically minimal non-empty suffix of $T[i..j]$ efficiently.

First, build the suffix array SA of T. As indicated in Section 2, SA occupies $O(n)$ memory and can be built in $O(n)$ time. Then compute ISA array by inverting SA. We preprocess ISA so as to answer *range minimum* queries over it in constant time. The answer to a range minimum query on $ISA[i..j]$ is the lexicographically minimal suffix among suffixes starting between positions i through j (inclusively). The preprocessing takes linear time and space, see e.g. [3].

Using the range minimum data structure on ISA, we can find the lexicographically minimal suffix $T[m..]$, $m \in [i, j]$, among $Suf[i, j] := \{T[i..], T[i + 1..], \ldots, T[j..]\}$ in $O(1)$ time. Let $T[\mu..j]$ be the requested lexicographically minimal suffix of $T[i..j]$.

Lemma 1. *If $T[m..j]$ is border-free then $T[\mu..j] = T[m..j]$. Otherwise $T[\mu..j]$ is the shortest non-empty border of $T[m..j]$.*

Proof. We first show that $T[\mu..j]$ is both a prefix and a suffix of $T[m..j]$. If $T[m..j] = T[\mu..j]$ then we are done otherwise $T[\mu..j] \prec T[m..j]$. By the definition of the lexicographic order, either (1) $T[\mu..j]$ is a prefix of $T[m..j]$, or (2) there exists $\ell < \min(|T[\mu..j]|, |T[m..j]|)$ such that $T[\mu..\mu + \ell] = T[m..m + \ell]$, and $T[\mu + \ell + 1] < T[m + \ell + 1]$.

In Case (1) we have $m < \mu$ and thus $T[\mu..j]$ is a suffix of $T[m..j]$ as well. Let us show that Case (2) is impossible. Indeed, it follows that $T[\mu..] \prec T[m..]$, but the lexicographically smallest suffix in $Suf[i, j]$ is $T[m..]$.

Hence $T[\mu..j]$ is both a prefix and a suffix of $T[m..j]$. If $T[m..j]$ is border-free then $\mu = m$. Otherwise $T[\mu..j]$ is a border of $T[m..j]$. Suppose $T[\mu..j]$ is not the shortest non-empty border, then there exists a shorter border β of $T[m..j]$. By definition β is a prefix of $T[m..j]$ and thus is also a prefix of $T[\mu..j]$. Therefore $\beta \prec T[\mu..j]$, which is a contradiction.   □

Summing up these observations, to compute the lexicographically minimal suffix of $T[i..j]$ it suffices to find the shortest border of $T[m..j]$ or, equivalently, to find the longest period of $T[m..j]$. As shown in [10], for any fixed $\varepsilon > 0$, T can be turned into a data structure of size $O(n)$ capable of computing, for given (m, j), all periods of $T[m..j]$ in $O(\log^{1+\varepsilon} n)$ time. These periods are reported as a set of $O(\log n)$ arithmetic progressions. We scan over these progressions and find the shortest border of $T[m..j]$.

We conclude:

Theorem 1. *Given a string T of length n and fixed $\varepsilon > 0$, one can construct in $O(n)$ time a data structure of size $O(n)$ that enables finding the lexicographically minimal suffix of any substring of T in $O(\log^{1+\varepsilon} n)$ time.*

4   Computing Maximal Suffix

Now we switch to the problem of computing the lexicographically maximal suffix of a substring. Let $T[\mu..j]$ be the desired lexicographically maximal suffix of $T[i..j]$. As earlier, let $Suf[i,j] := \{T[i..], T[i+1..], \ldots, T[j..]\}$.

4.1   Naive Algorithm

The following simple observation is crucial:

Lemma 2. *Let $P = T[m..j]$ be a prefix of $T[\mu..j]$. If there are no suffixes in $Suf[i, m-1]$ starting with P; then $m = \mu$. Otherwise, let $T[m_1..]$ be the maximal suffix in $Suf[i, m-1]$ among those starting with P; then $P_1 = T[m_1..j]$ is a another prefix of $T[\mu..j]$ obeying $|P_1| > |P|$.*

Proof. Suppose that no suffix in $Suf[i, m-1]$ starts with P. Then $\mu \notin [i, m-1]$ (as $T[\mu..]$ does start with P) so $\mu \geq m$. Also $\mu \leq m$ since $|T[\mu..j]| \geq |P|$. Hence $\mu = m$ and we are done.

Now let $T[m_1..]$ be the lexicographically maximal suffix in $Suf[i, m-1]$ among those starting with P. If $m_1 = \mu$ then we are done. Otherwise $T[m_1..j] \prec T[\mu..j]$ by the definition of μ. Suppose that $P_1 = T[m_1..j]$ is not a prefix of $T[\mu..j]$. Then $T[m_1..m_1 + \ell] = T[\mu..\mu + \ell]$ and $T[m_1 + \ell + 1] < T[\mu + \ell + 1]$ for some ℓ with $|P| \leq \ell < j - \mu + 1$. Therefore $T[m_1..] \prec T[\mu..]$ and $T[m_1..]$ is not the lexicographically largest suffix in $Suf[i, m-1]$ starting with P, which is a contradiction. □

The above lemma leads to the following procedure for computing the lexicographically maximal suffix of $T[i..j]$. We maintain a certain prefix $P = T[m..j]$ of $T[\mu..j]$ (initially $m = j + 1$ so P is empty) and execute a series of iterations. On each iteration we compute the lexicographically maximal suffix $T[m_1..]$ in $Suf[i, m - 1]$ among those starting with P. We apply Lemma 2, reset $m := m_1$ and proceed to the next iteration. We call this transition a *jump from m to m_1*. The iterations terminate when no suffixes in $Suf[i, m - 1]$ starting with P remain; then the algorithm stops with $P = T[\mu..j]$.

Two issues remain open. First, we need an efficient way to compute the lexicographically maximal suffix in $Suf[i, m - 1]$ among those starting with a given prefix P. Second, the number of jumps in the above method can be linear. Indeed, for $T = a^{n-1}b$ and its substring $T[1..n - 1]$ the answer is a^{n-1}. However, each iteration increases the length of P by one. We will address these concerns in the upcoming subsections.

4.2   Data Structures

The following data structures are crucial to the success of our approach:

Suffix Array, *LCP*, and Maximum Ranks: We start by constructing the suffix array SA of T. As indicated in Section 2, this suffix array uses $O(n)$

space and can be built in $O(n)$ time. Then SA is inverted and gives rise to ISA array. We preprocess ISA so as to answer *range maximum* queries over it in constant time. The answer to a range maximum query on $ISA[i..j]$ is the lexicographically maximal suffix among suffixes starting between positions i through j. The preprocessing takes linear time and space, see e.g. [3]. Next, we construct the LCP array of length $(n-1)$, where $LCP[i]$ is equal to the length of the longest common prefix of suffixes $T[SA[i]..]$ and $T[SA[i+1]..]$. As shown in [9], LCP can be built in linear time as well. Finally, we build a *range minimum* query data structure on top of LCP (again using the construction from [3]). This enables us to find, for every pair of suffixes $T[i..]$ and $T[j..]$, the length of their longest common prefix (denoted by $lcp(i,j)$) in $O(1)$ time.

Suffix Array for Reversed Text and LCS: Similar to the above, we construct the suffix array SA_r for the reversed string $T_r = T[n]\,T[n-1]\,\ldots\,T[1]$ and construct the LCS array of length $(n-1)$, where $LCS[i]$ is equal to the length of the longest common prefix of suffixes $T_r[SA_r[i]..]$ and $T_r[SA_r[i+1]..]$ (i.e. the length of the longest common suffix of $T[..n - SA_r[i] + 1]$ and $T[..n - SA_r[i+1]+1]$). As above, we build a *range minimum* query data structure on top of LCS. This enables computing, for every pair of prefixes $T[..i]$ and $T[..j]$, the length of their longest common suffix (denoted by $lcs(i,j)$) in $O(1)$ time.

4.3   Improved Algorithm

To achieve the desired running time per query, we improve the above naive method as follows. As earlier, we maintain a certain current prefix $P = T[m..j]$ of $T[\mu..j]$. Instead of gradually jumping according to Lemma 2 we perform a series of improved iterations. At each such iteration the algorithm either finds out that $P = T[\mu..j]$ (in which case it stops) or replaces m by m' such that $P' = T[m'..j]$ is another prefix of $T[\mu..j]$ obeying $|P'| \geq \frac{3}{2}|P|$. As we will see, each iteration takes $O(1)$ time.

The following invariants are maintained throughout the algorithm:

(1) (a) $P = T[m..j]$ is a prefix of $T[\mu..j]$;
     (b) Every suffix in $Suf[i, m-1]$ that starts with $P = T[m..j]$ is less than $T[m..]$.

We will describe the structure of the algorithm postponing some technical proofs until the end of this section.

Startup: Due to invariant (1,b) we cannot start iterations with the empty prefix. Instead we compute the lexicographically largest suffix $T[m..]$ in $Suf[i,j]$. This is done by performing a single range maximum query on ISA array in $O(1)$ time.

Next we explain how an iteration works.

Iteration: Initial Jump: The iteration starts by performing a usual jump from m in accordance to Lemma 2. To this aim, let $T[m_1..]$ be the lexicographically largest suffix among $Suf[i, m-1]$. Such a suffix can be found by executing a range maximum query for ISA array. A somewhat unexpected fact is that m_1 defines the jump destination for m:

Lemma 3. *If $T[m_1..]$ starts with P then $T[m_1..]$ is the lexicographically largest suffix in $Suf[i, m-1]$ among those starting with P. Otherwise there is no suffix in $Suf[i, m-1]$ that starts with P.*

Checking if $T[m_1..]$ indeed starts with P amounts to validating if $lcp(m_1, m) \geq |P|$. If the latter is false then no suffix in $Suf[i, m-1]$ starts with P, so by Lemma 2 the algorithm terminates with $T[m..j]$ being the answer.

Now let us assume that $T[m_1..]$ is the desired lexicographically maximal suffix in $Suf[i, m-1]$ starting with P. From Lemma 2 it follows that invariant (1,a) holds for $m := m_1$, while invariant (1,b) holds for $m := m_1$ by construction. If $|T[m_1..j]| \geq \frac{3}{2}|T[m..j]|$, the iteration completes with $m' := m_1$.

Otherwise the initial jump was a short one. Note that P occurs both at positions m_1 and m and the distance between these two occurrences is $m - m_1 < \frac{1}{2}|P|$. This large overlap implies a certain periodicity:

Lemma 4. *$\beta := T[m_1..m-1]$ is the shortest period of $T[m..j]$, i.e. $T[m..j] = \beta^s \beta'$ where $s \geq 1$ and β' is a prefix of β; also β is the shortest string with such a property.*

Iteration: Fast-Forward: Taking the above periodicity into account the jump procedure can be refined as follows:

Lemma 5. *Let $T[m..j] = \beta^s \beta'$, where $s \geq 1$, β' is a prefix of β, and β is the shortest period. Assume that the jump from m leads to m_1 and $T[m_1..m-1] = \beta$. Suppose there is an additional match of β to the left of $T[m_1..j]$ inside $T[i..j]$, i.e. $T[m_1 - |\beta|..m_1 - 1] = \beta$ and $m_1 - |\beta| \geq i$. Then the jump from m_1 leads to $m_1 - |\beta|$.*

According to Lemma 5, if a copy of β exists to the left of $T[m_1..j]$, then jumping from m_1 one gets into $m_1 - |\beta|$, etc. This process can be fast-forwarded; namely, let us match as many copies of β to the left of $T[m_1..]$ (while remaining inside $T[i..j]$) as possible, i.e. find the minimum $m_2 \in [i, m_1]$ such that $T[m_2..m_1 - 1] = \beta^t$ for some $t \geq 0$. Then from repeated application of Lemma 5 it follows that the sequence of jumps ultimately leads from m_1 to $m_1 - t|\beta|$.

Recall that for positions p, p' in T we write $lcs(p, p')$ to denote the longest common suffix of prefixes $T[..p]$ and $T[..p']$. Then $t := \lfloor \min(lcs(m_1 - |\beta| - 1, m_1 - 1), m_1 - i)/|\beta| \rfloor$. Since $lcs(p, p')$ can be computed in constant time by the range minimum data structure, values t and $m_2 := m_1 - t|\beta|$ can be found in constant time.

Iteration: Final Jump: To finish the iteration we perform a jump from the position m_2 to a position m_3 (applying Lemma 3 for $m := m_2$, $P := T[m_2..j]$ and making a range maximum query to ISA array). If there is no matching suffix, the algorithm reports $T[m_2..]$ as the lexicographically maximal suffix and stops. Otherwise we claim that total increase of prefix length is sufficiently large, i.e. $|T[m_3..j]| > \frac{3}{2}|T[m..j]|$. Indeed, consider the first $(s + t + 1)|\beta|$ letters of $T[m_2..]$ and $T[m_3..]$; both of these two substrings are equal to β^{s+t+1}. Since there is no match of β to the left of $T[m_2..j]$ these substrings must have an overlap of length less than $|\beta|$, for otherwise β has a non-trivial occurrence in β^2 and thus β is not the shortest period of $T[m..j]$ (see Lemma 3.2.1 in [7]). Therefore $m_3 < m_2 - (s+t)|\beta| \le m_2 - s|\beta| \le m_2 - \frac{1}{2}|T[m..j]|$, hence $|T[m_3..j]| > \frac{3}{2}|T[m..j]|$, as required.

This completes the description of the algorithm. Summing up, we obtain the following

Theorem 2. *Given a string T of length n, one can construct in $O(n)$ time a data structure of size $O(n)$ that enables computing the lexicographically maximal suffix of any substring $T[i..j]$ of T in $O(\log(j - i + 1)) = O(\log n)$ time.*

Proof. At each iteration of the algorithm $|T[m..j]|$ increases by at least a factor of $\frac{3}{2}$. Therefore, the number of iterations is $O(\log(j - i + 1))$. Since each of them takes $O(1)$ time, each request in answered in $O(\log(j - i + 1))$ time in total. The linearity of needed time and space follows immediately from the description.   □

4.4   Proof of Lemma 3

Define $\ell := j - m + 1$ and consider the substring $Q := T[m_1..m_1 + \ell - 1]$ of length ℓ. We claim that $Q \preceq P$. Indeed, if $Q \succ P$ then since $T[\mu..j]$ starts with P (by invariant (1,a)) it follows that $T[m_1..j] \succ T[\mu..j]$, which is impossible as $T[\mu..j]$ is the maximal suffix of $T[i..j]$.

If $Q = P$ then $T[m_1..]$ starts with P and is the lexicographically largest suffix among all in $Suf[i, m - 1]$ (even those not starting with P), so the jump from m leads to m_1.

Finally let $Q \prec P$. Then no suffix in $Suf[i, m-1]$ can start with P for otherwise such a suffix would be larger than $T[m_1..]$.   □

4.5   Proof of Lemma 4

We rely on the following well-known facts:

Lemma 6 (see Lemma 3.2.3 in [7]). *Given strings α and β, assume that α occurs in β at positions p and p', $p < p'$, $0 < p' - p < |\alpha|/2$. Then α is periodic with period $\beta[p..p' - 1]$.*

Lemma 7 (see Lemma 3.2.4 in [7]). *If k and k' are both periods of a string α obeying $k + k' \le |\alpha|$ then $\gcd(k, k')$ is also a period of α.*

Since $m - m_1 = |T[m_1..j]| - |T[m..j]| < \frac{1}{2}|T[m..j]|$, Lemma 6 implies (for $\alpha := T[m..j]$ and $\beta := T$) that $\beta = T[m_1..m - 1]$ is a period of $T[m..j]$. It remains to prove that the latter period is the shortest possible.

Observe the following:

Lemma 8. *If $\alpha \succ \beta^k \alpha$ for some strings α, β and integer $k \geq 1$, then $\beta^{k-1}\alpha \succ \beta^k \alpha$.*

Proof. Suppose that $\alpha \prec \beta\alpha$. Then prepending both parts of the latter inequality by multiples of β gives $\beta\alpha \prec \beta^2\alpha$, $\beta^2\alpha \prec \beta^3\alpha$, ..., $\beta^{k-1}\alpha \prec \beta^k\alpha$. By transitivity this implies $\alpha \prec \beta^k\alpha$, which is a contradiction. Therefore $\alpha \succ \beta\alpha$ and consequently $\beta^{k-1}\alpha \succ \beta^k\alpha$.   □

Now we complete the proof of Lemma 4. Let γ be the shortest period of $T[m..j]$. Suppose $|\gamma| < |\beta|$. Then $|\gamma| + |\beta| < 2|\beta| \leq |T[m..j]|$, and by Lemma 7 substring $T[m..j]$ has another period $gcd(|\gamma|, |\beta|)$. Since γ is the shortest period, $|\beta|$ must be a multiple of $|\gamma|$, i.e., $\beta = \gamma^k$ for some $k \geq 2$.

By invariant (1,b) we have $T[m..] \succ T[m_1..] = \gamma^k T[m..]$. Now from Lemma 8 it follows that $T[m_1 + |\gamma|..] = \gamma^{k-1}T[m..] \succ \gamma^k T[m..] = T[m_1..]$. Therefore $T[m_1 + |\gamma|..]$, which starts with $T[m..j]$, is greater than $T[m_1..]$. This contradicts invariant (1,b).   □

4.6   Proof of Lemma 5

Clearly $T[m_1 - |\beta|..] = \beta^{s+2}\beta'$ starts with $T[m_1..j] = \beta^{s+1}\beta'$. It remains to prove that this is the lexicographically largest suffix in $Suf[i, m_1 - 1]$ with the given prefix. Assume the contrary, i.e. $T[x..]$ starts with $\beta^{s+1}\beta'$, $x \in [i, m_1 - 1]$ and $T[x..] \succ T[m_1 - |\beta|..]$. Note that $x \neq m_1 - |\beta|$ (otherwise one would get the equality). Also $x > m_1 - |\beta|$ is impossible (otherwise β would have a non-trivial occurrence within β^2, which is impossible due to its minimality). Therefore $x \in [i, m_1 - |\beta| - 1]$. Define $y := x + |\beta|$, notice that $y \in [i, m_1 - 1]$, and consider the suffix $T[y..]$. Since both $T[x..]$ and $T[m_1 - |\beta|..]$ start with $\beta^{s+1}\beta'$ and $T[x..] \succ T[m_1 - |\beta|..]$ it follows that both $T[y..]$ and $T[m_1..]$ start with $\beta^s\beta' = T[m..j]$ and $T[y..] \succ T[m_1..]$. The latter contradicts the choice of m_1.   □

5   Conclusions

We showed that a string T of length n can be preprocessed in linear time into a data structure that takes linear space and solves MINSUF within $O(\log^{1+\varepsilon} n)$ time per query and MAXSUF within $O(\log n)$ time per query.

One may wonder if the algorithm for MAXSUF presented in Subsection 4.3 generalizes to MINSUF. Indeed, suppose we are asked to compute the minimal suffix of $T[i..j]$. The first step would be to compute the minimal suffix $T[m..]$ among $Suf[i..j]$. Now suppose that $T[m..j]$ has no proper borders, then by Lemma 1 the latter substring is the answer. Otherwise let $T[k..j]$ be any (non necessarily minimal) non-empty border of $T[m..j]$. Then if $|T[k..j]| < \frac{1}{2}|T[m..j]|$ we can reset $m := k$ and continue (achieving a noticeable reduction of length). Otherwise $T[k..j]$ and $T[m..j]$ have a large overlap implying that $T[m..j]$ is periodic. Now we can reset $T[m..j]$ to its period and again achieve a significant length reduction.

A closer look, however, immediately reveals an obstacle. The above method is correct and solves MINSUF in $O(\log n)$ time per query but requires a subroutine for finding an arbitrary non-empty border of a given substring in $O(1)$ time. In case of MAXSUF we used simple range queries to perform such a lookup; namely we were looking for a maximal suffix in $Suf[i..m-1]$ among those starting with $T[m..j]$. This approach does not seem to apply to MINSUF. Indeed, what we need is a shorter string, so it is not clear what suffixes in $Suf[m+1..j]$ to consider. Also just picking the minimal suffix $T[k..]$ among $Suf[m+1..j]$ we may end up having a substring $T[k..j]$ that is not a border of $T[m..j]$.

Another (more conceptual) way of explaining the difference between MINSUF and MAXSUF if to say that in MAXSUF we are approximating the answer from below and extend the current candidate on each iteration. This helps us to guide the search by fixing a prefix of the new candidate. On the other hand, in MINSUF we are approximating the answer from above and have no obvious way of narrowing down the search.

Soon after submitting the paper to CPM 2013 we found a way to overcome the above issues. After a suitable linear-time preprocessing, the new algorithm solves MINSUF in $O(\log(j-i+1)) = O(\log n)$ time per query and does not rely on any sophisticated data structures for finding periodicities. This algorithm, however, is rather complicated so we decided to postpone it until the full version of the paper.

Acknowledgements. Tatiana Starikovskaya was partly supported by RFBR grant 10-01-93109-CNRS-a and Dynasty Foundation. Maxim Babenko was partly supported by RFBR grants 12-01-00864-a and 10-01-93109-CNRS-a. Ignat Kolesnichenko was partly supported by RFBR grant 10-01-93109-CNRS-a. Authors are thankful to anonymous reviewers for valuable comments and suggestions.

References

1. Aho, A.V., Ullman, J.D., Hopcroft, J.E.: The design and analysis of computer algorithms. Addison-Wesley, Reading (1974)
2. Apostolico, A., Crochemore, M.: Fast parallel Lyndon factorization with applications. Theory of Computing Systems 28, 89–108 (1995), 10.1007/BF01191471
3. Bender, M.A., Farach-Colton, M.: The LCA problem revisited. In: Gonnet, G.H., Panario, D., Viola, A. (eds.) LATIN 2000. LNCS, vol. 1776, pp. 88–94. Springer, Heidelberg (2000)
4. Crochemore, M., Iliopoulos, C., Kubica, M., Radoszewski, J., Rytter, W., Waleń, T.: Extracting powers and periods in a string from its runs structure. In: Chavez, E., Lonardi, S. (eds.) SPIRE 2010. LNCS, vol. 6393, pp. 258–269. Springer, Heidelberg (2010)
5. Crochemore, M., Rytter, W.: Text Algorithms. Oxford University Press (1994)
6. Duval, J.-P.: Factorizing words over an ordered alphabet. J. Algorithms 4(4), 363–381 (1983)
7. Gusfield, D.: Algorithms on strings, trees, and sequences: computer science and computational biology. Cambridge University Press, New York (1997)

8. Karhumäki, J., Lifshits, Y., Rytter, W.: Tiling Periodicity. Discrete Mathematics & Theoretical Computer Science 12, 237–248 (2010)
9. Kasai, T., Lee, G., Arimura, H., Arikawa, S., Park, K.: Linear-time longest-common-prefix computation in suffix arrays and its applications. In: Amir, A., Landau, G.M. (eds.) CPM 2001. LNCS, vol. 2089, pp. 181–192. Springer, Heidelberg (2001)
10. Kociumaka, T., Radoszewski, J., Rytter, W., Waleń, T.: Efficient data structures for the factor periodicity problem. In: Calderón-Benavides, L., González-Caro, C., Chávez, E., Ziviani, N. (eds.) SPIRE 2012. LNCS, vol. 7608, pp. 284–294. Springer, Heidelberg (2012)
11. Puglisi, S.J., Smyth, W.F., Turpin, A.H.: A taxonomy of suffix array construction algorithms. ACM Comput. Surv. 39(2), 4 (2007)

Converting SLP to LZ78 in almost Linear Time

Hideo Bannai[1], Paweł Gawrychowski[2],
Shunsuke Inenaga[1], and Masayuki Takeda[1]

[1] Department of Informatics, Kyushu University, Japan
{bannai,inenaga,takeda}@inf.kyushu-u.ac.jp
[2] Max-Planck-Institut für Informatik, Saarbrücken, Germany
gawry@cs.uni.wroc.pl

Abstract. Given a straight line program of size n, we are interested in constructing the LZ78 factorization of the corresponding text. We show how to perform such conversion in $\mathcal{O}(n + m \log m)$ time, where m is the number of LZ78 codewords. This improves on the previously known $\mathcal{O}(n\sqrt{N} + m \log N)$ solution [Bannai et al., SPIRE 2012]. The main tool in our algorithm is a data structure which allows us to efficiently operate on labels of the paths in a growing trie, and a certain method of recompressing the parse whenever it leads to decreasing its size.

1 Introduction

Large scale string data are commonly compressed before being stored or transmitted, in order to save storage and communication costs. Compressed string processing (CSP) is an approach for processing such data without explicitly decompressing them, therefore enabling us to process the data using less space *and* time. Efficient CSP algorithms have been proposed for exact pattern matching [7,10,17,18], and other string problems [8,6,9,22] and can outperform the straightforward decompress-then-process approach, both in theory and even in practice [8,21].

In this paper, we consider the following *re-compression* problem in the CSP setting. Given a string t represented as a Straight Line Program (SLP) [11], compute the LZ78 factorization [24] of t. An SLP is a context free grammar in Chomsky normal form, that derives a single string. Since outputs of various grammar based [12,19] and dictionary based [23,24] compression algorithms (including LZ78) can be considered as SLPs, or quickly transformed to an SLP [20], it is widely used as a model of compressed representations in CSP. The LZ78 compression algorithm compresses a given string based on a dynamic dictionary which is constructed by partitioning the input string, the process of which is called LZ78 factorization. Other than its obvious use for compression, the LZ78 factorization is an important concept used in various string processing algorithms and applications [4,14,15,16].

The significance of recompression can be seen in the following situation. Some CSP algorithms make use of properties of the compressed representation that are specific to a certain compression algorithm. In order to apply such CSP

J. Fischer and P. Sanders (Eds.): CPM 2013, LNCS 7922, pp. 38–49, 2013.

algorithms to a compressed representation that was generated by a different compression algorithm, we must somehow convert the given compressed representation of a string to another compressed representation that would have been generated by the specific compression algorithm given the uncompressed string. Furthermore, for the CSP approach to be meaningful, this conversion must be done without explicitly reconstructing the uncompressed string. Other examples of application of recompression are dynamic updates of compressed strings and efficient computation of normalized compression distance (NCD) [13] via the CSP approach (see [2] for details).

The contribution of this paper is as follows. Given an SLP of size n that represents a string t of length N, we present an $\mathcal{O}(n + m \log m)$ time and space algorithm for computing the LZ78 factorization of t, where m is the number of LZ78 factors. Then we improve the space complexity to linear in $n + m$ at the cost of increasing the time complexity to $\mathcal{O}((n + m) \log m)$.

This improves on the previous $\mathcal{O}(n\sqrt{N} + m \log N)$ time and $\mathcal{O}(n\sqrt{N} + m)$ space solution [2]. Since $m = \Omega(\sqrt{N})$, the second term is asymptotically $m \log m = O(m \log N)$ and differs only by a constant factor. However, the first term can be significantly smaller, since N can be as large as 2^{n-1}. The main tool in our solution is a data structure which allows us to efficiently operate on labels of the paths in a growing trie, see Theorem 1, which we believe to be of independent interest. Additionally, we apply a certain method of recompressing the parse whenever it leads to decreasing size. A similar idea was previously applied to solve the fully LZ78-compressed pattern matching problem [5,7], but here we need to find the means to work with both the SLP and LZ78 parse at the same time.

We start with some preliminaries, including presenting the tools that we are using, in Section 2. Then in Section 3 we develop a data structure which allows us to maintain a growing trie so that we can quickly lexicographically compare suffixes of any two paths starting at the root. In Section 4 we use this structure to construct a faster recompression algorithm.

2   Preliminaries

2.1   Strings

Let Σ be a finite *alphabet*. An element of Σ^* is called a *string*. The length of a string t is denoted by $|t|$. For a string $t = xyz$, x, y and z are called a *prefix*, *substring*, and *suffix* of t, respectively. The i-th letter of a string t is denoted by $t[i]$ for $1 \leq i \leq |t|$, and the substring of a string t that begins at position i and ends at position j is denoted by $t[i..j]$ for $1 \leq i \leq j \leq |t|$. For convenience, let $t[i..j] = \varepsilon$ if $j < i$.

Our model of computation is the word RAM: we shall assume that the word size is at least $\lceil \log_2 |t| \rceil$, and hence operations on values representing lengths and positions of string t can be manipulated in constant time. Space complexities will be determined by the number of computer words (not bits).

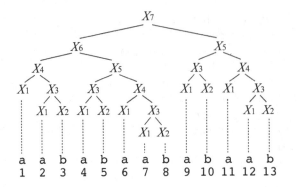

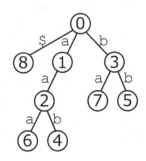

Fig. 1. The derivation tree of SLP $\mathcal{S} = \{X_1 \to$ a,
$X_2 \to$ b, $X_3 \to X_1 X_2$, $X_4 \to X_1 X_3$, $X_5 \to X_3 X_4$,
$X_6 \to X_4 X_5$, $X_7 \to X_6 X_5\}$ representing string
aababaababaab

Fig. 2. LZ78 trie for string
aaabaabbbaaaba\$. Node i represents f_i, e.g., $f_4 =$ aab.

2.2 Straight Line Programs

A *straight line program* (*SLP*) is a set of productions $\mathcal{T} = \{X_1 \to expr_1, X_2 \to expr_2, \dots, X_n \to expr_n\}$, where each X_i is a distinct non-terminal variable and each $expr_i$ is an expression that can be either $expr_i = a$ $(a \in \Sigma)$, or $expr_i = X_{\ell(i)} X_{r(i)}$ $(i > \ell(i), r(i))$. An SLP is essentially a context free grammar in the Chomsky normal form, that derives a single string. The *size* of the program $\mathcal{T}$ is the number n of assignments in $\mathcal{T}$.

The derivation tree of SLP $\mathcal{T}$ is a labeled ordered binary tree where each internal node is labeled with a non-terminal variable in $\{X_1, \dots, X_n\}$, and each leaf is labeled with a terminal letter in Σ. The root node has label X_n. Fig. 1 shows an example of such derivation tree.

2.3 LZ78 Parsing

Definition 1 (LZ78 parse). *The LZ78 parse of a string t is a partition of t into substrings $t = f_1 \cdots f_m$ of t, where (for convenience) $f_0 = \varepsilon$, and each LZ78 codeword $f_i \in \Sigma^+$ is defined as $f_i = t[p : p + |f_j|]$ where $p = |f_0 \cdots f_{i-1}| + 1$ and $f_j (0 \le j < i)$ is the longest previously used codeword which is a prefix of $t[p..|t|]$.*

The LZ78 parse can be encoded by a sequence of pairs, where the pair for f_i consists of the identifier j of the previous substring f_j and the new letter $t[|f_1 \cdots f_i|]$. Regarding this pair as a parent and edge label, the substrings can be represented as a trie, see Fig. 2.

We will need to operate on the trie representing the LZ78 codewords efficiently. For this we will apply two relatively well-known tools.

Lemma 1 (see [1]). *We can maintain a tree in linear space so that addition of a leaf and a level ancestor query are performed in worst-case $\mathcal{O}(1)$ time, where a level ancestor query gives us the k-th ancestor of a given node v.*

Lemma 2 (see [3]). *We can maintain a tree under adding new leaves and inserting new nodes on edges so that both updates and lowest common ancestor queries are performed in worst-case $\mathcal{O}(1)$ time.*

3   Dynamic LZ78 Trie for Constant-Time Chunk Comparison

Definition 2. *A chunk is a suffix of a codeword. We represent such suffix as a pair consisting of a node of the LZ78 trie and a number denoting the length.*

Given a chunk, using Lemma 1 we can either access any of its letters, or construct a chunk corresponding to any of its substrings. The goal of this section is to show how to maintain a data structure which allows lexicographical comparison of any two chunks for a growing trie. We will implement the comparison in $\mathcal{O}(1)$ time at the expense of adding new leaves in $\mathcal{O}(\log m)$ time. To gain some intuition, we begin with a static solution for the case when we are given the whole trie in the very beginning. Some of the bounds mentioned below are amortized, which is good enough for our purposes.

Lemma 3. *Given a static trie on m nodes, we can build in $\mathcal{O}(m \log m)$ time a structure of size $\mathcal{O}(m \log m)$ which allows lexicographical comparison of any two chunks in $\mathcal{O}(1)$ time.*

Proof. It is enough to show that it is possible to compare any two chunks whose length is a power of 2 in such complexity. Observe that for a fixed k, there are just m chunks of length 2^k (because for each node of the trie we get at most one chunk ending there). Hence we can afford to assign numbers to all of them in such a way that lexicographical comparisons can be performed by simply looking at the numbers. Observe that computing those numbers requires sorting, but the sorting can be actually implemented using radix sort, if we compute the numbers for increasing values of k. □

If the trie is dynamic, it is still enough to consider only chunks whose length is a power of 2, but now new chunks can arrive, thus destroying the numbering we have so far. We will use a different approach consisting of a number of steps building on each other. At a high level, we split all chunks into different types depending on their length and the value of $\log m$ (or, more precisely, the rounded down value of $\log m$) and deal with different types separately. The value of $\log m$ can (and will) actually change during the computation. To avoid the messy details, we apply the following reasoning: whenever the value of m doubles, we recompute all structures constructed so far with the new rounded down value of $\log m$. As the update time for a single codeword will be just $\mathcal{O}(\log m)$, the total construction time will be at most $\sum_i \mathcal{O}(\frac{m}{2^i} \log \frac{m}{2^i}) = \mathcal{O}(m \log m)$, where m is the final number of codewords, so $\mathcal{O}(\log m)$ amortized per codeword. Now our approach is as follows.

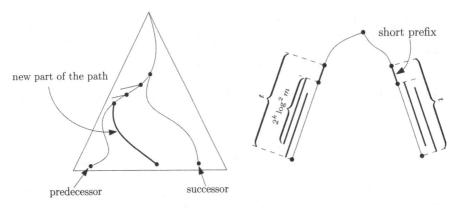

Fig. 3. Locating the existing prefix of the path   **Fig. 4.** Covering chunks with fragments of length $2^k \log^2 m$

1. We will show a structure which can be used to compute the longest common **suffix** of any two chunks.
2. We will construct a structure allowing us to compute the longest common **prefix** of any two chunks of length at most $\log m$, which we will call the *short chunks*.
3. We will extend the above structure so that it can be used to compute the longest common **prefix** of any two chunks of length at most $\log^2 m$ being a multiple of $\log m$.
4. We will choose and store in a compacted trie some of the longer chunks of lengths being powers of 2, which will allow computing the longest common **prefix** of any two of them. Updating this compacted trie will be costly, but we will make sure that it does not happen very often. More precisely, we will make sure that it happens at most twice for every addition, and takes just $\mathcal{O}(\log m)$ time.
5. Finally, we will show that even though we have chosen just some of the longer chunks, we can use the structure built for the short ones to fill in the missing information.

As soon as we know that the longest common prefix of two chunks is of length ℓ, we can access their $\ell + 1$-th letters and use them to determine their ordering.

Lemma 4. *We can maintain a structure of size $\mathcal{O}(m)$ for a growing trie so that each addition requires $\mathcal{O}(\log m)$ time and the longest common suffix of any two chunks can be computed $\mathcal{O}(1)$ time.*

Proof. It is enough to show that we are able to compute the longest common suffix of any two codewords. We store their reversals in a compacted trie. Together with a lowest common ancestor structure, this is enough to compute the longest common suffix in $\mathcal{O}(1)$ time. Adding a new codeword reduces to inserting one new path into the compacted trie, which requires locating its lowest already existing node. For this we maintain a lexicographically sorted list of all reversed codewords. Given a new codeword, we use binary search to locate its predecessor/successor on the sorted list, where each comparison reduces to computing the longest common suffix of two (already existing before the update) codewords. Then we compute the longest common prefix of the new reversed codeword and the predecessor/successor, which gives us the prefix of the path that already exists, see Fig. 3. Using yet another binary search on the sorted list we can retrieve the last already existing node on this prefix. Then we create the remaining part of the path, which requires either adding a new outgoing edge, or locating and splitting one of the already existing outgoing edges. If we store all outgoing edges in a balanced search tree, both can be done in $\mathcal{O}(\log m)$ time. Note that as new reversed codewords appear, the sorted list must be actually implemented as a (any) balanced search tree.   □

Lemma 5. *We can maintain a structure of size $\mathcal{O}(m \log m)$ for a growing trie so that each addition requires $\mathcal{O}(\log m)$ time and the longest common prefix of any two short chunks can be computed in $\mathcal{O}(1)$ time.*

Proof. We maintain a *short trie* containing all short chunks with a lowest common ancestor structure, which allows computing the longest common prefix for any two short chunks. Consider the total number of nodes in this trie: each edge of the original trie contributes at most $\log m$ to this number, hence the size of the short trie is at most $m \log m$. For each node in the original trie we store up to $\log m$ pointers to the nodes of the short trie corresponding to all short chunks ending there. In contrast to the previous proof, now whenever we add a new codeword, up to $\log m$ new short chunks must be inserted into the short trie (less if the depth of the new leaf is smaller than that, exactly $\log m$ otherwise, hence the situation is more complex. More precisely, adding a codeword ca, where c is an existing codeword and a is a letter, requires iterating through all sufficiently short (of length less than $\log m$) suffixes c' of c and inserting a as a child of the node of the short trie corresponding to c'. If the alphabet is of constant size, inserting a child takes constant time. If this is not the case, it could take as much as $\log m$ as we should check if such child already exists, and the total complexity would be $\mathcal{O}(\log^2 m)$, hence we need a slightly different approach.

Consider all sufficiently short suffixes $c[|c|..|c|], c[|c|-1..|c|], \ldots, c[|c|-\log m + 1..|c|]$ of c. If the node corresponding to $c[i..|c|]$ in the short trie has an outgoing edge labeled by a, so does the node corresponding to any shorter suffix. Hence we should identify the smallest such i, and then simply add a new outgoing edge labeled by a for each node corresponding to a longer suffix, which can be done in constant time per suffix, as there is no need to keep the children sorted. To identify this smallest i, we can apply binary search to locate the predecessor/successor of $(ca)^R$ in the sorted list of all reversed codewords. The list is implemented, as in Lemma 4, as a balanced search tree. Then we compute the longest common suffix of ca and the predecessor/successor, and the maximum of those two values gives us the value of i. The total complexity is $\mathcal{O}(\log m)$.   □

Now, observe that the above idea can be actually used to store all chunks of length $\alpha \log m$, where $\alpha = 1, 2, \ldots, \log m$. Indeed, because the method worked for any alphabet size, we can treat fragments of length $\log m$ as single letters which we can compare in constant time with Lemma 4. More precisely, we construct a new trie, where the parent of a node is its $\log m$-th ancestor in the original trie (or the root, if there is no such ancestor), and the labels of the edges are words of length up to $\log m$. Then by applying Lemma 5 to this new trie we get the following result.

Lemma 6. *We can maintain a structure of size $\mathcal{O}(m \log m)$ for a growing trie so that each addition requires $\mathcal{O}(\log m)$ time and the longest common prefix of any two chunks of length $\alpha \log m$ with $\alpha \leq \log m$ can be computed in in $\mathcal{O}(1)$ time.*

By first using Lemma 5 and then Lemma 6, we can actually compare any two chunks of length at most $\log^2 m$ in constant time. To deal with the longer chunks, we will store for each k a sorted list of chunks of length $2^k \log^2 m$. We would like to actually store all chunks of length 2^k, but this seems infeasible: insertion takes constant time when we know the place we want to insert at, and computing this place seems to require logarithmic time. Hence we will make sure that the total number of chunks we store is $\mathcal{O}(m)$, choosing them carefully. For each node we will store at most two chunks ending there. More precisely, let the *level* of a node be its distance from the root. Then if the level is ℓ, and $\ell = k_1 \log m + k_2$ (mod $\log^2 m$) where $0 \leq k_1, k_2 < \log m$, we store *selected chunks* of length $2^{k_1} \log^2 m$ and $2^{k_2} \log^2 m$ ending there. Selected chunks are used to show the following lemma:

Lemma 7. *Assume that we can compare any two selected chunks in constant time. Then, we can maintain a structure of size $\mathcal{O}(m \log m)$ for a growing trie so that each addition requires $\mathcal{O}(\log m)$ time and the longest common prefix of any two chunks can be computed in $\mathcal{O}(1)$ time.*

Proof. Let t be the length of the shorter chunk. First we use the method developed for short chunks to compare their prefixes of length $\log^2 m + (t \bmod \log^2 m)$ (if t is smaller, we are already done). Then cut this prefix and consider the resulting two chunks, which are of length $\alpha \log^2 m$. Choose k such that $2^k \leq \alpha < 2^{k+1}$ and cover both chunks with fragments of length $2^k \log^2 m$, see Fig. 4. To finish the computation, we only need to show how to compare the corresponding fragments of such length. Note that given such pair of fragments, we are actually allowed to move both of them up by at most $\log^2 m$. This is because we already compared the prefixes of length at least $\log^2 m$ of the original chunks, and realized that they are the same, and because we can compare any two other short chunks in constant time (so after the shift we can compare the uncovered suffix parts, if necessary). Hence the situation is actually simpler than comparing any two chunks of such length, and this is where the selected chunks come into the play. As long as at least one of the two fragments does not correspond to a selected chunk, we move both of them up (simultaneously) by one. Note that because of the way selection works, we never have to move up more than $\log^2 m$. More precisely, imagine that we first move up as long as the first one is not a selected chunk, which happens after at most $\log m$ moves. Then we move up skipping $\log m$ edges in a single step as long as the second one is not a selected chunk, which happens after at most $\log m$ steps consisting of $\log m$ edges. Moreover, how much we need to move up can actually be computed using simple arithmetic operations operating on the levels. Then we use the structure to compare the selected chunks we got. Repeating this at most twice (and comparing the remaining short parts, if necessary) gives us the final result. □

The only remaining part is comparing the selected chunks. The difficult part is updating the structure, and for this we will actually use the above lemma, making sure that this is not a circular reference.

Lemma 8. *We can maintain a structure of size $\mathcal{O}(m \log m)$ for a growing trie so that each addition requires $\mathcal{O}(\log m)$ time and the longest common prefix of any two selected chunks of the same length can be computed in $\mathcal{O}(1)$ time.*

Proof. For each k we maintain a separate compacted trie storing all selected chunks of length $2^k \log^2 m$. Together with the lowest common ancestor structure, this gives us enough information to compute the longest common prefix. To update the compacted trie, we maintain a sorted list of the corresponding chunks, which is stored in a balanced search tree. Each elements of this list keeps a bidirectional link to the corresponding leaf in the compacted trie.

After adding a new leaf, we need to insert two selected chunks ending there into their compacted tries. We first insert the shorter and then the longer. Each insertion requires locating the new selected chunk ca in the sorted list of all chunks of given length, which is stored in a balanced search tree. Navigating the tree can be done in $\mathcal{O}(\log m)$ time assuming we can compare the current chunk with any already existing one in constant time. We can use the existing data to first compare c (which is a chunk, although not necessarily selected) and then (if required) look at the next letter of the other chunk and compare it with a, hence each comparison takes $\mathcal{O}(1)$ time. As soon as we know the predecessor and successor of ca on the list, we can compute the longest prefix ca which already exists in the compacted trie, and then using another binary search retrieve the last node on this prefix. Then we either add a new outgoing edge, or perform a binary search among the already existing outgoing edges and split one of them. Finally, we add the new chunk to the sorted list. □

Theorem 1. *We can maintain a structure of size $\mathcal{O}(m \log m)$ for a growing trie so that each addition requires $\mathcal{O}(\log m)$ time and the longest common prefix of any two suffixes of words corresponding to the nodes can be computed in $\mathcal{O}(1)$ time.*

To decrease the space complexity of the final algorithm, we will also need the following consequence of the above theorem.

Lemma 9. *We can maintain a structure of size $\mathcal{O}(m)$ for a growing trie so that each addition requires $\mathcal{O}(\log m)$ time and the longest common prefix of any two*

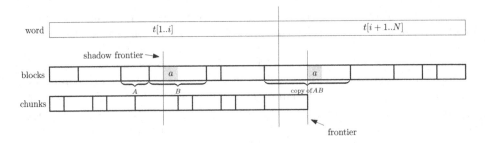

Fig. 5. Word, blocks list and chunks list

chunks can be approximately computed in $\mathcal{O}(1)$ time. If the length of the longest common prefix is ℓ, the approximation returns $\left\lfloor \frac{\ell}{\log^2 m} \right\rfloor$.

4   Algorithm

First we transform the SLP. We want to have a description of the text which consists of a number of *blocks*. Each block is either a single letter, or a concatenation of a consecutive range of blocks on the left. We can get such description consisting of $\mathcal{O}(n)$ blocks from the SLP in $\mathcal{O}(n)$ time [20]. We will store all blocks in a doubly linked list.

We process the text from left to right. Assuming we have already processed its prefix $t[1..i]$, we need to compute the longest prefix of $t[i + 1..N]$ which is a codeword. We would like to apply binary search here, so we keep all codewords sorted in the lexicographical order. As we will need to add new codewords, we actually store them in a balanced search tree. So, to compute the longest prefix of $t[i + 1..N]$ which is a codeword, we traverse the balanced search tree, and at each node we compare its codeword to $t[i + 1..N]$. Depending on the result, we go left or right. We need to execute each of those comparisons in amortized constant time.

We store the representation of the already processed prefix $t[1..i]$ as a concatenation of chunks, maintained as a doubly linked list. Additionally, for each boundary between two elements of the block list we maintain a link to the corresponding position on this chunks list. We will make sure that the only operations on the chunks list is appending a new element and merging two adjacent elements, hence the links can be easily maintained using a union-find data structure. Such structure can be implemented with $\mathcal{O}(1)$ time for find and $\mathcal{O}(\log U)$ time for union, where $U = \mathcal{O}(m)$ is the universe size, by the standard relabel-the-smaller-part technique. We require that $t[1..i]$ is a prefix of the concatenation of all chunks, but it might (and will) happen that this concatenation is actually a longer prefix of the whole word. It will be the case, though, that concatenating all but the last chunk gives us a prefix of $t[1..i]$, which will be called the *sticking out invariant*. Let the *frontier* be the position in the text just after the concatenation of all chunks. We maintain its corresponding block. If the frontier lies inside a block which is a copy of a range of blocks on the left, we also need the *shadow frontier*, which is the position in the text corresponding to the place we

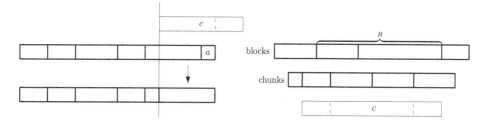

Fig. 6. Fixing the sticking out invariant   **Fig. 7.** Merging two adjacent chunks

should copy the letter at the frontier from. For the shadow frontier we maintain its corresponding chunk. See Fig. 5 which depicts all data we maintain.

Now consider comparing $t[i+1..N]$ with some codeword c. Imagine that we have access to the representation of this suffix as a concatenation of already defined codewords. Then we can simply go through those codewords one-by-one and compare each of them with the corresponding substring of c as long as they are equal, and the first mismatch allows us to determine the result. The total cost is proportional to the number of codewords we access and we must somehow amortize this quantity. This is the first difficulty. The second difficulty is that we do not really have the whole representation yet. Nevertheless, we will see how to be always able to extract the next element using the shadow frontier.

Lemma 10. *Given a codeword c which should be compared to $t[i+1..N]$ we can execute a constant time procedure which achieves at least one of the following goals:*

1. *deciding if $t[i+1..N]$ is lexicographically smaller or bigger than c,*
2. *moving the frontier to the next block,*
3. *merging two adjacent chunks.*

Proof. First we compare the suffix of the last chunk with the corresponding prefix of c. If they are not equal, we are done. Otherwise consider the next letter a of the text. If the corresponding block consists of just this letter, we compare a with the corresponding letter of c. If they are not equal, we are done. Otherwise we can move the frontier one position to the right and append a new chunk to the chunks list. Note that as a result we might break the sticking out invariant, but in such case before appending we shorten the last element and append the corresponding substring of c, see Fig. 6.

If the corresponding block is a copy of a range R of blocks on the left, we can use the shadow frontier to extract a chunk describing a prefix of the suffix of the text starting at the frontier. Again, we compare this chunk with the corresponding fragment of c, and if they are not equal, we are done. If they are equal, we look at the next chunk, and if it still intersects R, we use it to compare with c. We repeat this twice, stopping whenever we have inequality or move beyond R. Now, if we have inequality, we are done. If we move beyond R, we can move the frontier to the beginning of the next block and append at most 2 chunks to the chunks list. As before, this might break the sticking out invariant, but we can fix this using the same trick. We also have to update the shadow frontier by using the link computed for the next block in the very beginning. Hence the only remaining case is that we have equality, we are still inside R, and we repeated 2 times. Hence the situation looks like on Fig. 7, namely two chunks can be actually replaced with the corresponding fragment of c.   □

The above lemma actually allows us to compare $t[i+1..N]$ with any c in amortized constant time, as we either get the comparison result, consume one block, or decrease the number of chunks. Hence in amortized $\mathcal{O}(\log m)$ time we are able to find the longest codeword which is a prefix of $t[i + 1..N]$ by simulating a binary search for $t[i + 1..N]$ on the sorted list of all codewords and returning the codeword corresponding to the longest common prefix of $t[i + 1..N]$ and its predecessor on this list (because the the set of words on the list is prefix-closed, there is no need to compute the longest common prefix with the successor, as it cannot be larger). Then we create the next codeword and increase i. This requires possibly updating both frontier and shadow frontier. The former is easy to perform in amortized constant time, as we just need to skip a number of blocks, so we can amortize using the number of blocks n. The latter seems to be more complicated. Nevertheless, we can simply move to the right on the chunks list, and observe that whenever we go through at least two whole chunks, we can replace them with a corresponding fragment of the new codeword. Hence, again, we can amortize using the total number of codewords m.

Theorem 2. *Given a SLP on n productions, we can build a LZ78 parse describing the same text in $\mathcal{O}(n + m \log m)$ time and space.*

Using Lemma 9 and two levels of binary search, we can decrease the space usage at the cost of a slight increase in the time complexity.

Theorem 3. *Given a SLP on n productions, we can build a LZ78 parse describing the same text in $\mathcal{O}((n + m) \log m)$ time and $\mathcal{O}(n + m)$ space.*

5   Open Problems

The following question seems interesting: is it possible to implement the structure from Theorem 1 in linear space and/or with better update time?

References

1. Alstrup, S., Holm, J.: Improved algorithms for finding level ancestors in dynamic trees. In: Welzl, E., Montanari, U., Rolim, J.D.P. (eds.) ICALP 2000. LNCS, vol. 1853, pp. 73–84. Springer, Heidelberg (2000)
2. Bannai, H., Inenaga, S., Takeda, M.: Efficient LZ78 factorization of grammar compressed text. In: Calderón-Benavides, L., González-Caro, C., Chávez, E., Ziviani, N. (eds.) SPIRE 2012. LNCS, vol. 7608, pp. 86–98. Springer, Heidelberg (2012)
3. Cole, R., Hariharan, R.: Dynamic LCA queries on trees. In: Proc. SODA 1999, pp. 235–244 (1999)
4. Crochemore, M., Landau, G.M., Ziv-Ukelson, M.: A subquadratic sequence alignment algorithm for unrestricted scoring matrices. SIAM J. Comput. 32(6), 1654–1673 (2003)

5. Gasieniec, L., Rytter, W.: Almost optimal fully LZW-compressed pattern matching. In: DCC 1999: Proceedings of the Conference on Data Compression, p. 316. IEEE Computer Society, Washington, DC (1999)
6. Gawrychowski, P.: Faster algorithm for computing the edit distance between SLP-compressed strings. In: Calderón-Benavides, L., González-Caro, C., Chávez, E., Ziviani, N. (eds.) SPIRE 2012. LNCS, vol. 7608, pp. 229–236. Springer, Heidelberg (2012)
7. Gawrychowski, P.: Tying up the loose ends in fully LZW-compressed pattern matching. In: Proc. STACS 2012, pp. 624–635 (2012)
8. Goto, K., Bannai, H., Inenaga, S., Takeda, M.: Fast q-gram mining on SLP compressed strings. Journal of Discrete Algorithms 18, 89–99 (2013)
9. Hermelin, D., Landau, G.M., Landau, S., Weimann, O.: A unified algorithm for accelerating edit-distance computation via text-compression. In: Proc. STACS 2009, pp. 529–540 (2009)
10. Jeż, A.: Faster fully compressed pattern matching by recompression. In: Czumaj, A., Mehlhorn, K., Pitts, A., Wattenhofer, R. (eds.) ICALP 2012, Part I. LNCS, vol. 7391, pp. 533–544. Springer, Heidelberg (2012)
11. Karpinski, M., Rytter, W., Shinohara, A.: An efficient pattern-matching algorithm for strings with short descriptions. Nordic Journal of Computing 4, 172–186 (1997)
12. Larsson, N.J., Moffat, A.: Off-line dictionary-based compression. Proceedings of the IEEE 88(11), 1722–1732 (2000)
13. Li, M., Chen, X., Li, X., Ma, B., Vitányi, P.M.B.: The similarity metric. IEEE Transactions on Information Theory 50(12), 3250–3264 (2004)
14. Li, M., Sleep, R.: Genre classification via an LZ78-based string kernel. In: Proc. ISMIR 2005, pp. 252–259 (2005)
15. Li, M., Sleep, R.: An LZ78 based string kernel. In: Li, X., Wang, S., Dong, Z.Y. (eds.) ADMA 2005. LNCS (LNAI), vol. 3584, pp. 678–689. Springer, Heidelberg (2005)
16. Li, M., Zhu, Y.: Image classification via LZ78 based string kernel: A comparative study. In: Ng, W.-K., Kitsuregawa, M., Li, J., Chang, K. (eds.) PAKDD 2006. LNCS (LNAI), vol. 3918, pp. 704–712. Springer, Heidelberg (2006)
17. Lifshits, Y.: Processing compressed texts: A tractability border. In: Ma, B., Zhang, K. (eds.) CPM 2007. LNCS, vol. 4580, pp. 228–240. Springer, Heidelberg (2007)
18. Miyazaki, M., Shinohara, A., Takeda, M.: An improved pattern matching algorithm for strings in terms of straight-line programs. Journal of Discrete Algorithms 1(1), 187–204 (2000)
19. Nevill-Manning, C.G., Witten, I.H., Maulsby, D.L.: Compression by induction of hierarchical grammars. In: Proc. DCC 1994. pp. 244–253 (1994)
20. Rytter, W.: Application of Lempel-Ziv factorization to the approximation of grammar-based compression. Theoretical Computer Science 302(1-3), 211–222 (2003)
21. Takeda, M., Shibata, Y., Matsumoto, T., Kida, T., Shinohara, A., Fukamachi, S., Shinohara, T., Arikawa, S.: Speeding up string pattern matching by text compression: The dawn of a new era. Transactions of Information Processing Society of Japan 42(3), 370–384 (2001)
22. Yamamoto, T., Bannai, H., Inenaga, S., Takeda, M.: Faster subsequence and don't-care pattern matching on compressed texts. In: Giancarlo, R., Manzini, G. (eds.) CPM 2011. LNCS, vol. 6661, pp. 309–322. Springer, Heidelberg (2011)
23. Ziv, J., Lempel, A.: A universal algorithm for sequential data compression. IEEE Transactions on Information Theory IT-23(3), 337–349 (1977)
24. Ziv, J., Lempel, A.: Compression of individual sequences via variable-length coding. IEEE Transactions on Information Theory 24(5), 530–536 (1978)

A Bit-Parallel, General Integer-Scoring Sequence Alignment Algorithm

Gary Benson[1,2,3,*], Yozen Hernandez[1,2], and Joshua Loving[1,2]

[1] Laboratory for Biocomputing and Informatics, Boston University, Boston, MA
[2] Graduate Program in Bioinformatics, Boston University, Boston, MA
[3] Department of Computer Science, Boston University, Boston, MA
{gbenson,yhernand,jloving}@bu.edu

Abstract. Mapping of next-generation sequencing data and other processor-intensive sequence comparison applications have motivated a continued search for high efficiency sequence alignment algorithms. In one approach, which exploits the inherent parallelism in computer logic calculations, individual cells in an alignment scoring matrix are represented as bits in a computer word and the calculation of scores is emulated by a series of bit operations comprised of AND, OR, XOR, complement, shift, and addition. Bit-parallelism has been successfully applied to the Longest Common Subsequence (LCS) and edit-distance problems, producing solutions which are significantly faster than standard implementations. But, the intensive mental effort required to produce these solutions, which are closely tied to special properties of the problems, has limited efforts to extend bit-parallelism to more general scoring schemes. In this paper, we give the **first bit-parallel solution for general, integer-scoring global alignment**. Integer-scoring schemes, which are widely used, assign integer weights for match, mismatch, and insertion/deletion or indel. Our method depends on structural properties of the relationship between adjacent scores in the scoring matrix. We utilize these properties to construct a class of efficient algorithms, each designed for a particular set of weights, and we introduce a standard for characterizing the efficiency in terms of the average number of bit-operations per cell of the original scoring matrix.

Keywords: bit-parallelism, global sequence alignment, integer weights.

1 Introduction

Sequence alignment algorithms are critical tools in the analysis of biological sequence data including DNA, RNA, and protein sequences. But, the demands placed on computational resources by high-throughput experiments such as next-generation sequencing require new, more efficient methodologies. While standard

* This work was supported in part by NSF grants IIS-1017621 and DRL-1020166 (GB), NIH grant 1UL1 RR025771 (GB), and NSF IGERT grant DGE-0654108 (JL and YH).

J. Fischer and P. Sanders (Eds.): CPM 2013, LNCS 7922, pp. 50–61, 2013.

implementations of the Smith-Waterman [11] and Needleman-Wunsch [10] algorithms calculate the score in each cell of the alignment scoring matrix sequentially, a newer technique called bit-parallelism adapts the inherent parallelism in computer logic calculations to the task of overcoming the limited dependencies between adjacent scores in order to achieve much higher efficiencies.

Bit-parallel algorithms use computer words to represent multiple adjacent cells in the scoring matrix, and bit operations to mimic the result of dynamic programming. Bit-parallel methods have been successfully applied to the longest common subsequence (LCS) [1,3,5] and unit-cost edit distance (Levenshtein) [6,12,8] problems. These algorithms focus on computing the alignment score, delinking that computation from the traceback which produces the final alignment. In the LCS scoring matrix, scores are monotonically non-decreasing in the rows and columns and the bit-parallel implementations use bits to represent the cells where an increase occurs. In the edit distance scoring matrix, adjacent scores can differ by at most one, and the binary representation stores the locations of (two of the three) possible differences, $+1, -1$, and zero. These algorithms are adhoc in their approach, relying on specific properties of the underlying problems, making it difficult to directly adapt them to other alignment scoring schemes.

Bit-parallel algorithms have also been developed for the approximate string matching problem in which a pattern and text are given and occurrences of the pattern with at most k differences are sought in the text [13,12,2,9]. For example, the Wu and Manber algorithm [12] finds approximate matches to a pattern or regular expression where the number of differences between the pattern and the text is at most k. This algorithm is implemented as the Unix command *agrep*. The Navarro algorithm for approximate regular expressions [9] allows arbitrary integer weights for match, mismatch, and insertion or deletion and finds occurrences of the pattern where the *sum* of the edit weights is at most k. In these algorithms, the complexity (and computation time) increases with increasing k. By contrast, in our algorithm discussed below, the complexity depends on the edit weights, not the ultimate score of the alignment.

In this paper, we describe, to our knowledge, the first generalization of the bit-parallel method to integer-scoring similarity and distance based global alignment. Integer-scoring schemes, which are widely used, assign integer weights for match, mismatch, and indel operations. Our new contribution is an observation of the regularity of the relationship between adjacent scores in the scoring matrix when using general integer scoring (Section 3) and the design of an efficient series of bit operations to exploit that regularity (Section 4). We show how to construct a class of efficient algorithms, each designed for a particular set of weights. The method works, as described below, for general alphabets, but our interest derives from frequent use of DNA alignment when analyzing next-generation sequencing data to detect genetic variation. The remainder of the paper is organized as follows. In Section 2 we give a formal presentation of the problem, in Section 5 we compare the performance of our algorithm with five related algorithms, and in Section 6 we discuss future work.

2   Problem Description

We state our problem in terms of similarity scoring, but the technique can be used for distance scoring as well.

Problem: *Given two sequences X and Y, of length n and m respectively, and a similarity scoring function S defined by three* **integer** *weights M, I, G (match, mismatch, indel or gap), calculate the global alignment similarity score for X and Y using bit operations with computer words of length w in time $O(nm/w)$, and more specifically, such that the actual average count of bit operations* **per cell** *of the alignment scoring matrix, p/w, is $\leq e$ for some small number e, where p is the number of operations to complete the calculation for w cells.*

We will say that an algorithm (or program) that accomplishes the task has a **per-cell bit operation cost** of at most e. For example, in the case of the LCS, the per cell bit operation cost is $p/w = 1/16$ (that is, there are $p = 4$ bit operations per word of length $w = 64$) [5]. For the edit distance problem, $p/w = 15/64 < 1/4$ (15 bit operations per word, unpublished, improved from [8,6]). Note that in these examples we have counted only bit operations and not storage of computed values in program variables. Adding store operations is more accurate and increases the numbers here, but stores are difficult to count because they depend on specifics of the compiler and the level of optimization.

We require that the alignment method be global, but do not restrict the initializations in the first row or column of the alignment scoring matrix. Typical initializations require 1) a gap weight to be added successively to every cell (global alignment from the beginning of a sequence), and 2) a zero in every cell (global alignment where an initial gap has no penalty).

We assume that match scores are positive, $M > 0$, mismatch and gap scores are negative, $I, G < 0$ and that the use of mismatch is possible, meaning that its penalty is no worse than the penalty for two adjacent gaps, one in each sequence, $I \geq 2G$. While other weightings are possible, they either reduce to simpler problems from a bit-parallel perspective (i.e., Longest Common Subsequence has $G = 0$, $I = -\infty$, $M = 1$) or require more complicated structures than detailed here (protein alignment using PAM or BLOSUM style amino acid substitution tables).

We stress that we do not claim to have an optimal solution for any particular instance of the alignment weights. As can be seen in the cases of the LCS and edit-distance, extreme efficiency can be obtained by exploiting specific problem properties. Instead, we give a general framework for efficient bit-parallel implementation of alignment which works across a wide spectrum of weights.

3   Function Tables

Let S be a recursively-defined, similarity scoring function for computing the global alignment score between two sequences X and Y:

$$S[i,j] = \max \begin{cases} S[i-1,j-1] + M & \text{if } X_i = Y_j \\ S[i-1,j-1] + I & \text{if } X_i \neq Y_j \\ S[i-1,j] + G & \text{delete } X_i \\ S[i,j-1] + G & \text{delete } Y_j \end{cases}$$

We assume the convention that S is computed in an alignment scoring matrix. Suppose that instead of knowing the actual value in a cell $S[i,j]$ we know only the *difference*, ΔV, between that cell and the cell above, and the *difference*, ΔH, between that cell and the cell to it's left:

$$\Delta V[i,j] = S[i,j] - S[i-1,j]$$
$$\Delta H[i,j] = S[i,j] - S[i,j-1].$$

Lemma 1 defines the minimum and maximum values of ΔV and ΔH and Lemma 2 gives their recursive definitions. Proofs are omitted.

Lemma 1. *Given S, X, and Y as described above where match score $M > 0$, mismatch score $I < 0$ and gap (indel) score $G < 0$, the minimum and maximum differences between adjacent values in the same row (i.e., $\Delta H[i,j]$) or column (i.e., $\Delta V[i,j]$) are G and $M - G$.*

Lemma 2. *The values for ΔV are shown below and the values for ΔH are the transpose, that is $\Delta H[i,j] = \Delta V[j,i]$.*

$$\Delta V[i,j] = \begin{cases} \text{Score comes diagonally from a match:} \\ M - \Delta H[i-1,j] \qquad \text{if } X_i = Y_j \\ \\ \text{Score comes diagonally from a mismatch:} \\ I - \Delta H[i-1,j] \qquad \text{if } I - G \geq \begin{cases} \Delta H[i-1,j] \\ \Delta V[i,j-1] \end{cases} \\ \text{Score comes from the cell above:} \\ G \qquad \qquad \qquad \text{if } \Delta H[i-1,j] \geq \begin{cases} I - G \\ \Delta V[i,j-1] \end{cases} \\ \text{Score comes from the cell to the left:} \\ \Delta V[i,j-1] + G - \Delta H[i-1,j] \quad \text{if } \Delta V[i,j-1] \geq \begin{cases} I - G \\ \Delta H[i-1,j] \end{cases} \end{cases}$$

$$\left(V[0,j] = G \text{ or } V[0,j] = 0 \right) \text{ and } \left(H[i,0] = G \text{ or } H[i,0] = 0 \right).$$

The recursion for the ΔV values can be summarized in a Function Table (Figure 1). Note the key value $I - G$ from the recursion and the relation $\Delta H = \Delta V$. They set the boundaries for the marked zones in the table. These zones comprise $(\Delta V, \Delta H)$ value pairs which determine how the best score of a cell in S is obtained in the absence of a match, either as an indel from the left (Zones A and B), a mismatch (Zone C), or an indel from above (Zone D). Borders between zones, indicated by dotted lines, yield ties for the best score. Figure 2 shows how the relative size of the Zones changes with changes in I and G.

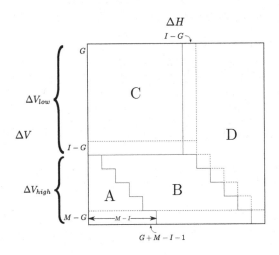

Fig. 1. Zones in the Function Table for ΔV

We use the following (see Figure 1):

Definitions:
$\Delta V_{\min} = \Delta H_{\min} = G$
$\Delta V_{\max} = \Delta H_{\max} = M - G$
$\Delta V_{low}, \Delta H_{low} \in [G, I - G]$
$\Delta V_{high}, \Delta H_{high} \in [I - G + 1, M - G]$

Observations:
Zone A – All value are in V_{high}
Zone B – All values are in V_{low}
Zone C – All values are in V_{low}. Values depend only on ΔH.
Zone D – All values are G
Last Row – Values from this row also apply when there is a Match.
First Column – Identity column for values in V_{high}

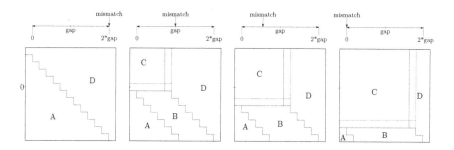

Fig. 2. Relative size of Zones as I (mismatch penalty) decreases from $2G$ (twice gap penalty) where there is no preference for mismatches, to zero, where mismatches are free and gaps are introduced only to obtain matches

4   Bit-Parallel Alignment

Our goal is to develop an algorithm for calculating the ΔH values in row i from:

- the ΔH values in row $i - 1$,
- the initial ΔV value in row i
- the Match positions in row i.

What follows is a description of the simplest case where the length of the first sequence, n, is less than the computer word size w. Longer sequences can be handled in "chunks," where each chunk has size w. The Match positions for every row are computed prior to the calculation of the row values as is done for the LCS and edit-distance problems. Details are given at the end. For the remainder of the paper, we will use the following set of scoring weights for illustration:

$$M = 2, \; I = -3, \; G = -5.$$

The ΔV Function Table for these weights is shown in Figure 3.

Representation of ΔH and ΔV Values. In the bit-parallel framework, we use one computer word (sometimes referred to later as a vector) to represent each possible value of ΔH and ΔV. Bit i in a word refers to column i in the alignment scoring matrix. With the weights used for illustration, there are 13 values, each, for ΔH and ΔV.

Algorithm Outline. We first calculate all ΔV values for row i and then use them to calculate the ΔH values. Close inspection of the Function Table (Figure 3) reveals that the values in Zone A, which are all in ΔV_{high}, are interdependent, and require computing in order from high to low. The other complication is the identity column ΔH_{min} which requires carrying a ΔV input value through runs of ΔH_{min}. Values in Zones B, C, and D, which are all in ΔV_{low}, can be computed once the values from Zone A are obtained. The main work is combining outputs from $(\Delta V, \Delta H)$ pairs which intersect along the same diagonal. Once all ΔV values are available, the ΔH values can be computed in any order.

$$\Delta H$$

		-5	-4	-3	-2	-1	0	1	2	3	4	5	6	7
	-5...2	2	1	0	-1	-2	-3	-4	-5	-5	-5	-5	-5	-5
	3	3	2	1	0	-1	-2	-3	-4	-5	-5	-5	-5	-5
ΔV	4	4	3	2	1	0	-1	-2	-3	-4	-5	-5	-5	-5
	5	5	4	3	2	1	0	-1	-2	-3	-4	-5	-5	-5
	6	6	5	4	3	2	1	0	-1	-2	-3	-4	-5	-5
7 and match		7	6	5	4	3	2	1	0	-1	-2	-3	-4	-5

Fig. 3. The ΔV Function Table for the weights $M = 2$, $I = -3$, $G = -5$. Note that $\Delta V_{high}, \Delta H_{high} \in [3,7]$; $\Delta V_{low}, \Delta H_{low} \in [-5,2]$; $\Delta V_{min} = \Delta H_{min} = -5$; $\Delta V_{max} = \Delta H_{max} = 7$.

4.1   Computing ΔV

We present the following without theorems or proofs for compactness.

Zone A – Dependencies. To compute its output value, each cell needs to know its ΔH and ΔV input values. The ΔH values are already known, as is the input ΔV value for the first cell. As in standard left to right processing, the output ΔV value from one cell becomes the input value for the cell to its right. The arrangement of the ΔV values in Zone A of Figure 3 indicates a chain of dependency:

$$Matches \to 7 \to 6 \to \ldots \to 3.$$

Additionally, the identity column ΔH_{min} ($= -5$) indicates that a ΔV_{high} value that is fed into a ΔH run of -5s will yield the identical output for every cell in the run. Therefore, to know where the 7s are (for ΔV) first requires knowing where the Matches are and then which of the 7s from Matches carry through runs of -5; to know where the 6s are first requires knowing where the Matches and 7s are and then which of those 6s carry through runs of -5, etc. The "carry through runs of ΔH_{min}" is really the only obstacle, but can be accomplished with an addition (+) as seen below. Addition is also used to solve similar left to right dependency problems in the LCS and edit-distance bit-parallel algorithms.

Zone A – Finding ΔV_{max}. The vector is calculated with four operations (Figure 4). The result stores locations of ΔV_{max} shifted one position to the right for input to subsequent calculations. The operations are 1) an AND to find the 7s from Matches, 2) an addition (+) to carry through adjacent runs of ΔH_{min} and into the position following a run (and causing erroneous internal bit flips if there are multiple Matches in the same run), 3) an XOR to complement the bits within the runs, and 4) an XOR to correct any erroneous bits and accomplish the shift by removing the leading 1 in a run.

```
        1       1     1   1 1   1      Matches
AND    1110    1110      111110  1110  ΔHmin
     ----------------------------------------
        0100    1000     010100  0000  ΔVmax (initial)
 +     1110    1110      111110  1110  ΔHmin
     ----------------------------------------
        1001    0001     100X01  1110
XOR    1110    1110      111110  1110  ΔHmin
     ----------------------------------------
        0111    1111     011011  0000
XOR    0100    1000     010100  0000  ΔVmax (initial)
     ----------------------------------------
        0011    0111     001111  0000  >> ΔVmax (final and shifted)

Example Code:
INITpos7 = DHneg5 & Matches;
DVpos7shift = ((INITpos7 + DHneg5) ^ DHneg5) ^ INITpos7;
```

Fig. 4. Finding $\Delta V_{\max}$. Each line represents a computer word with low order bit, corresponding to the first position in a sequence, on the left. 1s are shown explicitly, 0s are only shown to fill runs of $\Delta H_{\min}$ and the first position to the right of each run. Symbol $>>$ indicates that the final $\Delta V_{\max}$ values are shifted to the right one position. Erroneous bit set by the ADD (+) is marked X.

Zone A – Others. Remaining ΔV_{high} vectors are calculated, in descending order as discussed above. First, initial vectors are computed by AND of appropriate $(\Delta V, \Delta H)$ pairs (which intersect along a common diagonal in the Function Table) and collected together with ORs. Second, the intial vectors are shifted right one position for subsequent calculations. Third, the carry through runs of $\Delta H_{\min}$ is computed in two operations (Figure 5), an addition (+) as before and an XOR to complement the bits within the runs. Before the carry operation, those $\Delta H_{\min}$ positions that have already output a $\Delta V_{\max}$ value must be removed. Note that initial ΔV values, when shifted to the right, can only occur at the leftmost position of a $\Delta H_{\min}$ run, and not at the single bit between adjacent runs.

Zones B and C and D. (Figure 6). At this point, all the ΔV_{high} input values for Zone B have been computed, remaining output values are all ΔV_{low}, and Zone C output depends only on ΔH values. Each output vector is an OR combination of 1) Zone B – the AND of appropriate $(\Delta V, \Delta H)$ pairs, which intersect along a common diagonal, and 2) Zone C – the AND of the appropriate ΔH vector and all positions without a ΔV_{high} output. The result is shifted one position to the right for subsequent calculations. Zone D has only one output value, ΔV_{min}. It is assigned to all remaining positions as well as the first position if gap penalty in the first column is being used.

4.2 Computing ΔH

After the ΔV values are computed, there are no longer any dependencies. All the new vectors for ΔH can be immediately computed. Again, each vector is an OR combination of the AND of appropriate pairs of ΔH and ΔV values.

```
          1110    1110 11101110  ΔH_min (remaining)
    +      1    1  1    1  X      >> ΔV (initial shifted)
          ------------------------
          0001 1  0001 00011110
    XOR   1110    1110 11101110  ΔH_min (remaining)
          ------------------------
          1111 1  1111 11110000  >> ΔV (final and shifted)

Example Code:
RemainDHneg5 = DHneg5 ^ (DVpos7shift >> 1);
INITpos3s = (DHneg1 & DVpos7shiftorMatch)|(DHneg2 & DVpos6shiftNotMatch)|
            (DHneg3 & DVpos5shiftNotMatch)|(DHneg4 & DVpos4shiftNotMatch);
DVpos3shift = ((INITpos3s << 1) + RemainDHneg5) ^ RemainDHneg5;
DVpos3shiftNotMatch = DVpos3shift & NotMatches;
```

Fig. 5. Carry through runs of $\Delta H_{\min}$ for remaining values in ΔV_{high}. Symbol X marks a single position between runs which cannot be 1 in the initial shifted values.

4.3 Other Tasks

Determining Matches. Prior to computing any ΔH or ΔV, the position of the matches are determined for each character σ in the sequence alphabet Σ. A bit vector $Match_\sigma$ records those positions in sequence X where σ occurs. Filling all the $Match_\sigma$ simultaneously can be accomplished efficiently in a single pass through X. For row i of the ΔH and ΔV calculations, the $Match$ vector for character Y_i is used.

Decoding the Alignment Score. The score in the last column of the last row of the alignment scoring matrix can be obtained by calculating the score in the zero column $(= m * G)$ and then adding the number of 1 bits in each of the ΔH vectors multiplied by the value of the vector. Using the method described in [7], this takes $O(n + M - 2G)$ operations with a small constant:

$$S[m, n] = m * G + \sum_{i=G}^{M-G} \text{bits}_i * i.$$

```
Example Code Zones B and C:

DVnot7to3shiftorMatch = ~ (DVpos7shiftorMatch|DVpos6shift|DVpos5shift|DVpos4shift|
            DVpos3shift);
DVpos2shift = ((DHzero & DVpos7shiftorMatch)|(DHneg1 & DVpos6shiftNotMatch)|
            (DHneg2 & DVpos5shiftNotMatch)|(DHneg3 & DVpos4shiftNotMatch)|
            (DHneg4 & DVpos3shiftNotMatch)|(DHneg5 & DVnot7to3shiftorMatch)) << 1;

Example Code Zone D:

DVneg5shift = all_ones ^ (DVpos7shift|DVpos6shift|DVpos5shift|DVpos4shift|DVpos3shift|
            DVpos2shift|DVpos1shift|DVzeroshift|DVneg1shift|DVneg2shift|DVneg3shift|
            DVneg4shift);
```

Fig. 6. Code for Zones B and C and D

Several methods can be used to efficiently find all scores in the last row. Discussion of these is omitted due to space limitations.

4.4   Complexity and Number of Operations

The time complexity of our algorithm is $O(znm/w)$ where z depends on the combined size of the Zones A, B, and C (the latter is reduced to a single row as in Figure 3) in the Function Table. This in turn depends on the alignment weights M, I, and G:

$$z = \frac{(M - 2G + 1)^2 - (I - 2G)^2}{2}$$

and the constant hidden in the big O notation is approximately 4 (dominated by two operations per cell of Zones A, B, and C for ΔV and separately for ΔH). For the example weights used in this paper, the number of bit operations per w cells of the scoring matrix is 265 yielding a per-cell bit operation cost of $265/64 \approx 4.2$.

5   Experimental Results

We compared running times for several related bit-parallel algorithms: 1) BHL – our new algorithm with 5 sets of alignment weights to show the effect of the weights on the running time, 2) NW – the classical Needleman-Wunch [10] dynamic programming alignment algorithm, 3) LCS – the bit-parallel LCS algorithm of [5], 4) ED – a bit-parallel, unit-cost edit-distance algorithm, improved from [8,6], 5) WM – the unit-cost Wu-Manber approximate pattern matching algorithm [12], and 6) N – the Navarro, general integer scoring, approximate regular expression matching algorithm [9]. We implemented BHL, NW, LCS, ED, and WM. N was graciously provided by Gonzalo Navarro.

For all experiments, we used human DNA and ran 100 pattern sequences against 250,000 text sequences for a total of 25 million alignments. (Pattern and text distinctions are irrelevant for BHL, NW, LCS, and ED.) All sequences were 63 characters long. For WM we varied k, the maximum number of allowed errors, from 1 to 15. For N, we varied k from 1 to 12. (Additionally, we selected the smallest values for *tables* and *mask length*, internal parameters to the N method, for which the program could be successfully run. For $k = 1$, we used *tables* = 8 and *mask length* = 4. For $k = 2$ through 5, *tables* = 9 and *mask length* = 5. For $k = 6$ through 12, *tables* = 13 and *mask length* = 7.)

All programs were compiled with GCC using optimization level O3 and were run on an Intel Core 2 Duo E8400 3.0 GHz CPU running Ubuntu Linux 12.10. Results are shown in Figure 7.

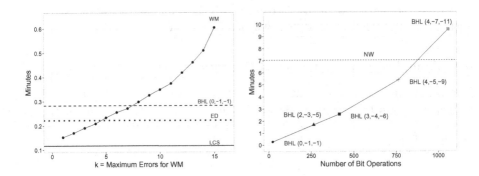

Fig. 7. Running times. For BHL, alignment weights (M, I, G) are shown in parenthesis. All times are averages of three runs. **Left.** Unit-cost BHL, unit-cost WM, LCS, and ED. k is the maximum number of errors allowed for WM. k is not a parameter for the other algorithms and their times are shown as horizontal lines. LCS uses 4 bit operations per w cells, ED uses 15 bit operations, BHL (0, -1, -1) uses 23 bit operations. For $k = 7$, the times for BHL and WM are nearly the same. By $k = 15$, BHL is approximately twice as fast. Results for N are not shown on the graph due to the much longer running time. For N, we ran 2 sequences against 250,000 and multiplied by 50 to find comparable times. For values of k from 1 to 12, the performance of N varied from 20 to 50 times slower than BHL (2, -3, -5). **Right.** Variants of BHL and NW. For BHL, time is approximately linearly proportional to the number of bit operations (and z) as explained in Section 4.4. For NW, the number of bit operations is not available. Its time is shown as a horizontal line. BHL (2, -3, -5) is approximately 4.2 times faster than NW and BHL (0,-1,-1) is approximately 24.9 times faster.

6   Discussion

The algorithm outlined above can be extended in several ways. Computers now in common usage have a word size of $w = 64$ bits. A straightforward extension is to use the 128 bit SIMD registers (Single Instruction, Multiple Data). This essentially halves the number of operations per cell (with the addition of several bookkeeping operations) and doubles the speed of computation. Details of the method will be given in the journal version of this abstract. Another extension is due to the unexploited parallelism of the operations. There are no dependencies on prior computations after the ΔV vectors in Zone A are computed. This means that all the computations in Zones B, C, and D for ΔV and all the subsequent computations for ΔH can be computed simultaneously, an ideal situation for the use of general purpose graphical processing units (GPGPU). CUDA programming (Compute Unified Device Architecture) for this method will be presented in a separate paper.

Because a different program has to be created for each unique set of weights for M, I, and G, adoption of this algorithm would require a complete understanding of the program structure. To simplify usage, we are constructing a web site that will generate C computer code given the weights as input.

This method has already been used to accelerate software for detecting tandem repeat variants in next-generation sequencing reads [4] and is well suited to other DNA sequence comparison tasks that involve computing many alignments.

References

1. Allison, L., Dix, T.I.: A bit-string longest-common-subsequence algorithm. Information Processing Letters 23(5), 305–310 (1986)
2. Bergeron, A., Hamel, S.: Vector algorithms for approximate string matching. International Journal of Foundations of Computer Science 13(01), 53–65 (2002)
3. Crochemore, M., Iliopoulos, C.S., Pinzon, Y.J., Reid, J.F.: A fast and practical bit-vector algorithm for the longest common subsequence problem. Information Processing Letters 80(6), 279–285 (2001)
4. Gelfand, Y., Loving, J., Hernandez, Y., Benson, G.: VNTRseek – A Computational Pipeline to Detect Tandem Repeat Variants in Next-Generation Sequencing Data: Analysis of the 454 Watson Genome. In: Proc. of RECOMB-seq: The Third Annual RECOMB Satellite Workshop on Massively Parallel Sequencing (to appear, 2013)
5. Hyyrö, H.: Bit-parallel LCS-length computation revisited. In: Proc. 15th Australasian Workshop on Combinatorial Algorithms, AWOCA 2004 (2004)
6. Hyyrö, H., Fredriksson, K., Navarro, G.: Increased bit-parallelism for approximate and multiple string matching. Journal of Experimental Algorithmics (JEA) 10, 2–6 (2005)
7. Kernighan, B.W., Ritchie, D.M.: The C programming language, 2nd edn. Prentice Hall (1988)
8. Myers, G.: A fast bit-vector algorithm for approximate string matching based on dynamic programming. Journal of the ACM (JACM) 46(3), 395–415 (1999)
9. Navarro, G.: Approximate regular expression searching with arbitrary integer weights. Nordic Journal of Computing 11(4), 356–373 (2004)
10. Needleman, S., Wunch, C.: A general method applicable to the search for similarities in the amino acid sequence of two proteins. J. Mol. Biol. 48, 443–453 (1970)
11. Smith, T.F., Waterman, M.S.: Identification of common molecular subsequences. Journal of Molecular Biology 147(1), 195–197 (1981)
12. Wu, S., Manber, U.: Fast text searching: allowing errors. Communications of the ACM 35(10), 83–91 (1992)
13. Wu, S., Manber, U., Myers, G.: A subquadratic algorithm for approximate limited expression matching. Algorithmica 15(1), 50–67 (1996)

Compact q-Gram Profiling
of Compressed Strings

Philip Bille, Patrick Hagge Cording, and Inge Li Gørtz

DTU Compute, Technical University of Denmark
{phbi,phaco,inge}@dtu.dk

Abstract. We consider the problem of computing the q-gram profile of a string T of size N compressed by a context-free grammar with n production rules. We present an algorithm that runs in $O(N - \alpha)$ expected time and uses $O(n + k_{T,q})$ space, where $N - \alpha \leq qn$ is the exact number of characters decompressed by the algorithm and $k_{T,q} \leq N - \alpha$ is the number of distinct q-grams in T. This simultaneously matches the current best known time bound and improves the best known space bound. Our space bound is asymptotically optimal in the sense that any algorithm storing the grammar and the q-gram profile must use $\Omega(n + k_{T,q})$ space. To achieve this we introduce the q-gram graph that space-efficiently captures the structure of a string with respect to its q-grams, and show how to construct it from a grammar.

1 Introduction

Given a string T, the q-gram profile of T is a data structure that can answer substring frequency queries for substrings of length q (q-grams) in $O(q)$ time. We study the problem of computing the q-gram profile from a string T of size N compressed by a context-free grammar with n production rules.

The generalization of string algorithms to grammar-based compressed text is currently an active area of research. Grammar-based compression is studied because it offers a simple and strict setting and is capable of modelling many commonly used compression schemes, such as those in the Lempel-Ziv family [15,16], with little expansion [1,12]. The problem of computing the q-gram profile has its applications in bioinformatics, data mining, and machine learning [3,9,11]. All are fields where handling large amount of data effectively is crucial. Also, the q-gram distance can be computed from the q-gram profiles of two strings and used for filtering in string matching [14].

Recently the first dedicated solution to computing the q-gram profile from a grammar-based compressed string was proposed by Goto et al. [5]. Their algorithm runs in $O(qn)$ expected time[1] and uses $O(qn)$ space. This was later improved by the same authors [6] to an algorithm that takes $O(N - \alpha)$ expected

---

[1] The bound in [5] is stated as worst-case since they assume integer alphabets for fast suffix sorting. We make no such assumptions and without it hashing can be used to obtain the same bound in expectation.

J. Fischer and P. Sanders (Eds.): CPM 2013, LNCS 7922, pp. 62–73, 2013.

time and uses $O(N - \alpha)$ space, where N is the size of the uncompressed string, and α is a parameter depending on how well T is compressed with respect to its q-grams. $N - \alpha \leq \min(qn, N)$ is in fact the exact number of characters decompressed by the algorithm in order to compute the q-gram profile, meaning that the latter algorithm excels in avoiding decompressing the same character more than once. These algorithms, as well as the one presented in this paper, assume the RAM model of computation with a word size of $\log N$ bits.

We present a Las Vegas-type randomized algorithm that gives Theorem 1.

Theorem 1. *Let T be a string of size N compressed by a grammar of size n. The q-gram profile can be computed in $O(N - \alpha)$ expected time and $O(n + k_{T,q})$ space, where $k_{T,q} \leq N - \alpha$ is the number of distinct q-grams in T.*

Hence, our algorithm simultaneously matches the current best known time bound and improves the best known space bound. Our space bound is asymptotically optimal in the sense that any algorithm storing the grammar and the q-gram profile must use $\Omega(n + k_{T,q})$ space.

A straightforward approach to computing the q-gram profile is to first decompress the string and then use an algorithm for computing the profile from a string. For instance, we could construct a compact trie of the q-grams using an algorithm similar to a suffix tree construction algorithm as mentioned in [7], or use Rabin-Karp fingerprints to obtain a randomized algorithm [14]. However, both approaches are impractical because the time and space usage associated with a complete decompression of T is linear in its size $N = O(2^n)$. To achieve our bounds we introduce the q-gram graph, a data structure that space efficiently captures the structure of a string in terms of its q-grams, and show how to compute the graph from a grammar. We then transform the graph to a suffix tree containing the q-grams of T. Because our algorithm uses randomization to construct the q-gram graph, the answer to a query may be incorrect. However, as a final step of our algorithm, we show how to use the suffix tree to verify that the fingerprint function is collision free and thereby obtain Theorem 1.

2   Preliminaries and Notation

2.1   Strings and Suffix Trees

Let T be a string of length $|T|$ consisting of characters from the alphabet Σ. We use $T[i : j]$, $0 \leq i \leq j < |T|$, to denote the substring starting in position i of T and ending in position j of T. We define $socc(s, T)$ to be the number of occurrences of the string s in T.

The suffix tree of T is a compact trie containing all suffixes of T. That is, it is a trie containing the strings $T[i : |T| - 1]$ for $i = 0..|T| - 1$. The suffix tree of T can be constructed in $O(|T|)$ time and uses $O(|T|)$ space for integer alphabets [2]. The generalized suffix tree is the suffix tree for a set of strings. It can be constructed using time and space linear in the sum of the lengths of the strings in the set. The set of strings may be compactly represented as a common suffix tree (CS-tree). The CS-tree has the characters of the strings on its edges, and the strings start in the leaves and end in the root. If two strings have some suffix in common, the suffixes are merged to one path. In other words, the CS-tree is a trie of the reversed strings, and is not to be confused with the suffix tree. For CS-trees, the following is known.

Lemma 1 (Shibuya [13]). *Given a set of strings represented by a CS-tree of size n and comprised of characters from an integer alphabet, the generalized suffix tree of the set of strings can be constructed in $O(n)$ time using $O(n)$ space.*

For a node v in a suffix tree, the string depth $sd(v)$ is the sum of the lengths of the labels on the edges from the root to v. We use $parent(v)$ to get the parent of v, and $nca(v, u)$ is the nearest common ancestor of the nodes v and u.

2.2 Straight Line Programs

A Straight Line Program (SLP) is a context-free grammar in Chomsky normal form that unambigously derives a string T of length N over the alphabet Σ. In other words, an SLP S is a set of n production rules of the form $X_i = X_l X_r$ or $X_i = a$, where a is a character from the alphabet Σ, and each rule is reachable from the start rule X_n. Our algorithm assumes without loss of generality that the compressed string given as input is compressed by an SLP.

It is convenient to view an SLP as a directed acyclic graph (DAG) in which each node represents a production rule. Consequently, nodes in the DAG have exactly two outgoing edges. An example of an SLP is seen in Figure 1(a). When a string is decompressed we get a derivation tree which corresponds to the depth-first traversal of the DAG.

We denote by t_{X_i} the string derived from production rule X_i, so $T = t_{X_n}$. For convenience we say that $|X_i|$ is the length of the string derived from X_i, and these values can be computed in linear time in a bottom-up fashion using the following recursion. For each $X_i = X_l X_r$ in S,

$$|X_i| = \begin{cases} |X_l| + |X_r| & \text{if } X_i \text{ is a nonterminal,} \\ 1 & \text{otherwise.} \end{cases}$$

Finally, we denote by $occ(X_i)$ the number of times the production rule X_i occurs in the derivation tree. We can compute the occurrences using the following linear time and space algorithm due to Goto et al. [5]. Set $occ(X_i) = 1$ for $i = 1..n$. For each production rule of the form $X_i = X_l X_r$, in decreasing order of i, we set $occ(X_l) = occ(X_l) + occ(X_i)$ and similarly for $occ(X_r)$.

2.3 Fingerprints

A Rabin-Karp fingerprint function ϕ takes a string as input and produces a value small enough to let us determine with high probability whether two strings match in constant time. Let s be a substring of T, c be some constant, $2N^{c+4} < p \le 4N^{c+4}$ be a prime, and choose $b \in \mathbb{Z}_p$ uniformly at random. Then,

$$\phi(s) = \sum_{k=1}^{|s|} s[k] \cdot b^k \quad \mod p.$$

Lemma 2 (Rabin and Karp [8]). *Let ϕ be defined as above. Then, for all $0 \le i, j \le |T| - q$,*

$$\phi(T[i : i+q]) = \phi(T[j : j+q]) \quad iff \quad T[i : i+q] = T[j : j+q] \quad w.h.p.$$

We denote the case when $T[i : i+q] \ne T[j : j+q]$ and $\phi(T[i : i+q]) = \phi(T[j : j+q])$ for some i and j a collision, and say that ϕ is collision free on substrings of length q in T if $\phi(T[i : i+q]) = \phi(T[j : j+q])$ iff $T[i : i+q] = T[j : j+q]$ for all i and j, $0 \le i, j < |T| - q$.

Besides Lemma 2, fingerprints exhibit the useful property that once we have computed $\phi(T[i : i+q])$ we can compute the fingerprint $\phi(T[i+1 : i+q+1])$ in constant time using the update function,

$$\phi(T[i+1 : i+q+1]) = \phi(T[i : i+q])/b - T[i] + T[i+q+1] \cdot b^q \quad \mod p.$$

3 Key Concepts

3.1 Relevant Substrings

Consider a production rule $X_i = X_l X_r$ that derives the string $t_{X_i} = t_{X_l} t_{X_r}$. Assume that we have counted the number of occurrences of q-grams in t_{X_l} and t_{X_r} separately. Then the relevant substring r_{X_i} is the smallest substring of t_{X_i} that is necessary and sufficient to process in order to detect and count q-grams that have not already been counted. In other words, r_{X_i} is the substring that contains q-grams that start in t_{X_l} and end in t_{X_r}. Formally, for a production rule $X_i = X_l X_r$, the relevant substring is $r_{X_i} = t_{X_i}[\max(0, |X_l| - q + 1) : \min(|X_l| + q - 2, |X_i| - 1)]$. We want the relevant substrings to contain at least one q-gram, so we say that a production rule X_i only has a relevant substring if $|X_i| \ge q$. From this definition we see that the size of a relevant substring is $q \le |r_{X_i}| \le 2(q-1)$.

The concept of relevant substrings is the backbone of our algorithm because of the following. If X_i occurs $occ(X_i)$ times in the derivation tree for $\mathcal{S}$, then the substring t_{X_i} occurs at least $occ(X_i)$ times in T. It follows that if a q-gram s occurs $socc(s, t_{X_i})$ times in some substring t_{X_i} then we know that it occurs at least $socc(s, t_{X_i}) \cdot occ(X_i)$ times in T. Using our description of relevant substrings we can rewrite the latter statement to $socc(s, t_{X_i}) \cdot occ(X_i) = socc(s, t_{X_l}) \cdot occ(X_l) + socc(s, t_{X_r}) \cdot occ(X_r) + socc(s, r_{X_i}) \cdot occ(X_i)$ for the production rule $X_i = X_l X_r$. By applying this recursively to the root X_n of the SLP we get the following Lemma, which is implicit in [5].

Lemma 3. *Let $\mathcal{S}_q = \{X_i \mid X_i \in \mathcal{S}$ and $|X_i| \ge q\}$ be the set of production rules that have a relevant substring, and let s be some q-gram. Then,*

$$socc(s, T) = \sum_{X_i \in \mathcal{S}_q} socc(s, r_{X_i}) \cdot occ(X_i).$$

3.2   Prefix and Suffix Decompression

The following Lemma states a result that is crucial to the algorithm presented in this paper.

Lemma 4 (Gąsieniec et al. [4]). *An SLP S of size n can be preprocessed in $O(n)$ time using $O(n)$ extra space such that, given a pointer to a variable X_i in S, the j length prefix and suffix of t_{X_i} can be decompressed in $O(j)$ time.*

Gąsieniec et al. give a data structure that supports linear time decompression of prefixes, but it is easy to extend the result to also hold for suffixes. Let s be some string and s^R the reversed string. If we reverse the prefix of length j of s^R this corresponds to the suffix of length j of s. To obtain an SLP for the reversed string we swap the two variables on the right-hand side of each nonterminal production rule. The reversed SLP S' contains n production rules and the transformation ensures that $t_{X_{i'}} = t^R_{X_i}$ for each production rule $X_{i'}$ in S'. A proof of this can be found in [10]. Producing the reversed SLP takes linear time and in the process we create pointers from each variable to its corresponding variable in the reversed SLP. After both SLP's are preprocessed for linear time prefix decompression, a query for the j length suffix of t_{X_i} is handled by following the pointer from X_i to its counterpart in the reversed SLP, decompressing the j length prefix of this, and reversing the prefix.

3.3   The q-Gram Graph

We now describe a data structure that we call the q-gram graph. It too will play an important role in our algorithm. The q-gram graph $G_q(T)$ captures the structure of a string T in terms of its q-grams. In fact, it is a subgraph of the De Bruijn graph over Σ^q with a few augmentations to give it some useful properties. We will show that its size is linear in the number of distinct q-grams in T, and we give a randomized algorithm to construct the graph in linear time in N.

A node in the graph represents a distinct $(q-1)$-gram, and the label on the node is the fingerprint of the respective $(q-1)$-gram. The graph has a special node that represents the first $(q-1)$-gram of T and which we will denote the start node. Let x and y be characters and α a string such that $|\alpha| = q - 2$. There is an edge between two nodes with labels $\phi(x\alpha)$ and $\phi(\alpha y)$ if $x\alpha y$ is a substring of T. The graph may contain self-loops. Each edge has a label and a counter. The label of the edge $\{\phi(x\alpha), \phi(\alpha y)\}$ is y, and its counter indicates the number of times the substring $x\alpha y$ occurs in T. Since $|x\alpha y| = q$ this data structure contains information about the frequencies of q-grams in T.

Lemma 5. *The q-gram graph of T, $G_q(T)$, has $O(k_{T,q})$ nodes and $O(k_{T,q})$ edges.*

Proof. Each node represents a distinct $(q-1)$-gram, and because of the way we construct the graph, its outgoing edges have unique labels. The combination of a node and an outgoing edge thus represents a distinct q-gram, and therefore there can be at most $k_{T,q}$ edges in the graph. For each new q-gram the algorithm adds an edge from an existing node to a new node, so the graph is connected. Therefore, it has at most $k_{T,q} + 1$ nodes.   □

The graph can be constructed using the following online algorithm which takes a string T, an integer q, and a fingerprint function ϕ as input. Let the start node of the graph have the fingerprint $\phi(T[0 : (q-1)-1])$. Assume that we have built the graph $G_q(T[0 : k + (q-1) - 1])$ and that we keep its nodes and edges in two dictionaries implemented using hashing. We then compute the fingerprint $\phi(T[k+1 : k+(q-1)])$ for the $(q-1)$-gram starting in position $k+1$ in T. Recall that since this is the next successive q-gram, this computation takes constant time. If a node with label $\phi(T[k + 1 : k + (q - 1)])$ already exists we check if there is an edge from $\phi(T[k : k+(q-1)-1])$ to $\phi(T[k+1 : k+(q-1)])$. If such an edge exists we increment its counter by one. If it does not exist we create it and set its counter to 1. If a node with label $\phi(T[k + 1 : k + (q - 1)])$ does not exist we create it along with an edge from $\phi(T[k : k + (q - 1) - 1])$ to it.

Lemma 6. *For a string T of length N, the algorithm is a Monte-Carlo type randomized algorithm that builds the q-gram graph $G_q(T)$ in $O(N)$ expected time.*

4 Algorithm

Our main algorithm is comprised of four steps: preparing the SLP, constructing the q-gram graph from the SLP, turning it into a CS-tree, and computing the suffix tree of the CS-tree. Ultimately the algorithm produces a suffix tree containing the reversed q-grams of T, so to answer a query for a q-gram s we will have to lookup s^R in the suffix tree. Below we will describe the algorithm and we will show that it runs in $O(qn)$ expected time while using $O(n + k_{T,q})$ space; an improvement over the best known algorithm in terms of space usage. The catch is that a frequency query to the resulting data structure may yield incorrect results due to randomization. However, we show how to turn the algorithm from a Monte Carlo to a Las Vegas-type randomized algorithm with constant overhead. Finally, we show that by decompressing substrings of T in a specific order, we can construct the q-gram graph by decompressing exactly the same number of characters as decompressed by the best known algorithm.

The algorithm is as follows. Figure 1 shows an example of the data structures after each step of the algorithm.

Preprocessing. As the first step of our algorithm we preprocess the SLP such that we know the size of the string derived from a production rule, $|X_i|$, and the number of occurrences in the derivation tree, $occ(X_i)$. We also prepare the SLP for linear time prefix and suffix decompressions using Lemma 4.

Computing the q-gram graph. In this step we construct the q-gram graph $G_q(T)$ from the SLP $\mathcal{S}$. Initially we choose a suitable fingerprint function for the q-gram graph construction algorithm and proceed as follows. For each production rule

$X_i = X_l X_r$ in $\mathcal{S}$, such that $|X_i| \geq q$, we decompress its relevant substring r_{X_i}. Recall from the definition of relevant substrings that r_{X_i} is the concatenation of the $q-1$ length suffix of t_{X_l} and the $q-1$ length prefix of t_{X_r}. If $|X_l| \leq q-1$ we decompress the entire string t_{X_l}, and similarly for t_{X_r}. Given r_{X_i} we compute the fingerprint of the first $(q-1)$-gram, $\phi(r_{X_i}[0 : (q-1)-1])$, and find the node in $G_q(T)$ with this fingerprint as its label. The node is created if it does not exist. Now the construction of $G_q(T)$ can continue from this node, albeit with the following change to the construction algorithm. When incrementing the counter of an edge we increment it by $occ(X_i)$ instead of 1.

The q-gram graph now contains the information needed for the q-gram profile; namely the frequencies of the q-grams in T. The purpose of the next two steps is to restructure the graph to a data structure that supports frequency queries in $O(q)$ time.

Transforming the q-gram graph to a CS-tree. The CS-tree that we want to create is basically the depth-first tree of $G_q(T)$ with the extension that all edges in $G_q(T)$ are also in the tree. We create it as follows. Let the start node of $G_q(T)$ be the node whose label match the fingerprint of the first $q-1$ characters of T. Do a depth-first traversal of $G_q(T)$ starting from the start node. For a previously unvisited node, create a node in the CS-tree with an incoming edge from its predecessor. When reaching a previously visited node, create a new leaf in the CS-tree with an incoming edge from its predecessor. Labels on nodes and edges are copied from their corresponding labels in $G_q(T)$. We now create a path of length $q-1$ with the first $q-1$ characters of T as labels on its edges. We set the last node on this path to be the root of the depth-first tree. The first node on the path is the root of the final CS-tree.

Computing the suffix tree of the CS-tree. Recall that a suffix in the CS-tree starts in a node and ends in the root of the tree. Usually we store a pointer from a leaf in the suffix tree to the node in the CS-tree from which the particular suffix starts. However, when we construct the suffix tree, we store the value of the counter of the first edge in the suffix as well as the label of the first node on the path of the suffix.

4.1   Correctness

Before showing that our algorithm is correct, we will prove some crucial properties of the q-gram graph, the CS-tree, and the suffix tree of the CS-tree subsequent to their construction in the algorithm.

Lemma 7. *The q-gram graph $G_q(T)$ constructed from the SLP is connected.*

Proof. Omitted due to lack of space. $\square$

Lemma 8. *Assuming that we are given a fingerprint function ϕ that is collision free for substrings of length $q-1$ in T, then the extended CS-tree contains each distinct q-gram in T exactly once.*

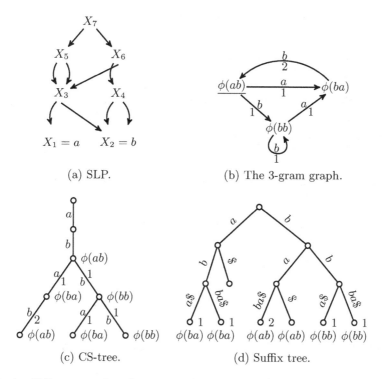

Fig. 1. An SLP compressing the string **ababbbab**, and the data structures after each step of the algorithm executed with $q = 3$

Proof. Let v be a node with an outgoing edge e in $G_q(T)$. The combination of the label of v followed by the character on e is a distinct q-gram and occurs only once in $G_q(T)$ due to the way we construct it. There may be several paths of length $q - 1$ ending in v spelling the same string s, and because the fingerprint function is deterministic, there can not be a path spelling s ending in some other node. Since the depth-first traversal of $G_q(T)$ only visits e once, the resulting CS-tree will only contain the combination of the labels on v and e once.   □

Lemma 9. *Assuming that we are given a fingerprint function ϕ that is collision free for substrings of length $q - 1$ in T, then any node v in the suffix tree of the CS-tree with $sd(v) \geq q$ is a leaf.*

Proof. Each suffix of length $\geq q$ in the CS-tree has a distinct q length prefix (Lemma 8), so therefore each node in the suffix tree with string depth $\geq q$ is a leaf.   □

We have now established the necessary properties to prove that our algorithm is correct.

Lemma 10. *Assuming that we are given a fingerprint function ϕ that is collision free on all substrings of length $q-1$ of T, our algorithm correctly computes a q-gram profile for T.*

Proof. Our algorithm inserts each relevant substring r_{X_i} exactly once, and if a q-gram s occurs $socc(s, r_{X_i})$ times in r_{X_i}, the counter on the edge representing s is incremented by exactly $socc(s, r_{X_i}) \cdot occ(X_i)$. From Lemma 3 we then know that when $G_q(T)$ is fully constructed, the counters on its edges correspond to the frequencies of the q-grams in T. Since $G_q(T)$ is connected (Lemma 7) the modified depth-first traversal will correctly produce a CS-tree, and it contains each q-gram from $G_q(T)$ exactly once (Lemma 8). Finally, we know from Lemma 9 that a node v with $sd(v) \geq q$ in the suffix tree is a leaf, so searching for a string of length q in the suffix tree will yield a unique result and can be done in $O(q)$ time. $\qquad\square$

4.2 Analysis

Theorem 2. *The algorithm runs in $O(qn)$ expected time and uses $O(n + k_{T,q})$ space.*

Proof. Let $\mathcal{S}_q = \{X_i \mid X_i \in \mathcal{S} \text{ and } |X_i| \geq q\}$ be the set of production rules that have a relevant substring. For each production rule $X_i = X_l X_r \in \mathcal{S}_q$ we decompress its relevant substring of size $|r_{X_i}|$ and insert it into the q-gram graph. Since r_{X_i} is comprised of the suffix of t_{X_l} and the prefix of t_{X_r} we know from Lemma 4 that r_{X_i} can be decompressed in $O(|r_{X_i}|)$ time. Inserting r_{X_i} into the q-gram graph can be done in $O(|r_{X_i}|)$ expected time (Lemma 6). Since $|\mathcal{S}_q| = O(n)$ and $q \leq |r_{X_i}| \leq 2(q-1)$ this step of the algorithm takes $O(qn)$ time. From Lemma 5 we know that $G_q(T)$ has size $O(k_{T,q})$, so transforming the graph to the CS-tree takes $O(k_{T,q})$ expected time. We also know that constructing the suffix tree takes expected linear time in the size of the CS-tree if we hash the characters of the alphabet to a polynomial range first (Lemma 1). Finally, observe that since our algorithm is correct, it detects all q-grams in T and therefore there can be at most $k_{T,q} \leq \sum_{X_i \in \mathcal{S}_q} |r_{X_i}| = O(qn)$ distinct q-grams in T. Thus, the expected running time of our algorithm is $O(qn)$.

In the preprocessing step of our algorithm we use $O(n)$ space to store the size of the derived substrings and the number of occurrences in the derivation tree as well as the data structure needed for linear time prefix and suffix decompressions (Lemma 4). The space used by the q-gram graph is $O(k_{T,q})$, and after the transformation and augmentation, the CS-tree and suffix tree uses $O(k_{T,q} + q)$ space. In total our algorithm uses $O(n + k_{T,q})$ space. $\qquad\square$

4.3 Verifying the Fingerprint Function

Until now we have assumed that the fingerprints used as labels for the nodes in the q-gram graph are collision free. In this section we describe an algorithm that verifies if the chosen fingerprint function is collision free using the suffix tree resultant from our algorithm.

If there is a collision among fingerprints, the q-gram graph construction algorithm will add an edge such that there are two paths of length $q - 1$ ending in the same node while spelling two different strings. This observation is formalized in the next Lemma.

Lemma 11. *For each node v in $G_q(T)$, if every path of length $q - 1$ ending in v spell the same string, then the fingerprint function used to construct $G_q(T)$ is collision free for all $(q - 1)$-grams in T.*

Proof. From the q-gram graph construction algorithm we know that we create a path of characters in the same order as we read them from T. This means that every path of length $q - 1$ ending in a node v represents the $q - 1$ characters generating the fingerprint stored in v, regardless of what comes before those $q-1$ characters. If all the paths of length $q - 1$ ending in v spell the same string s, then we know that there is no other substring $s' \neq s$ of length $q - 1$ in T that yields the fingerprint $\phi(s)$. □

It is not straightforward to check Lemma 11 directly on the q-gram graph without using too much time. However, the error introduced by a collision naturally propagates to the CS-tree and the suffix tree of the CS-tree, and as we shall now see, the suffix tree offers a clever way to check for collisions. First, recall that in a leaf v in the suffix tree, we store the fingerprint of the reversed prefix of length $q - 1$ of the suffix ending in v. Now consider the following property of the suffix tree.

Lemma 12. *Let v_ϕ be the fingerprint stored in a leaf v in the suffix tree. The fingerprint function ϕ is collision free for $(q - 1)$-grams in T if $v_\phi \neq u_\phi$ or $sd(nca(v, u)) \geq q - 1$ for all pairs v, u of leaves in the suffix tree.*

Proof. Consider the contrapositive statement: If ϕ is not collision free on T then there exists some pair v, u for which $v_\phi = u_\phi$ and $sd(nca(v, u)) < q - 1$. Assume that there is a collision. Then at least two paths of length $q - 1$ spelling the same string end in the same node in $G_q(T)$. Regardless of the order of the nodes in the depth-first traversal of $G_q(T)$, the CS-tree will have two paths of length $q - 1$ spelling different strings and yet starting in nodes storing the same fingerprint. Therefore, the suffix tree contains two suffixes that differ by at least one character in their $q - 1$ length prefix while ending in leaves storing the same fingerprint, which is what we want to show. □

Checking if there exists a pair of leaves where $v_\phi = u_\phi$ and $sd(nca(v, u)) < q - 1$ is straightforward. For each leaf we store a pointer to its ancestor w that satisfies $sd(w) \geq q-1$ and $sd(parent(w)) < q-1$. Then we visit each leaf v again and store v_ϕ in a dictionary along with the ancestor pointer just defined. If the dictionary already contains v_ϕ and the ancestor pointer points to a different node, then it means that $v_\phi = u_\phi$ and $sd(nca(v, u)) < q - 1$ for some two leaves.

The algorithm does two passes of the suffix tree which has size $O(k_{T,q} + q)$. Using a hashing scheme for the dictionary we obtain an algorithm that runs in $O(k_{T,q} + q)$ expected time.

4.4 Eliminating Redundant Decompressions

We now present an alternative approach to constructing the q-gram graph from the SLP. The resulting algorithm decompresses fewer characters.

In our first algorithm for constructing the q-gram graph we did not specify in which order to insert the relevant substrings into the graph. For that reason we do not know from which node to resume construction of the graph when inserting a new relevant substring. So to determine the node to continue from, we need to compute the fingerprint of the first $(q-1)$-gram of each relevant substring. In other words, the relevant substrings are overlapping, and consequently some characters are decompressed more than once. Our improved algorithm is based on the following observation. Consider a production rule $X_i = X_l X_r$. If all relevant substrings of production rules reachable from X_l (including r_{X_i}) have been inserted into the graph, then we know that all q-grams in t_{X_l} are in the graph. Since the $q-1$ length prefix of r_{X_i} is also a suffix of t_{X_l}, then we know that a node with the label $\phi(r_{X_i}[0:(q-1)-1])$ is already in the graph. Hence, after inserting all relevant substrings of production rules reachable from X_l we can proceed to insert r_{X_i} without having to decompress $r_{X_i}[0:(q-1)-1]$.

Algorithm. First we compute and store the size of the relevant substring $|r_{X_i}| = \min(q-1, |X_l|) + \min(q-1, |X_r|)$ for each production rule $X_i = X_l X_r$ in the subset $\mathcal{S}_q = \{X_i \mid X_i \in \mathcal{S} \text{ and } X_i \geq q\}$ of the production rules in the SLP. We maintain a linked list L with a pointer to its head and tail, denoted by $head(L)$ and $tail(L)$. The list initially contains the leftmost node in $\mathcal{S}_q$, say X_k, from the root of the SLP and the sentinel value $|X_k|$. We now start decompressing T by traversing the SLP depth-first, left-to-right. When following a pointer from X_i to a right child, and $X_i \in \mathcal{S}_q$, we add X_i and the sentinel value $|r_{X_i}| - (q-1)$ to the back of L. As characters are decompressed they are fed to the q-gram graph construction algorithm, and when a counter on an edge in $G_q(T)$ is incremented, we increment it by $occ(head(L))$. For each character we decompress, we decrement the sentinel value for $head(L)$, and if this value becomes 0 we remove the head of the list and set $head(L)$ to be the next production rule in the list. When leaving a node X_i we mark it as visited and store a pointer from X_i to the node with label $\phi(t_{X_i}[|X_i| - (q-1) : |X_i| - 1])$ in $G_q(T)$. If we encounter a node that has already been visited, we decompress its $q-1$ length prefix, set the node with label $\phi(t_{X_i}[|X_i| - (q-1) : |X_i| - 1])$ to be the node from where construction of the q-gram graph should continue, and do not proceed to visit its children nor add it to L.

Analysis. Assume without loss of generality that the algorithm is at a production rule deriving the string $t_{X_i} = t_{X_l} t_{X_r}$ and all q-grams in t_{X_l} are in $G_q(T)$. Since we start by decompressing the leftmost production rule in $\mathcal{S}_q$ there is always such a rule. We then decompress $|r_{X_i}| - (q-1)$ characters before X_i is removed from the list. We only add a production once to the list, so the total number of characters decompressed is $(q-1) + \sum_{X_i \in \mathcal{S}_q} |r_{X_i}| - (q-1) = O(N - \alpha)$, and we hereby obtain our result from Theorem 1. This is fewer characters than our first algorithm that require $\sum_{X_i \in \mathcal{S}_q} |r_{X_i}|$ characters to be decompressed. Furthermore, it is exactly the same number of characters decompressed by the fastest known algorithm due to Goto et al. [6].

References

1. Charikar, M., Lehman, E., Liu, D., Panigrahy, R., Prabhakaran, M., Sahai, A., Shelat, A.: The smallest grammar problem. IEEE Trans. Inf. Theory 51(7), 2554–2576 (2005)
2. Farach, M.: Optimal suffix tree construction with large alphabets. In: Proc. 38th FOCS, pp. 137–143 (1997)
3. Gärtner, T.: A survey of kernels for structured data. ACM SIGKDD Explorations Newsletter 5(1), 49–58 (2003)
4. Gąsieniec, L., Kolpakov, R., Potapov, I., Sant, P.: Real-time traversal in grammar-based compressed files. In: Proc. 15th DCC, p. 458 (2005)
5. Goto, K., Bannai, H., Inenaga, S., Takeda, M.: Fast q-gram mining on SLP compressed strings. In: Grossi, R., Sebastiani, F., Silvestri, F. (eds.) SPIRE 2011. LNCS, vol. 7024, pp. 278–289. Springer, Heidelberg (2011)
6. Goto, K., Bannai, H., Inenaga, S., Takeda, M.: Speeding up q-gram mining on grammar-based compressed texts. In: Kärkkäinen, J., Stoye, J. (eds.) CPM 2012. LNCS, vol. 7354, pp. 220–231. Springer, Heidelberg (2012)
7. Kärkkäinen, J., Sutinen, E.: Lempel–Ziv index for q-grams. Algorithmica 21(1), 137–154 (1998)
8. Karp, R.M., Rabin, M.O.: Efficient randomized pattern-matching algorithms. IBM J. Res. Dev. 31(2), 249–260 (1987)
9. Leslie, C., Eskin, E., Noble, W.S.: The spectrum kernel: A string kernel for SVM protein classification. In: Proc. PSB, vol. 7, pp. 566–575 (2002)
10. Matsubara, W., Inenaga, S., Ishino, A., Shinohara, A., Nakamura, T., Hashimoto, K.: Efficient algorithms to compute compressed longest common substrings and compressed palindromes. Theoret. Comput. Sci. 410(8), 900–913 (2009)
11. Paaß, G., Leopold, E., Larson, M., Kindermann, J., Eickeler, S.: SVM classification using sequences of phonemes and syllables. In: Elomaa, T., Mannila, H., Toivonen, H. (eds.) PKDD 2002. LNCS (LNAI), vol. 2431, p. 373. Springer, Heidelberg (2002)
12. Rytter, W.: Application of Lempel–Ziv factorization to the approximation of grammar-based compression. Theoret. Comput. Sci. 302(1), 211–222 (2003)
13. Shibuya, T.: Constructing the suffix tree of a tree with a large alphabet. IEICE Trans. Fundamentals 86(5), 1061–1066 (2003)
14. Ukkonen, E.: Approximate string-matching with q-grams and maximal matches. Theoret. Comput. Sci. 92(1), 191–211 (1992)
15. Ziv, J., Lempel, A.: A universal algorithm for sequential data compression. IEEE Trans. Inf. Theory 23(3), 337–343 (1977)
16. Ziv, J., Lempel, A.: Compression of individual sequences via variable-rate coding. IEEE Trans. Inf. Theory 24(5), 530–536 (1978)

A Constant-Space Comparison-Based Algorithm for Computing the Burrows–Wheeler Transform*

Maxime Crochemore[1], Roberto Grossi[2],
Juha Kärkkäinen[3], and Gad M. Landau[4]

[1] King's College London, UK
Maxime.Crochemore@kcl.ac.uk
[2] Dipartimento di Informatica, Università di Pisa, Italy
grossi@di.unipi.it
[3] Department of Computer Science, University of Helsinki, Finland
Juha.Karkkainen@cs.helsinki.fi
[4] Department of Computer Science, University of Haifa, Israel, and
Department of Computer Science and Engineering, NYU-Poly, Brooklyn NY, USA
landau@cs.haifa.ac.il

Abstract. We introduce the problem of computing the Burrows–Wheeler Transform (BWT) using just $O(1)$ additional space. Our in-place algorithm does not need the explicit storage for the suffix sort array and the output array, as typically required in previous work. It relies on the combinatorial properties of the BWT, and runs in $O(n^2)$ time in the comparison model using $O(1)$ extra memory cells, apart from the array of n cells storing the n characters of the input text. We also discuss some time-space trade-offs for the inverse algorithm to obtain the text from the given BWT.

1 Introduction

The Burrows–Wheeler Transform [2] (known as BWT) of a text string is at the heart of the `bzip2` family of text compressors, and finds also applications in text indexing and sequence processing. Consider an input text string $T \equiv T[0 \, . . \, n-1]$ and the set of its suffixes $T_i \equiv T[i \, . . \, n - 1]$ $(0 \le i < n)$ under the lexicographic order, where $T[n-1]$ is an endmarker symbol \$ smaller than any other symbol in T. The alphabet Σ from which the symbols in T are drawn can be unbounded.

A classical way to define the BWT uses the n circular shifts of the text $T = $ `mississippi$` as shown in the first column of Table 1. We perform a lexicographic sort of these shifts, as shown in the second column: if we mark the last symbol from each of the circular shifts in this order, we obtain a sequence L of n symbols that is called the BWT of T. Its relation with suffix sort is well

---

* The work of the third author has been supported by the Academy of Finland grant 118653 (ALGODAN). The work of the fourth author has been partially supported by the National Science Foundation Award 0904246, Israel Science Foundation grant 347/09, Yahoo, Grant No. 2008217 from the United States-Israel Binational Science Foundation (BSF) and DFG.

J. Fischer and P. Sanders (Eds.): CPM 2013, LNCS 7922, pp. 74–82, 2013.

Table 1. BWT L for the text $T = $ mississippi$ and its relation with suffix sort

cyclic shifts	sorted cyclic shifts	suffixes
	L	i T_i
mississippi$	$mississipp i	11 $
$mississippi	i$mississip p	10 i$
i$mississipp	ippi$missis s	7 ippi$
pi$mississip	issippi$mis s	4 issippi$
ppi$mississi	issississippi$ m	1 ississippi$
ippi$mississ	mississippi $	0 mississippi$
sippi$missis	pi$mississi p	9 pi$
ssippi$missi	ppi$mississ i	8 ppi$
issippi$miss	sippi$missi s	6 sippi$
sissippi$mis	sissippi$mi s	3 sissippi$
ssissippi$mi	ssippi$miss i	5 ssippi$
ississippi$m	ssissippi$m i	2 ssissippi$

known, as illustrated in the third column: the rth symbol in L is $T[j-1]$ if and only if T_j is the rth suffix in the sort (except the borderline case $j = 0$, for which we take $T[n-1]$),

As it can be seen in the example of Table 1, the BWT produces a text of the same length as the input text T. The transform is reversible since it is a one-to-one function when the input text is terminated by an endmarker $. Thus, not only we can recover T from L alone, but typically L is more compressible than T itself using 0th-order compressors [16]. There are now efficient methods that convert T to L and vice versa, taking $O(n \log n)$ time for unbounded alphabets in the worst case [1].[1] The BWT is also a key element of some compressed text indexing implementations due to the small amount of space it requires: some examples are the solution by Ferragina and Manzini [5] or that by Grossi et al. [7], where the transform is associated with the techniques of wavelet trees and of succinct data structures using rank-select queries on binary sequences [17].

One of the prominent applications of the BWT is for software dealing with Next Generation Sequencing, where millions of short strings, called reads, are mapped onto a reference genome. Typical and popular software of this type are Bowtie [13], BWA [14] and SOAP2 [12]. Here it is crucial that the genome is indexed in a compact manner to get reasonable running time. Space issues for computing the BWT are thus relevant: it is not rare the case when the input data is so large that the input text T stays in main memory but any additional data structure of similar size cannot fit in the rest of the main memory [10].

All the previous work for computing the BWT of T relies on the fact that (a) we need first to *store* the suffix sorting of T (also known as suffix array [15]), thus occupying n memory cells for storing integers, and (b) we need to output the BWT in *another* array storing n characters. Motivated by these observations,

---

[1] As is standard in many string algorithms, we assume that any two symbols in Σ can only be compared and this takes $O(1)$ time. Hence, comparing symbolwise any two suffixes may require $O(n)$ time in the worst case.

we want to study the case in which (a) and (b) are avoided, thus saving over the space occupied by them.

In this paper, our goal is to obtain the BWT by directly permuting T and using just $O(1)$ memory cells, i.e., we aim at an *in-place* algorithm for computing the BWT. We consider the model in which the text T is stored as an array of n entries, where each entry stores exactly one character of T. Note that storing an integer usually takes more space than a character, so we assume that only the characters of T can be kept in the array T. Moreover, T is not read-only but it can be modified at any time, and just $O(1)$ additional memory cells (besides T) can be kept for storing auxiliary information.[2]

Note that our model represents some realistic situations in which one has to handle large text collections, or large genomic sequences, without relying on extra memory for (a) and (b). Hence it is crucial to maximize the amount of data that can fit into main memory: not storing explicitly (a) and (b) permits to save space, which is typically regarded as taking more than half of the total space required. For instance, DNA sequences are stored by using 2 bits per character and machine integers take 64 bits. Here we just need $2n$ bits to store the (genomic) text and save the $64n$ bits for storing the intermediate suffix sorting in (a) and the $2n$ bits for storing the output of BWT in another array in (b): this means that during the BWT construction, we can fit almost 33 times more text using the same main memory size, thus eliminating the usage of the slower external memory for this time-consuming task in these cases.

From the combinatorial point of view, the in-place BWT is an interesting question to solve on strings. There are space saving approaches storing the suffix sorting in compressed form [9,19,10,20] or only partially at a time [11], but none of them provides an in-place algorithm. In-place selection and sorting does not seem to help either [4,6,8,18,21]. It well known that in-place sorting requires the same comparison cost of $\Theta(n \log n)$ as in standard sorting. But for the BWT, we only know its comparison cost of $\Theta(n \log n)$ for the standard construction. As far as we know, no result is known for the in-place construction of BWT: a naive solution is not that simple, even allowing for exponential time. Indeed, any movement of a character $T[j]$ to another position inside T at least changes the content of its suffixes T_i for $0 \leq i \leq j$, making the algorithmic flavor of this problem different from that of in-place sorting n elements.

The above discussion suggests that a careful orchestration of the movement of the characters inside T is needed to avoid losing the content of some suffixes before they contribute to the BWT. Our idea is to define a sequence of transformations $B_0, B_1, \ldots, B_n$, where B_0 is the input text T and B_n is the final BWT of T. For $1 \leq k \leq n$, we have that B_k is the BWT of the last k characters in T and is computed from B_{k-1} (re)using just $O(1)$ extra memory cells. We think that this sequence of transformations could be of independent interest for the community of string algorithms, and some of the combinatorial properties can be found in [22].

---

[2] In C code, we would declare T as `unsigned char T[n]` and use this storage plus $O(1)$ local variables of constant size.

In this paper we propose an $O(n^2)$-time approach that builds the above sequence of transformations using four integer variables and one character variable, taking $O(n)$ time per transformation in the worst case. The resulting in-place algorithm is simple and can be easily encoded in few lines of C code or similar programming languages. However we do not claim any practicality of our solution due to its quadratic cost. Our contribution is that it could lay out the path towards faster methods for the space-efficient computation of the BWT: any method to compute B_k from B_{k-1} in $t(s)$ time (re)using $s(n)$ space, would lead to a construction of the BWT in $O(n \cdot t(n))$ time using $O(1 + s(n))$ space. To this end, it is worth noting that the inputs for BWT are typically large and a fast algorithm that is in-place or uses very low additional memory, would be relevant in practice.

The paper is organized as follows. We describe how to perform the in-place BWT in Section 2. We then discuss how to invert the BWT, so as to obtain the original text T, in Section 3. Finally, we draw some conclusions in Section 4.

2   In-Place BWT

Given the input text $T = T[0 \dots n-1]$ where $T[n-1] = \$$, moving a single character inside T can change the content of many suffixes. The idea to circumvent this difficulty without using storage for the suffix sort is to proceed by induction from right to left in T, while maintaining the BWT of the current suffix T_s, denoted by $\mathrm{BWT}(T_s)$. We assume $0 \leq s \leq n-3$, since the last two suffixes of T are equal to their respective BWT.

To compute $\mathrm{BWT}(T_s)$, suppose that $\mathrm{BWT}(T_{s+1})$ has been already computed and stored in the last positions of T, i.e. $T[s+1 \dots n-1]$. Consider the current symbol $c = T[s]$: if we look at the content of $T[s \dots n-1]$, we no longer find T_s, but the symbol c followed by the permutation $\mathrm{BWT}(T_{s+1})$ of T_{s+1}. Nevertheless, we still have enough information as we will show in the proof of Theorem 1 that the position of $\$$ inside $\mathrm{BWT}(T_{s+1})$ is related to the rank of T_{s+1} among the suffixes $T_{s+1}, \dots, T_{n-1}$. We exploit this fact in the following steps.

1. Find the position p of the $\$$ in $T[s+1 \dots n-1]$: note that $p - s$ is the (local) rank of the suffix T_{s+1} that originally was starting at position $s+1$.
2. Find the rank r of the suffix T_s (originally in position s) using just symbol c. To this end, scan $T[s+1 \dots n-1]$ and count how many symbols are strictly smaller than c and how many occurrences of c appear in $T[s+1 \dots p]$ (and add s as an offset to obtain r).
3. Store c into $T[p]$ (thus replacing the $\$$).
4. Insert the symbol $\$$ in $T[r]$ by shifting $T[s+1 \dots r]$ by one position left.

The simple and short C code reported in Fig. 1 implements Steps 1–4 above, where END_MARKER denotes $\$$. For example, consider $T = \mathtt{mississippi\$}$ and $s = 4$, where we use capital letters to denote the BWT partially built on the last positions of T. Suppose that we have already computed the BWT for the last 7 characters in T, namely, we have $\mathtt{missiIPSPIS\$}$. We then have $p = 11$

```
void inplaceBWT( unsigned char T[ ], int n ){
 int i, p, r, s;
 unsigned char c;

 for ( s = n-3;  s >= 0; s-- ){
   c = T[ s ];

   /* steps 1 and 2 */
   r = s;
   for ( i = s+1; T[ i ] != END_MARKER; i++ )
     if ( T[ i ] <= c ) r++;
   p = i;
   while ( i < n )
     if ( T[ i++ ] < c ) r++;

   /* step 3 */
   T[ p ] = c;

   /* step 4 */
   for ( i = s; i < r; i++ )
     T[ i ] = T[ i+1 ];
   T[ r ] = END_MARKER;
 }
}
```

Fig. 1. In-place construction of BWT

and, since there is one symbol ($) smaller than $c = $ i, and two symbols that are equal to c and occur before position p, we have $r = s + 3 = 7$. This means that we have to replace $ by c and shift IPS by one position left so as to insert $ in position r. The next configuration is missIPS$PISI, which is the BWT of T_s.

Theorem 1. *Given a text T of n symbols, we can compute its Burrows–Wheeler Transform (BWT) in $O(n^2)$ time in the comparison model using $O(1)$ additional memory cells.*

Proof. We prove first the correctness. Let T be the input text and T' be its modification at a generic iteration s, where $0 \leq s \leq n - 3$. Note that $T'[0..s] = T[0..s]$ while $T'[s + 1..n - 1] = \text{BWT}(T[s + 1..n - 1])$. By induction, the position p of $ in $T'[s + 1..n - 1]$ indicates the rank $p - s$ of T_{s+1} among the suffixes in $\{T_{s+1}, T_{s+2}, \ldots, T_{n-1}\}$. The base case for T_{n-2} and T_{n-1} is trivially satisfied. Hence, we show how to preserve this property for $0 \leq s \leq n - 3$.

First note that the symbol $c = T[s]$ goes in position p, since it precedes T_{s+1} inside T. Next, we have to find the new position r for T_s, so that $r - (s - 1)$ is its rank among the suffixes in $S = \{T_s, T_{s+1}, \ldots, T_{n-1}\}$. First count how many symbols smaller than c occur in $T'[s + 1..n - 1]$: there are as many suffixes in S that are smaller than T_s since their first character is smaller than c.

To this quantity, add the number of occurrences of c in $T'[s+1..p]$: these are also smaller since they start with c but have rank smaller than p, i.e. the rank of T_{s+1}. In this way, we discover how many suffixes are smaller than T_s in S: inserting $ in the corresponding location of T', by shifting some of the characters in $T'[s..n-1]$, we maintain the induction. Hence, $T'[0..s-1] = T[0..s-1]$ and $T'[s..n-1] = \mathrm{BWT}(T[s..n-1])$. When $s = 0$, we obtain the BWT of T.

As for the complexity, note that each of the $n-2$ iterations requires $O(n)$ time, since it can be implemented by $O(1)$ scans of $T'[s..n-1]$. This gives a total cost of $O(n^2)$. We use four integer variables (i, p, r, s) and one character variable (c) in the C code shown in Fig. 1, and thus we need $O(1)$ memory cells for the local variables.   □

3   Inverting the BWT

Reversing the permutation performed by the in-place BWT is called *inverting* the BWT. Initially we have the BWT of the original input text T, denoted $\mathrm{BWT}(T)$. We want to invert the latter by permuting its symbols. Thus we reverse the approach described in Section 2. We maintain the invariant that there is a pointer L to a certain position in the input buffer storing $\mathrm{BWT}(T)$ so that, at any time, (a) the prefix of the buffer to the left of L stores the prefix of T obtained so far by the inverting process and (b) the remaining suffix of the buffer (pointed by L till the end of the input buffer) stores the portion of the BWT still to be inverted. For the sake as discussion, we identify L with the entire suffix of the input buffer that still has to be inverted.

Under this invariant, which is initially true by setting L to the beginning of the input buffer for $\mathrm{BWT}(T)$, we proceed as follows. We find the position p of $ in L, and then select the pth symbol in the multiset given by the symbols of L. Stability is needed, since equal symbols should be considered in the order of their appearance in L, as detailed below.

1. Find the position p of the $ in L, and increment p (since array indexing starts from 0).
2. Let `select` be a selection algorithm that works on read-only input, i.e., it does not move elements around while finding the pth smallest element. Using `select` on L, select the pth symbol c in the multiset of the symbols of L
3. Let f denote the fth occurrence of c, which we hit in a stable fashion when finding c in L, and let q be the position of this occurrence inside L.
4. Replace the occurrence of c at position q by $, and remove the old occurrence of $ by shifting to the right the symbols in L.
5. At this point, the first position in L is free: store the symbol c in it, and shorten L by one symbol at the beginning (i.e. advance the pointer L by one position towards the end of the input buffer).

The C code in Fig. 2 implements Steps 1–5 above, where `END_MARKER` denotes $. Note that it is a bit longer than the code for the in-place BWT in Fig. 1. The proof of correctness proceeds along the same lines as in the proof of Theorem 1,

```
void inplaceIBWT( unsigned char L[ ], int n ){
  int f, i, p, q, count;
  unsigned char c;

  /* step 1 */
  p = 0;
  while( L[ p ] != END_MARKER )
    p++;
  p++;

  while ( n > 2 ){
    /* step 2 */
    c = select( L, p );
    count = 0;
    for ( i = 0; i < n; i++ ){
      if (T[i] < c) count++;
    }
    /* step 3 */
    f = p - count;
    q = -1;
    while ( f > 0 ){
      q++;
      if ( L[ q ] == c ) f--;
    }
    /* step 4 */
    L[ q ] = END_MARKER;
    for ( i = p-1; i > 0; i-- ){
      L[i] = L[i-1];
    }
    /* step 5 */
    L[0] = c;
    L++; n--;
    /* step 1 */
    if (p-1 > q)
      p = q+1; /* also the new END_MARKER has been shifted */
    else
      p = q;
  }
}
```

Fig. 2. Reverting the permutation of the inverse BWT

since we are reversing the procedure described there. As for the complexity, each of the $n - 2$ iterations is dominated by the cost of select. We obtain the following result.

Theorem 2. *Let $t_s(n)$ be the time complexity in the comparison model and $s_s(n)$ be the space complexity required by* select. *Given the BWT of a text T of n symbols, we can recover T by permuting the BWT (also known as inverse*

BWT$)$ in $O(n \cdot t_s(n))$ time in the comparison model using $O(1+s_s(n))$ additional memory cells.

We give some examples of the bounds that can be attained with Theorem 2. Using the result in [18], where $t_s(n)$ is $O(n^{1+\epsilon})$ in the worst case for any fixed small constant $\epsilon > 0$, and $O(n \log \log n)$ on the average (which meets the randomized lower bound in [3]), with $s_s(n) = O(1)$, we have the following.

Corollary 1. *Given the* BWT *of a text* T *of* n *symbols, we can recover* T *by inverting the* BWT *in* $O(n^{5/2})$ *time in the worst case, or* $O(n^2 \log \log n)$ *time on the average, in the comparison model using* $O(1)$ *additional memory cells.*

Using slightly more additional space than a constant—literally speaking, the algorithm is no more in-place—and the result in [21], where $t_s(n) = O(n(\log n)^2)$ and $s_s(n) = O(\log n)$, we derive the following.

Corollary 2. *Given the* BWT *of a text* T *of* n *symbols, we can recover* T *by inverting the* BWT *in* $O((n \log n)^2)$ *time in the comparison model using* $O(\log n)$ *additional memory cells.*

Finally, for the special case in which the alphabet of the distinct symbols in T is of constant size (as in DNA and ASCII texts), we obtain an improved bound since select can be immediately implemented by a simple scheme that employs $O(|\Sigma|) = O(1)$ counters.

Corollary 3. *Given the* BWT *of a text* T *of* n *symbols drawn from a constant-size alphabet, we can recover* T *by inverting the* BWT *in* $O(n^2)$ *time in the comparison model using* $O(1)$ *additional memory cells.*

4   Conclusions

We presented an in-place BWT construction taking $O(n^2)$ time in the comparison model. It would be interesting to improve this bound. Note that the while loop in our in-place BWT can be avoided using $O(\Sigma)$ space, where Σ is the alphabet of symbols occurring in T. Unfortunately, some experiments on DNA sequences and on the Calgary corpus show that the overall number of steps of the inner for loop (preceding the while loop) is roughly $n^2/4$ and, with some trickery, it can be reduced to $n^2/10$. In any case, the cost of the in-place BWT in practical situations is actually $\Theta(n^2)$. Time can be further reduced to $O(n^2/\log_\sigma n)$ by packing symbols but still not useful for large text collections.

We do not know whether a lower bound better than $\Omega(n \log n)$ holds for the problem in the comparison model since the space is very constrained. This is an interesting question to investigate.

Acknowledgments. The second author is grateful to Gianni Franceschini for some preliminary discussions and to Venkatesh Raman for pointing out the results in [18,21]. We also thank some anonymous reviewers for their suggestions to make explicit some observations in the introduction and to improve the written text.

References

1. Adjeroh, D., Bell, T., Mukherjee, A.: The Burrows–Wheeler Transform: Data Compression, Suffix Arrays, and Pattern Matching. Springer (2008)
2. Burrows, M., Wheeler, D.J.: A block-sorting lossless data compression algorithm. Research Report 124, Digital SRC, Palo Alto, CA, USA (May 1994)
3. Chan, T.M.: Comparison-based time-space lower bounds for selection. ACM Trans. Algorithms 6(2), 1–16 (2010)
4. Dobkin, D.J., Ian Munro, J.: Optimal time minimal space selection algorithms. Journal of the ACM 28(3), 454–461 (1981)
5. Ferragina, P., Manzini, G.: Indexing compressed text. J. ACM 52(4), 552–581 (2005)
6. Franceschini, G., Muthukrishnan, S.: In-Place Suffix Sorting. In: Arge, L., Cachin, C., Jurdziński, T., Tarlecki, A. (eds.) ICALP 2007. LNCS, vol. 4596, pp. 533–545. Springer, Heidelberg (2007)
7. Grossi, R., Gupta, A., Vitter, J.S.: High-order entropy-compressed text indexes. In: SODA, pp. 841–850 (2003)
8. Hoare, C.A.R.: Algorithm 65: Find. Communications of the ACM 4(7), 321–322 (1961)
9. Hon, W.-K., Lam, T.W., Sadakane, K., Sung, W.-K., Yiu, S.-M.: A space and time efficient algorithm for constructing compressed suffix arrays. Algorithmica 48(1), 23–36 (2007)
10. Hon, W.-K., Sadakane, K., Sung, W.-K.: Breaking a time-and-space barrier in constructing full-text indices. SIAM J. Comput. 38(6), 2162–2178 (2009)
11. Kärkkäinen, J.: Fast BWT in small space by blockwise suffix sorting. Theor. Comput. Sci. 387(3), 249–257 (2007)
12. Lam, T.W., Li, R., Tam, A., Wong, S., Wu, E., Yiu, S.M.: High Throughput Short Read Alignment via Bi-directional BWT. In: IEEE International Conference on Bioinformatics and Biomedicine, pp. 31–36 (2009)
13. Langmead, B., Trapnell, C., Pop, M., Salzberg, S.L.: Ultrafast and memory-efficient alignment of short DNA sequences to the human genome. Genome Biology 10(3), R25 (2009)
14. Li, H., Durbin, R.: Fast and accurate long-read alignment with Burrows-Wheeler transform. Bioinformatics 26(5), 589–595 (2010)
15. Manber, U., Myers, G.: Suffix arrays: A new method for on-line string searches. SIAM Journal on Computing 22(5), 935–948 (1993)
16. Manzini, G.: An analysis of the Burrows-Wheeler transform. J. ACM 48(3), 407–430 (2001)
17. Ian Munro, J.: Tables. In: Chandru, V., Vinay, V. (eds.) FSTTCS 1996. LNCS, vol. 1180, pp. 37–42. Springer, Heidelberg (1996)
18. Ian Munro, J., Raman, V.: Selection from read-only memory and sorting with minimum data movement. Theoretical Computer Science 165(2), 311–323 (1996)
19. Na, J.C., Park, K.: Alphabet-independent linear-time construction of compressed suffix arrays using o(nlogn)-bit working space. Theor. Comput. Sci. 385(1-3), 127–136 (2007)
20. Okanohara, D., Sadakane, K.: A linear-time Burrows-Wheeler transform using induced sorting. In: Karlgren, J., Tarhio, J., Hyyrö, H. (eds.) SPIRE 2009. LNCS, vol. 5721, pp. 90–101. Springer, Heidelberg (2009)
21. Raman, V., Ramnath, S.: Improved Upper Bounds for Time-Space Trade-offs for Selection. Nordic J. Computing 6(2), 162–180 (1999)
22. Salson, M., Lecroq, T., Léonard, M., Mouchard, L.: A four-stage algorithm for updating a Burrows–Wheeler Transform. Theor. Comput. Sci. 410(43), 4350–4359 (2009)

Pattern Matching with Variables: A Multivariate Complexity Analysis

(Extended Abstract)

Henning Fernau and Markus L. Schmid[*]

Fachbereich 4 – Abteilung Informatik, Universität Trier, D-54286 Trier, Germany
{Fernau,MSchmid}@uni-trier.de

Abstract. In the context of this paper, a pattern is a string that contains variables and terminals. A pattern α matches a terminal word w if w can be obtained by uniformly substituting the variables of α by terminal words. It is a well-known fact that deciding whether a given terminal word matches a given pattern is an NP-complete problem. In this work, we consider numerous parameters of this problem and for all possible combinations of these parameters, we investigate the question whether or not the variant obtained by bounding these parameters by constants can be solved efficiently.

Keywords: Parameterised Pattern Matching, Function Matching, NP-Completeness, Membership Problem for Pattern Languages, Morphisms.

1 Introduction

In the present work, a detailed complexity analysis of a computationally hard pattern matching problem is provided. The *patterns* considered in this context are strings containing *variables* from $\{x_1, x_2, x_3, \ldots\}$ and *terminal symbols* from a finite alphabet Σ, e.g., $\alpha := x_1\,\mathsf{a}\,x_1\,\mathsf{b}\,x_2\,x_2$ is a pattern, where $\mathsf{a}, \mathsf{b} \in \Sigma$. We say that a word w over Σ *matches* a pattern α if and only if w can be derived from α by uniformly substituting the variables in α by terminal words. The respective pattern matching problem is then to decide for a given pattern and a given word, whether or not the word matches the pattern. For example, the pattern α from above is matched by the word $u := \mathsf{bacaabacabbaba}$, since substituting x_1 and x_2 in α by baca and ba, respectively, yields u. On the other hand, α is not matched by the word $v := \mathsf{cbcabbcbbccbc}$, since v cannot be obtained by substituting the variables of α by some words.

To the knowledge of the authors, this kind of pattern matching problem first appeared in the literature in 1979 in form of the membership problem for Angluin's *pattern languages* [3, 4] (i.e., the set of all words that match a certain pattern) and, independently, it has been studied by Ehrenfeucht and Rozenberg in [9], where they investigate the more general problem of deciding on the

---

[*] Corresponding author.

J. Fischer and P. Sanders (Eds.): CPM 2013, LNCS 7922, pp. 83–94, 2013.
© Springer-Verlag Berlin Heidelberg 2013

existence of a morphism between two given words (which is equivalent to the above pattern matching problem, if the patterns are *terminal-free*, i. e., they only contain variables).

Since their introduction by Angluin, pattern languages have been intensely studied in the learning theory community in the context of inductive inference (see, e. g., Angluin [4], Shinohara [28], Reidenbach [22, 23] and, for a survey, Ng and Shinohara [20]) and, furthermore, their language theoretical aspects have been investigated (see, e. g., Angluin [4], Jiang et al. [17], Ohlebusch and Ukkonen [21], Freydenberger and Reidenbach [10], Bremer and Freydenberger [6]). However, a detailed investigation of the complexity of their membership problem, i. e., the above described pattern matching problem, has been somewhat neglected. Some of the early work that is worth mentioning in this regard is by Ibarra et al. [15], who provide a more thorough worst case complexity analysis, and by Shinohara [29], who shows that matching patterns with variables can be done in polynomial time for certain special classes of patterns. Recently, Reidenbach and Schmid [24, 25] identify complicated structural parameters of patterns that, if bounded by a constant, allow the corresponding matching problem to be performed in polynomial time (see also Schmid [27]).

In the pattern matching community, independent from Angluin's work, the above described pattern matching problem has been rediscovered by a series of papers. This development starts with [5] in which Baker introduces so-called *parameterised pattern matching*, where a text is not searched for all occurrences of a specific factor, but for all occurrences of factors that satisfy a given pattern with parameters (i. e., variables). In the original version of parameterised pattern matching, the variables in the pattern can only be substituted by single symbols and, furthermore, the substitution must be injective, i. e., different variables cannot be substituted by the same symbol. Amir et al. [1] generalise this problem to *function matching* by dropping the injectivity condition and in [2], Amir and Nor introduce *generalized function matching*, where variables can be substituted by words instead of single symbols and "don't care" symbols can be used in addition to variables. In 2009, Clifford et al. [8] considered generalised function matching as introduced by Amir and Nor, but without "don't care" symbols, which leads to patterns as introduced by Angluin.

In [2], motivations for this kind of pattern matching can be found from such diverse areas as software engineering, image searching, DNA analysis, poetry and music analysis, or author validation. Another motivation arises from the observation that the problem of matching patterns with variables constitutes a special case of the matchtest for *regular expressions with backreferences* (see, e. g., Câmpeanu et al. [7]), which nowadays are a standard element of most text editors and programming languages (cf. Friedl [12]). Due to its simple definition, the above described pattern matching paradigm also has connections to numerous other areas of theoretical computer science and discrete mathematics, such as (un-)avoidable patterns (cf. Jiang et al. [16]), word equations (cf. Mateescu and Salomaa [19]), the ambiguity of morphisms (cf. Freydenberger et al. [11]) and equality sets (cf. Harju and Karhumäki [14]).

It is a well-known fact that – in its general sense – pattern matching with variables is an NP-complete problem; a result that has been independently reported several times (cf. Angluin [4], Ehrenfeucht and Rozenberg [9], Clifford et al. [8]). However, there are many different versions of the problem, tailored to different aspects and research questions. For example, in Angluin's original definition, variables can only be substituted by non-empty words and Shinohara soon afterwards complemented this definition in [28] by including the empty word as well. This marginal difference, as pointed out by numerous results, can have a substantial impact on learnability and decidability questions of the corresponding classes of *nonerasing* pattern languages on the one hand and *erasing* pattern languages on the other. If we turn from the languages point of view of patterns to the respective pattern matching task, then, at a first glance, this difference whether or not variables can be erased seems negligible. However, in the context of pattern matching, other aspects are relevant, which for pattern languages are only of secondary importance. For example, requiring variables to be substituted in an *injective* way is a natural assumption for most pattern matching tasks and bounding the maximal length of these terminal words by a constant (which would turn pattern languages into finite languages) makes sense for special applications (cf. Baker [5]). Hence, there are many variants of the above described pattern matching problem, each with its individual motivation, and the computational hardness of all these variants cannot directly be concluded from the existing NP-completeness results.

For a systematic investigation, we consider the following natural parameters: the number of different variables in the pattern, the maximal number of occurrences of the same variable in the pattern, the length of the terminal word, the maximum length of the substitution words for variables and the cardinality of the terminal alphabet. For all combinations of these parameters, we answer the question whether or not the parameters can be bounded by (preferably small) constants such that the resulting variant of the pattern matching problem is still NP-complete. In addition to this, we also study the differences between the erasing and nonerasing case, between the injective and non-injective case and between the case where patterns may contain terminal symbols and the terminal-free case.

Due to space constraints, the formal proofs for most of the results presented in this paper are omitted.

2   Definitions

Let $\mathbb{N} := \{1, 2, 3, \ldots\}$. For an arbitrary alphabet A, a *word* (*over* A) is a finite sequence of symbols from A, and ε is the *empty word*. The notation A^+ denotes the set of all non-empty words over A, and $A^* := A^+ \cup \{\varepsilon\}$. For the *concatenation* of two words w_1, w_2 we write $w_1 w_2$. We say that a word $v \in A^*$ is a *factor* of a word $w \in A^*$ if there are $u_1, u_2 \in A^*$ such that $w = u_1 v u_2$. The notation $|K|$ stands for the size of a set K or the length of a word K.

Let $X := \{x_1, x_2, x_3, \ldots\}$ and every $x \in X$ is a *variable*. Let Σ be a finite alphabet of *terminals*. Every $\alpha \in (X \cup \Sigma)^+$ is a *pattern* and every $w \in \Sigma^*$ is a (*terminal*) *word*. For any pattern α, we refer to the set of variables in α as var(α) and, for any variable $x \in$ var(α), $|\alpha|_x$ denotes the number of occurrences of x in α.

Let α be a pattern. A *substitution* (*for* α) is a mapping $h : \text{var}(\alpha) \to \Sigma^*$. For every $x \in \text{var}(\alpha)$, we say that x *is substituted by* $h(x)$ and $h(\alpha)$ denotes the word obtained by substituting every occurrence of a variable x in α by $h(x)$ and leaving the terminals unchanged. If, for every $x \in \text{var}(\alpha)$, $h(x) \neq \varepsilon$, then h is *nonerasing* (h is also called *erasing* if it is not non-erasing). If, for all $x, y \in \text{var}(\alpha)$, $x \neq y$ and $h(x) \neq \varepsilon \neq h(y)$ implies $h(x) \neq h(y)$, then h is E-*injective*[1] and h is called *injective* if it is E-injective and, for at most one $x \in \text{var}(\alpha)$, $h(x) = \varepsilon$.

Example 1. Let $\beta := x_1 \, \mathsf{a} \, x_2 \, \mathsf{b} \, x_2 \, x_1 \, x_2$ be a pattern, let $u := \mathsf{bacbabbacb}$ and let $v := \mathsf{abaabbababab}$. It can be verified that $h(\beta) = u$, where $h(x_1) = \mathsf{bacb}$, $h(x_2) = \varepsilon$ and $g(\beta) = v$, where $g(x_1) = g(x_2) = \mathsf{ab}$. Furthermore, β cannot be mapped to u by a nonerasing substitution and β cannot be mapped to v by an injective substitution.

If the type of substitution is clear from the context, then we simply say that a word w *matches* α to denote that there exists such a substitution h with $h(\alpha) = w$. We can now formally define the *pattern matching problem with variables*, denoted by PMV, which has informally been described in Section 1:

> PMV
> *Instance*: A pattern α and a word $w \in \Sigma^*$.
> *Question*: Does there exist a substitution h with $h(\alpha) = w$?

As explained in Section 1, the above problem exists in various contexts with individual terminology. Since we consider the problem in a broader sense, we term it pattern matching problem with variables in order to distinguish it – and all its variants to be investigated in this paper – from the classical pattern matching paradigm without variables.

Next, we define several parameters of PMV. To this end, let α be a pattern, let w be a word and let h be a substitution for α.

- $\rho_{|\text{var}(\alpha)|} := |\text{var}(\alpha)|$,
- $\rho_{|\alpha|_x} := \max\{|\alpha|_x \mid x \in \text{var}(\alpha)\}$,
- $\rho_{|w|} := |w|$,
- $\rho_{|\Sigma|} := |\Sigma|$,
- $\rho_{|h(x)|} := \max\{|h(x)| \mid x \in \text{var}(\alpha)\}$.

The restricted versions of the problem PMV are now defined by P-$[Z, I, T]$-PMV, where P is a list of parameters that are bounded, $Z \in \{\text{E}, \text{NE}\}$ denotes whether we are considering the erasing or nonerasing case, $T \in \{\text{tf}, \text{n-tf}\}$ denotes whether or not we require the patterns to be terminal-free and $I \in \{\text{inj}, \text{n-inj}\}$ denotes whether or not we require the substitution to be injective (more precisely, if $Z = \text{NE}$, then $I = \text{inj}$ denotes injectivity, but if $Z = \text{E}$, then $I = \text{inj}$ denotes E-injectivity). Hence, $[\rho_{|\alpha|_x}^{c_1}, \rho_{|\Sigma|}^{c_2}, \rho_{|h(x)|}^{c_3}]$-[NE, n-inj, tf]- PMV denotes the problem to decide for a given *terminal-free* pattern α and a given word $w \in \Sigma^*$ with $\max\{|\alpha|_x \mid x \in \text{var}(\alpha)\} \leq c_1$ and $|\Sigma| \leq c_2$, whether or not there exists a

---

[1] We use E-injectivity, since if an erasing substitution is injective in the classical sense, then it is "almost" nonerasing, i.e., only one variable can be erased.

nonerasing substitution h (possibly *non-injective*) that satisfies $\max\{|h(x)| \mid x \in \text{var}(\alpha)\} \leq c_3$ and $h(\alpha) = w$, where c_1, c_2 and c_3 are some constants.

The contribution of this paper is to show for each of the 256 individual problems $P\text{-}[Z, I, T]\text{-PMV}$ whether or not there exist constants such that if the parameters in P are bounded by these constants, this version of PMV is still NP-complete or whether it can be solved in polynomial time. To this end, we first summarise all the respective known results from the literature and then we close the remaining gaps.

3   Known Results and Preliminary Observations

In this section, we briefly summarise those variants of PMV, for which NP-completeness or membership in P has already been established. To this end, we first informally describe an obvious and simple brute-force algorithm that solves the pattern matching problem with variables. For some instance (α, w) of PMV with $m := |\text{var}(\alpha)|$, we simply enumerate all tuples $(u_1, u_2, \ldots, u_m)$, where, for every i, $1 \leq i \leq m$, u_i is a factor of w. Then, for each such tuple $(u_1, u_2, \ldots, u_m)$, we check whether $h(\alpha) = w$, where h is defined by $h(x_i) := u_i$, $1 \leq i \leq m$. This procedure can be performed in time exponential only in m and, furthermore, it is generic in that it works for any variant of PMV. This particularly implies that every version of PMV, for which $\rho_{|\text{var}(\alpha)|}$ is restricted, can be solved in polynomial time.

Next, we note that in the nonerasing case, a restriction of $\rho_{|w|}$ implicitly bounds $\rho_{|\text{var}(\alpha)|}$ as well and, thus, all the corresponding versions of the pattern matching problem with variables can be solved efficiently. Moreover, in [13], Geilke and Zilles note that if $\rho_{|w|} \leq c$, for some constant c, then this particularly implies that the number of variables that are *not* erased is bounded by c as well. As demonstrated in [13], this means that also for the erasing case PMV can be solved in polynomial time if the length of the input word is bounded by a constant. Consequently, every version of PMV, for which $\rho_{|\text{var}(\alpha)|}$ or $\rho_{|w|}$ is restricted, can be solved in polynomial time; thus, in the following, we shall neglect these two parameters and focus on the remaining 3 parameters $\rho_{|\alpha|_x}$, $\rho_{|\Sigma|}$ and $\rho_{|h(x)|}$.

In the next table, we briefly summarise those variants of PMV, for which NP-completeness has already been established. A numerical entry denotes the constant bound of a parameter and "–" means that a parameter is unrestricted.

	E / NE	inj / n-inj	tf / n-tf	$|h(x)|$	$|\alpha|_x$	$|\Sigma|$	Complexity
1	NE	n-inj	n-tf	3	–	2	NP-C [4]
2	E, NE	n-inj	tf	3	–	2	NP-C [9]
3	NE	n-inj	tf	2	–	2	NP-C [8]
4	NE	inj	tf	–	–	2	NP-C [8]
5	NE	inj	tf	2	–	–	NP-C [8]
6	E	n-inj	n-tf	–	2	2	NP-C [26]

The main contribution of this paper is to close the gaps that are left open in the above table. Before we present our main results in this regard, we conclude this section by taking a closer look at the parameters $\rho_{|\alpha|_x}$ and $\rho_{|\Sigma|}$. As indicated by rows 1 to 4 and row 6, restricting $\rho_{|\Sigma|}$ does not seem to help to solve PMV efficiently. In [26] it is shown that even if we additionally require the number of occurrences per variable to be bounded by 2, then PMV is still NP-complete. However, regarding these two parameters, we seem to have reached the boundary between P and NP-completeness, since it can be easily shown that PMV can be solved in polynomial time if parameter ρ_Σ or $\rho_{|\alpha|_x}$ is bounded by 1 (see, e.g., Schmid [27]).

4   Main Results

In this section, we investigate the complexity of all the variants of the pattern matching problem with variables that are not already covered by the table presented in the previous section. Most of these variants turn out to be NP-complete. The general proof technique used to establish these results is illustrated in Section 5. We shall now first consider the non-injective case, i.e., we consider the problems P-$[Z, \text{n-inj}, T]$-PMV first and the problems P-$[Z, \text{inj}, T]$-PMV later on.

4.1   The Non-injective Case

All the results of this section are first presented in a table of the form already used in the previous section and then we discuss them in more detail.

E / NE	inj / n-inj	tf / n-tf	$\|h(x)\|$	$\|\alpha\|_x$	$\|\Sigma\|$	Complexity
E	n-inj	n-tf	1	2	2	**NP-C**
NE	n-inj	n-tf	3	2	2	**NP-C**
E	n-inj	tf	1	8	2	**NP-C**
NE	n-inj	tf	3	3	4	**NP-C**

As mentioned in Section 3, Clifford et al. show in [8] that the nonerasing, terminal-free and non-injective case of the pattern matching problem with variables is NP-complete, even if additionally the parameters $\rho_{|\Sigma|}$ and $\rho_{|h(x)|}$ are bounded. By the rows 2 and 4 of the above table, we strengthen this result by stating that the NP-completeness is preserved, even if in addition also $\rho_{|\alpha|_x}$ is bounded and this holds both for the terminal-free and non-terminal-free case. However, we are only able to prove that these results hold if the parameter $\rho_{|\alpha|_x}$ is bounded by 3 and the case where $\rho_{|\alpha|_x}$ is bounded by 2 is left open.

With respect to the erasing case, i.e., rows 1 and 3 of the above table, we observe a surprising situation that deserves to be discussed in a bit more detail. To this end, we introduce a special case of a substitution. A substitution h (for a pattern α) is called a *renaming* if every variable of α is either erased by h or substituted by a single symbol, i.e., for every $x \in \text{var}(\alpha)$, $|h(x)| \leq 1$. Now row 1 shows that the erasing, non-injective and non-terminal-free version of the pattern matching problem with variables remains NP-complete, even if both $\rho_{|\alpha|_x}$ and $\rho_{|\Sigma|}$ are bounded by 2 and the substitution needs to be a renaming. This is a very restricted version of the pattern matching problem with variables, which seems to be located directly on the border between NP-completeness and polynomial time solvability, since the nonerasing version of this problem is trivially solvable in linear time, the parameter $\rho_{|h(x)|}$ is already bounded in the strongest possible sense, if $\rho_{|\alpha|_x}$ or $\rho_{|\Sigma|}$ is bounded by 1 instead of 2, then, as mentioned in Section 3, the problem becomes polynomial time solvable and, in the next section, we shall see that the injective version is in P as well.

With respect to the terminal-free case (row 3 of the table), we are only able to show NP-completeness if the parameter $\rho_{|\alpha|_x}$ is bounded by 8 instead of 2. This version of the pattern matching problem with variables can be rephrased as a more general problem on strings: given two strings u and v, can u be transformed into v in such a way that every symbol of w is either erased, substituted by a or substituted by b? This problem is NP-complete, even if every symbol in u occurs at most 8 times. It is open, however, whether it is still NP-complete if at most two occurrences per symbol are allowed.

We conclude this section by pointing out that the pattern matching problem that Baker considers in [5], and for which she presents efficient algorithms, in fact relies on the problem of finding a renaming between two words. However, in [5] only nonerasing and injective renamings are considered and with our above result we can conclude that Baker's pattern matching problem most likely cannot be solved in polynomial time if it is generalised to erasing and non-injective renamings.

4.2   The Injective Case

A main difference between the complexity of the injective and non-injective cases is that in the injective case, bounding $\rho_{|\Sigma|}$ and $\rho_{|h(x)|}$ already yields polynomial time solvability (see Theorem 1 below), whereas the non-injective case remains NP-complete, even if we additionally bound $\rho_{|\alpha|_x}$ (as stated in Section 4.1). Informally speaking, this is due to the fact that if $\rho_{|\Sigma|}$ and $\rho_{|h(x)|}$ are bounded by some constants, then the number of words variables can be substituted with is bounded by some constant, say c, as well. Now if we additionally require injectivity, then the number of variables that are substituted with non-empty words is bounded by c, too, which directly implies the polynomial time solvability for the nonerasing case. In order to extend this result to the erasing case, we apply a technique similar to the one used by Geilke and Zilles in [13].

Theorem 1. *Let* $k_1, k_2 \in \mathbb{N}$, *let* $Z \in \{E, NE\}$ *and let* $T \in \{tf, n\text{-}tf\}$. *The problem* $[\rho_{|\Sigma|}^{k_1}, \rho_{|h(x)|}^{k_2}]\text{-}[Z, inj, T]\text{-}PMV$ *is in P.*

Proof. Since the case $Z = E$ implies the case $Z = NE$, we shall only prove the former.

Let α be a pattern and let w be a word over $\Sigma := \{a_1, a_2, \ldots, a_{k_1}\}$. Let S be an arbitrary subset of $var(\alpha)$. We say that S satisfies condition $(*)$ if and only if if there exists an E-injective substitution h with $h(\alpha) = w$, $1 \leq |h(x)| \leq k_2$, for every $x \in S$, and $h(x) = \varepsilon$, for every $x \in \text{var}(\alpha) \setminus S$. For any set $S \subseteq \text{var}(\alpha)$, it can be checked in time exponential in $|S|$, whether S satisfies condition $(*)$. More precisely, this can be done in the following way. First, we obtain a pattern β from α by erasing all variables in $\text{var}(\alpha) \setminus S$. Then we use a brute-force algorithm to check whether or not there exists an injective nonerasing substitution h with $h(\beta) = w$ and $1 \leq |h(x)| \leq k_2$, $x \in \text{var}(\beta)$, which can be done in time $O(k_2^{|S|})$.

For the sake of convenience, we define $k' := k_2 \times k_1^{k_2}$. We observe that there are $O(k')$ non-empty words over $\{a_1, a_2, \ldots, a_{k_1}\}$ of length at most k_2. This implies that every substitution h that maps more than k' variables to non-empty words of length at most k_2 is necessarily not E-injective. So, for every set $S \subseteq \text{var}(\alpha)$, if $|S| > k'$, then S does not satisfy condition $(*)$. Consequently, there exists an E-injective substitution h with $h(\alpha) = w$, $|h(x)| \leq k_2$, for every $x \in \text{var}(\alpha)$, if and only if there exists a set $S \subseteq \text{var}(\alpha)$ with $|S| \leq k'$ and S satisfies the condition $(*)$.

We conclude that we can solve the problem stated in the theorem by enumerating all possible sets $S \subseteq \text{var}(\alpha)$ with $|S| \leq k'$ and, for each of these sets, checking whether they satisfy condition $(*)$. Since the number of sets $S \subseteq \text{var}(\alpha)$ with $|S| \leq k'$ is

$$\sum_{i=0}^{k'} \binom{|\text{var}(\alpha)|}{i} \leq \sum_{i=0}^{k'} |\text{var}(\alpha)|^i \leq (k'+1) \text{var}(\alpha)^{k'} = O(|\text{var}(\alpha)|^{k'}),$$

the runtime of this procedure is exponential only in k'; thus, since k' is a constant, it is polynomial. □

On the other hand, as pointed out by the following table, for all other possibilities to bound some of the parameters $\rho_{|\Sigma|}$, $\rho_{|\alpha|_x}$ and $\rho_{|h(x)|}$, without bounding both $\rho_{|\Sigma|}$ and $\rho_{|h(x)|}$ at the same time, we can show NP-completeness:

E / NE	inj / n-inj	tf / n-tf	$|h(x)|$	$|\alpha|_x$	$|\Sigma|$	Complexity
E	inj	n-tf	5	2	–	NP-C
NE	inj	n-tf	19	2	–	NP-C
E, NE	inj	n-tf	–	2	2	NP-C
E, NE	inj	tf	19	4	–	NP-C
E, NE	inj	tf	–	9	5	NP-C

With respect to the injective case (and in contrast to the non-injective case), we are not able to conclude any results about renamings. In particular, the most interesting open question in this regard is whether or not the following problem is NP-complete:

Instance: A pattern α and a word $w \in \Sigma^*$.
Question: Does there exist an E-injective renaming h with $h(\alpha) = w$?

We conjecture that this question can be answered in the affirmative.

In order to conclude this section, we wish to point out that for every variant of the pattern matching problem with variables that is not explicitly mentioned in the above tables, either NP-completeness or membership in P is directly implied by one of the results presented in this section or Section 3.

5   Proof Techniques

In this section, we give a sketch of the main proof technique for the hardness results presented in Section 4. To this end, we first define a graph problem, which is particularly suitable for our purposes.

Let $\mathcal{G} = (V, E)$ be a graph with $V := \{v_1, v_2, \ldots, v_n\}$. The *neighbourhood* of a vertex $v \in V$ is the set $N_{\mathcal{G}}(v) := \{u \mid \{v, u\} \in E\}$ and $N_{\mathcal{G}}[v] := N_{\mathcal{G}}(v) \cup \{v\}$ is called the *closed neighbourhood* of v. If, for some $k \in \mathbb{N}$, $|N_{\mathcal{G}}(v)| = k$, for every $v \in V$, then $\mathcal{G}$ is *k-regular*. A *perfect code* for $\mathcal{G}$ is a subset $C \subseteq V$ with the property that, for every $v \in V$, $|N_{\mathcal{G}}[v] \cap C| = 1$. Next, we define the problem to decide whether or not a given 3-regular graph has a perfect code:

3R-PERFECT-CODE
Instance: A 3-regular graph $\mathcal{G}$.
Question: Does $\mathcal{G}$ contain a perfect code?

In [18], Kratochvíl and Křivánek prove the problem 3R-PERFECT-CODE to be NP-complete:

Theorem 2 (Kratochvíl and Křivánek [18]). 3R-PERFECT-CODE *is NP-complete.*

All the NP-completeness results of Section 4 can be proved by reducing 3R-PERFECT-CODE to the appropriate variant of PMV. However, these reductions must be individually tailored to these different variants. As an example, we give a reduction from 3R-PERFECT-CODE to $[\rho^5_{|h(x)|}, \rho^2_{|\alpha|_x}]$-[E, inj, n-tf]-PMV, which implies the result stated in row 1 of the table presented in Section 4.2.

Let $\mathcal{G} = (V, E)$ with $V := \{v_1, v_2, \ldots, v_n\}$ be a 3-regular graph and, for every i, $1 \leq i \leq n$, let N_i be the closed neighbourhood of v_i. We transform the graph $\mathcal{G}$ into a pattern α and a word w over $\Sigma := \{a_i, \mathfrak{c}_i, \#_j \mid 1 \leq i \leq n, 1 \leq j \leq 2n-1\}$, such that, for every $x \in \text{var}(\alpha)$, $|\alpha|_x \leq 2$. Now, for any i, $1 \leq i \leq n$, let $N_{j_1}, N_{j_2}, N_{j_3}, N_{j_4}$ be exactly the closed neighbourhoods that contain vertex v_i. We transform vertex v_i into the pattern variables $x_{i,j_1}, x_{i,j_2}, x_{i,j_3}, x_{i,j_4}$; thus, our interpretation shall be that variable $x_{i,j}$ refers to vertex v_i in the closed neighbourhood of vertex v_j.

For every i, $1 \leq i \leq n$, the closed neighbourhood $N_i := \{v_{j_1}, v_{j_2}, v_{j_3}, v_{j_4}\}$ is transformed into

$$\beta_i := x_{j_1,i}\, x_{j_2,i}\, x_{j_3,i}\, x_{j_4,i} \text{ and}$$
$$u_i := a_i\,.$$

Furthermore, for every i, $1 \leq i \leq n$, we define

$$\gamma_i := z_i \, \mathfrak{C}_i \, x_{i,j_1} \, x_{i,j_2} \, x_{i,j_3} \, x_{i,j_4} \, \mathfrak{C}_i \, z_i' \text{ and}$$

$$v_i := \mathfrak{C}_i \, \mathfrak{C}_i \, \mathsf{a}_{j_1} \, \mathsf{a}_{j_2} \, \mathsf{a}_{j_3} \, \mathsf{a}_{j_4} \, \mathfrak{C}_i \, ,$$

where $N_i = \{v_{j_1}, v_{j_2}, v_{j_3}, v_{j_4}\}$. Finally, we define

$$\alpha := \beta_1 \, \#_1 \, \beta_2 \, \#_2 \cdots \#_{n-1} \, \beta_n \, \#_n \, \gamma_1 \, \#_{n+1} \, \gamma_2 \, \#_{n+2} \cdots \#_{2n-1} \, \gamma_n \text{ and}$$

$$w := u_1 \, \#_1 \, u_2 \, \#_2 \cdots \#_{n-1} \, u_n \, \#_n \, v_1 \, \#_{n+1} \, v_2 \, \#_{n+2} \cdots \#_{2n-1} \, v_n \, .$$

Every variable z_i, z_i', $1 \leq i \leq n$, has only one occurrence in α. For every i, $1 \leq i \leq n$, and every j with $v_j \in N_i$, variable $x_{j,i}$ has exactly one occurrence in β_i and exactly one occurrence in γ_j. Thus, for every $x \in \mathrm{var}(\alpha)$, $|\alpha|_x \leq 2$.

In order to see that the existence of an E-injective substitution h for α with $h(\alpha) = w$ and $|h(x)| \leq 5$ implies the existence of a perfect code for $\mathcal{G}$, we first observe that, for any substitution h with $h(\alpha) = w$, $h(\beta_i) = u_i$ and $h(\gamma_i) = v_i$, $1 \leq i \leq n$, is satisfied. This implies that, for every i, $1 \leq i \leq n$, exactly one of the variables $x_{j_l,i}$, $1 \leq l \leq 4$, where $N_i = \{v_{j_1}, v_{j_2}, v_{j_3}, v_{j_4}\}$, is mapped to a_i and the other three variables are erased. Furthermore either each of the variables x_{i,j_l}, $1 \leq l \leq 4$, is mapped to a_{j_l} or all these variables are erased. This directly translates into the situation that it is possible to pick exactly one vertex from each neighbourhood. The converse statement, i.e., the existence of a perfect code implies the existence of such a substitution h, follows from the observation that for the variables $x_{i,j}$ we can define h as induced by the perfect code and, for every i, $1 \leq i \leq n$, either $h(z_i) := \mathfrak{C}_i$ and $h(z_i') := \varepsilon$ or $h(z_i) := \varepsilon$ and $h(z_i') := \mathsf{a}_{j_1} \, \mathsf{a}_{j_2} \, \mathsf{a}_{j_3} \, \mathsf{a}_{j_4} \, \mathfrak{C}_i$, depending on whether or not vertex v_i is a member of the perfect code.

We wish to point out that the above reduction strongly relies on the possibility to erase variables and to have terminal symbols in the pattern; thus, as pointed out by the following explanations, converting it to the nonerasing or the terminal-free case is non-trivial. The general idea of extending our reduction to the terminal-free case is that instead of using terminals $\#$ in the pattern, we use variables that are forced to be substituted by $\#$. Especially for the erasing case, this is technically challenging and, furthermore, if we use an unbounded number of occurrences of the same terminal symbol in the pattern, then it is difficult to maintain the restriction on the number of variable occurrences and injectivity at the same time. In the above reduction, we also use the possibility of having an unbounded number of terminal symbols. Hence, if parameter $\rho_{|\Sigma|}$ is bounded, then instead of using arbitrarily many different symbols $\mathsf{a}_1, \mathsf{a}_2, \ldots, \mathsf{a}_n$, we either have to use only one symbol a for different variables, which destroys the injectivity, or we have to encode a single symbol a_i by a string $\mathsf{ba}^i\mathsf{b}$, which breaks the bound on parameter $\rho_{|h(x)|}$.

6   Future Research Directions

In this paper, for every variant P-$[Z, I, T]$-PMV of the pattern matching problem with variables, we either show that bounding the parameters by *any* constants leads to polynomial time solvability or that the parameters can be bounded by *some* constants, such that P-$[Z, I, T]$-PMV is NP-complete. Although for the results of the latter type we are mostly able to present rather small constants, we do not provide a full dichotomy result for the class of problems P-$[Z, I, T]$-PMV.

As pointed out in Section 4.1, $[\rho^1_{|h(x)|}, \rho^2_{|\alpha|_x}, \rho^2_{|\Sigma|}]$-$[\text{E}, \text{n-inj}, \text{n-tf}]$-PMV is an example for an NP-complete version of the pattern matching problem with variables for which we provable know that any further restriction – except to the terminal-free case, which is open – makes the problem polynomial time solvable. On the other hand, we do not know when exactly the problem $[\rho^3_{|h(x)|}, \rho^3_{|\alpha|_x}, \rho^4_{|\Sigma|}]$-$[\text{NE}, \text{n-inj}, \text{tf}]$-PMV shifts from NP-completeness to polynomial time solvability when the constants are decreased.

Consequently, possible further research is to completely determine these borderlines between NP-completeness and P with respect to the pattern matching problem with variables.

References

1. Amir, A., Aumann, Y., Cole, R., Lewenstein, M., Porat, E.: Function matching: Algorithms, applications, and a lower bound. In: Baeten, J.C.M., Lenstra, J.K., Parrow, J., Woeginger, G.J. (eds.) ICALP 2003. LNCS, vol. 2719, pp. 929–942. Springer, Heidelberg (2003)
2. Amir, A., Nor, I.: Generalized function matching. Journal of Discrete Algorithms 5, 514–523 (2007)
3. Angluin, D.: Finding patterns common to a set of strings. In: Proc. 11th Annual ACM Symposium on Theory of Computing, STOC 1979, pp. 130–141 (1979)
4. Angluin, D.: Finding patterns common to a set of strings. Journal of Computer and System Sciences 21, 46–62 (1980)
5. Baker, B.S.: Parameterized pattern matching: Algorithms and applications. Journal of Computer and System Sciences 52, 28–42 (1996)
6. Bremer, J., Freydenberger, D.D.: Inclusion problems for patterns with a bounded number of variables. In: Gao, Y., Lu, H., Seki, S., Yu, S. (eds.) DLT 2010. LNCS, vol. 6224, pp. 100–111. Springer, Heidelberg (2010)
7. Câmpeanu, C., Salomaa, K., Yu, S.: A formal study of practical regular expressions. International Journal of Foundations of Computer Science 14, 1007–1018 (2003)
8. Clifford, R., Harrow, A.W., Popa, A., Sach, B.: Generalised matching. In: Karlgren, J., Tarhio, J., Hyyrö, H. (eds.) SPIRE 2009. LNCS, vol. 5721, pp. 295–301. Springer, Heidelberg (2009)
9. Ehrenfeucht, A., Rozenberg, G.: Finding a homomorphism between two words is NP-complete. Information Processing Letters 9, 86–88 (1979)
10. Freydenberger, D.D., Reidenbach, D.: Bad news on decision problems for patterns. Information and Computation 208, 83–96 (2010)
11. Freydenberger, D.D., Reidenbach, D., Schneider, J.C.: Unambiguous morphic images of strings. International Journal of Foundations of Computer Science 17, 601–628 (2006)
12. Friedl, J.E.F.: Mastering Regular Expressions, 3rd edn. O'Reilly, Sebastopol (2006)
13. Geilke, M., Zilles, S.: Learning relational patterns. In: Kivinen, J., Szepesvári, C., Ukkonen, E., Zeugmann, T. (eds.) ALT 2011. LNCS, vol. 6925, pp. 84–98. Springer, Heidelberg (2011)

14. Harju, T., Karhumäki, J.: Morphisms. In: Rozenberg, G., Salomaa, A. (eds.) Handbook of Formal Languages, vol. 1, ch. 7, pp. 439–510. Springer (1997)

15. Ibarra, O., Pong, T.-C., Sohn, S.: A note on parsing pattern languages. Pattern Recognition Letters 16, 179–182 (1995)

16. Jiang, T., Kinber, E., Salomaa, A., Salomaa, K., Yu, S.: Pattern languages with and without erasing. International Journal of Computer Mathematics 50, 147–163 (1994)

17. Jiang, T., Salomaa, A., Salomaa, K., Yu, S.: Decision problems for patterns. Journal of Computer and System Sciences 50, 53–63 (1995)

18. Kratochvíl, J., Křivánek, M.: On the computational complexity of codes in graphs. In: Koubek, V., Janiga, L., Chytil, M.P. (eds.) MFCS 1988. LNCS, vol. 324, pp. 396–404. Springer, Heidelberg (1988)

19. Mateescu, A., Salomaa, A.: Finite degrees of ambiguity in pattern languages. RAIRO Informatique Théoretique et Applications 28, 233–253 (1994)

20. Ng, Y.K., Shinohara, T.: Developments from enquiries into the learnability of the pattern languages from positive data. Theoretical Computer Science 397, 150–165 (2008)

21. Ohlebusch, E., Ukkonen, E.: On the equivalence problem for E-pattern languages. Theoretical Computer Science 186, 231–248 (1997)

22. Reidenbach, D.: A non-learnable class of E-pattern languages. Theoretical Computer Science 350, 91–102 (2006)

23. Reidenbach, D.: Discontinuities in pattern inference. Theoretical Computer Science 397, 166–193 (2008)

24. Reidenbach, D., Schmid, M.L.: A polynomial time match test for large classes of extended regular expressions. In: Domaratzki, M., Salomaa, K. (eds.) CIAA 2010. LNCS, vol. 6482, pp. 241–250. Springer, Heidelberg (2011)

25. Reidenbach, D., Schmid, M.L.: Patterns with bounded treewidth. In: Dediu, A.-H., Martín-Vide, C. (eds.) LATA 2012. LNCS, vol. 7183, pp. 468–479. Springer, Heidelberg (2012)

26. Schmid, M.L.: A note on the complexity of matching patterns with variables. Information Processing Letters (Submitted)

27. Schmid, M.L.: On the Membership Problem for Pattern Languages and Related Topics. PhD thesis, Department of Computer Science, Loughborough University (2012)

28. Shinohara, T.: Polynomial time inference of extended regular pattern languages. In: Goto, E., Furukawa, K., Nakajima, R., Nakata, I., Yonezawa, A. (eds.) RIMS 1982. LNCS, vol. 147, pp. 115–127. Springer, Heidelberg (1983)

29. Shinohara, T.: Polynomial time inference of pattern languages and its application. In: Proc. 7th IBM Symposium on Mathematical Foundations of Computer Science, pp. 191–209 (1982)

New Algorithms for Position Heaps

Travis Gagie[1], Wing-Kai Hon[2], and Tsung-Han Ku[2]

[1] HIIT and Department of Computer Science, University of Helsinki, Finland
[2] Department of Computer Science, National Tsing Hua University, Taiwan

Abstract. We present several results about position heaps, a relatively new alternative to suffix trees and suffix arrays. First, we show that if we limit the maximum length of patterns to be sought, then we can also easily limit the height of the heap and reduce the worst-case cost of insertions and deletions. Second, we show how to build a position heap in linear time independent of the size of the alphabet. Third, we show how to augment a position heap such that it supports access to the corresponding suffix array, and vice versa. Fourth, we introduce a variant of a position heap that can be simulated efficiently by a compressed suffix array with a linear number of extra bits.

1 Introduction

String-indexing data structure have played a central role in pattern matching at least since the introduction of suffix trees forty years ago, and their importance has only increased with the introduction of suffix arrays, compressed suffix arrays, FM-indexes, etc. There are still many open problems about them, however, such as how best to make them dynamic. There are now fairly practical dynamic versions of suffix arrays and FM-indexes but these have poor worst-case theoretical bounds for updates [10,11]. Relatively recently, Ehrenfeucht, McConnell, Osheim and Woo [6] introduced a new and simple indexing data structure, called a position-heap, and showed how it can easily be made dynamic (albeit with a logarithmic slowdown for searches and also with a poor worst-case bound for updates). Like suffix trees and suffix arrays, position heaps take linear space and support searching in time proportional to the length of the pattern plus the number of occurrences reported, which is optimal. Ehrenfeucht et al. gave a construction algorithm that works in linear time when the size of the alphabet is constant. Shortly thereafter, Kucherov [7] gave a simpler, online construction that also takes linear time when the alphabet size is constant. Ehrenfeucht et al.'s and Kucherov's constructions of position heaps are analogous to Weiner's and Ukkonen's construction of suffix trees, respectively, and Kucherov asked whether there is a construction that takes linear time independent of the alphabet size, analogous to Farach's construction of suffix trees. Kucherov also asked whether position heaps can be compressed, as can suffix trees, suffix arrays and FM-indexes. Most recently, Nakashima, I, Inenaga, Bannai and Takeda [8] showed how to build the position heap for a set of strings given as a trie in linear time when the alphabet size is constant.

J. Fischer and P. Sanders (Eds.): CPM 2013, LNCS 7922, pp. 95–106, 2013.
© Springer-Verlag Berlin Heidelberg 2013

In this paper we answer some of the open problems about position heaps. We show in Section 3 that, if we limit the maximum length of patterns to be sought, then we can use a position heap with limited height as an index, which reduces the maximum cost of updating the heap after we make insertions or deletions in the string. In many practical applications we are interested only in fairly short patterns anyway, so this seems like a reasonable tradeoff. We also note that, if we replace a splay tree by an AVL-tree in Ehrenfeucht et al.'s implementation of dynamic position heaps, then their time bounds become worst-case instead of amortized. In Section 4 we show how to turn a suffix tree into a position heap in linear time independent of the alphabet size, using a simple modification of a recent algorithm by Bannai, Inenaga and Takeda [1] for building the LZ78 parse from a straight-line program. Combined with Farach's algorithm for building suffix trees in linear time, this means we can build position heaps in linear time independent of the alphabet size, answering Kucherov's first question affirmatively. In Section 5 we show how to augment a position heap with $\mathcal{O}(n \log h)$ bits such that it supports $\mathcal{O}(1)$-time access to the corresponding suffix array and inverse suffix array, where n is the length of the string and h is the height of the heap. Ehrenfeucht et al. showed that, although h can be as large as n in the worst case, it is typically $\mathcal{O}(\log n)$. We also show how to augment a compressed suffix array with $\mathcal{O}(n \log h)$ bits such that it supports access to the position heap in the same time needed to access the suffix array and inverse suffix array. Finally, in Section 7 we introduce a variant of a position heap, which we call a suffix heap, that still supports indexed pattern matching but which can be simulated by a compressed suffix array with only a linear number of extra bits. This seems at least partly to answer Kucherov's second question affirmatively as well.

2   Position Heaps

Ehrenfeucht et al.'s position heap data structure is a modification of an older data structure by Coffman and Eve [5] for hashing. Kucherov gave a simplified definition according to which, for a string $S[1..n]$ terminated by a special symbol $S[n] = \$$, the position heap is the trie *Heap* in which

- the root is labelled 0 and the other nodes are labelled 1 to n such that parents' labels are smaller than their children's labels;
- for $1 \le i \le n$, the path label of the node labelled i is a prefix of $S[i..n]$;
- for $1 \le i \le n$, the node labelled i stores a pointer (called its maximal-reach pointer) to the deepest node whose path label is a prefix of $S[i..n]$.

For example, if $S = \text{abaababbabbab}\$$ then *Heap* is as shown in Figure 1 (except that maximal-reach pointers are omitted there when they point back to the nodes themselves), overlaid on the the suffix trie for S. One reason to label the root 0 is so that, for $1 \le i \le n$, $S[i]$ is equal to the first edge label on the path from the root to the node labelled i.

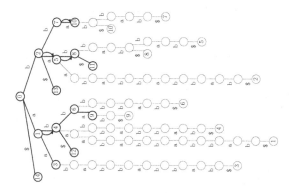

Fig. 1. The position heap *Heap* (in heavy lines) overlaid on the suffix trie for $S =$ abaababbabbab$

To be able to use *Heap* for indexed pattern matching in S we store, first, an array of pointers such that, given i, in $\mathcal{O}(1)$ time we can find the node labelled i; and second, a data structure such that, given i and j, in $\mathcal{O}(1)$ time we can determine whether the node labelled i is an ancestor of the node labelled j. The total space for *Heap* and these data structures is $\mathcal{O}(n \log n)$ bits, i.e., linear space on a word RAM.

To search for a pattern $P[1..m]$ in S, we start at the root and descend to the deepest node v whose path label is a prefix of P. If v's depth $d = m$, then we report the label of each node that is either in the subtree of v or on the path from the root to v with a maximal reach pointer into the subtree of v. Otherwise, we build a list containing the label of each node on the path from the root to v with a maximal reach pointer to v. We return to the root and descend to the deepest node v' whose path label is a prefix of $P[d + 1..m]$. If the depth d' of v' is $m - d$, then we report each label i in our list for which the node labelled $i + d$ is either in the subtree of v' or on the path from the root to v' with a maximal-reach pointer into the subtree of v'. Otherwise, we filter our list, keeping each label i only if the node labelled $i + d$ is on the path from the root to v' with a maximal reach pointer to v'. We return to the root and descend again, using d' in place of d, and keep repeatedly descending until we reach the end of P. By induction, this yields a list of the starting positions of the occurrences of P in S and, with the data structures mentioned above, takes time linear in m and the number of those occurrences.

For example, to search for $P = $ aabab in $S = $ abaababbabbab$, we start at the root and descend along two edges labelled $P[1] = P[2] = $ a to the node v labelled 3. Since v is at depth only $d = 2 \leq m = 5$, we check the nodes labelled 1 and 3 and then, since the former's maximal-reach pointer is not to the latter, build a list containing only 3. We return to the root and descend along edges labelled $P[3] = $ b, $P[4] = $ a and $P[5] = $ b to the node v' labelled 8. Since v' is at depth $3 = m - d$, we find the node labelled $3 + d = 5$ and, since it is on the path from the root to the v' and its maximal-reach pointer is into the subtree of v', we report position 3.

3 Limiting Length and Height

If we will never search for a pattern of length greater than M, then we can easily build a position heap of height $\mathcal{O}(M)$ that works as an index for S. To do this, we make two copies of S called S' and S''; insert a unique character after every $2M$ characters, counting from the first character of S' and the $(M + 1)$st character of S''; and build the position heap for $S'\,!\,S''$, where $!$ is another unique character. We refer to the inserted unique characters and $!$ as dividers and to the substrings of S' and S'' strictly between dividers as blocks. For example, if $S = \text{abaababbabbbab\$}$ and $M = 3$, then $S' = \text{abaaba}\,\#_1\,\text{bbabba}\,\#_2\,\text{b\$}$, $S'' = \text{ababba}\,\#_3\,\text{bbab\$}$ and we build the position heap for

$$S'\,!\,S'' = \text{abaaba}\,\#_1\,\text{bbabba}\,\#_2\,\text{b\$}\,!\,\text{ababba}\,\#_3\,\text{bbab\$}\,.$$

Notice that any substring of S with length at most M occurs in either S' or S'' or both. Moreover, given the endpoints of a substring in $S'\,!\,S''$, in $\mathcal{O}(1)$ time we can determine whether it contains any dividers and, if not, where it occurs in S. Therefore, we can use the position heap for $S'\,!\,S''$ as an index for S. The position heap for $S'\,!\,S''$ has height at most a factor of 2 larger than the height of the position heap for S and the dividers guarantee there are no common prefixes in S longer than $2M$, so the position heap for $S'\,!\,S''$ has height $\mathcal{O}(M)$.

If we insert or delete a substring in S, then we should update S' and S'' to maintain the invariants that every substring of S with length at most M occurs in either S' or S'' or both, and that the position heap for $S'\,!\,S''$ has height $\mathcal{O}(M)$. Consider first how we update S' when we insert a substring of length at most $4M$ into S. We insert that substring into the appropriate block of S'; if that block then has length more than $4M$, we split the block into two parts, each of length between $2M$ and $4M$, and insert a new divider between them. If we insert a substring with length greater than $4M$ into S, then we split that substring into blocks of length at most $2M$ separated by dividers, split the block of S' where the substring is to be inserted into two parts, concatenate the first part with the first block of the substring and concatenate the last block of the substring with the second part.

If we delete a substring of S, then we delete any blocks of S' completely contained in that substring and perform separate deletions from the blocks where the substring starts and finishes. To delete a substring from a single block of S', we delete that substring and then check whether the block still has length at least $2M$. If not, we remove the divider between that block and an adjacent one (assuming S is still long enough for there to be another block); if the resulting block then has length more than $4M$, we split it into two parts, each with length between $2M$ and $4M$, and insert a new divider between them.

Once we have updated S', we update S'' so that the blocks of S'' are again centered on the dividers in S' and have length exactly $2M$ (or less if they reach an end of S). Notice that inserting or deleting a substring of length ℓ into or from S requires inserting or deleting $\mathcal{O}(1)$ substrings of length $\mathcal{O}(\ell)$ into or from $S'\,!\,S''$. For example, if $S = \text{abaababbabbbab\$}$, $M = 3$ and we insert bba in position 5 to obtain $S = $ abaa*bba*babbabbab$, then we update S' to be abaa*bba*ba #$_1$ bbabba #$_2$ b$ and S'' to be *a*babba #$_3$ bbab$.

Ehrenfeucht et al.'s dynamic index has two parts, a dynamic position heap and the data structure for storing the dynamic string itself. They suggested using a splay tree to store the dynamic string but noted that this choice gives only amortized time bounds. If we use their dynamic position heap for $S'! S''$ exactly as they described but use an AVL-tree (which can also be split and joined in logarithmic time) instead of a splay tree to store $S'! S''$, then we obtain the following result with no amortization. We will give more details in the full version of this paper. Our use of dividers makes the alphabet size more than constant but, as we show in the next Section, it is still possible to build the position heap in linear time.

Theorem 1. *If we will never search for a pattern of length greater than M in a dynamic string S, then we can maintain a position heap that works as an index for S such that*

- *searching for a pattern of length $m \leq M$ takes $\mathcal{O}(m \log |S| + occ)$ time,*
- *inserting a substring of length ℓ takes $\mathcal{O}((M + \ell)M \log(|S| + \ell))$ time,*
- *deleting a substring of length ℓ takes $\mathcal{O}((M + \ell)M \log |S|)$ time.*

4   Turning a Suffix Tree into a Position Heap

Bannai et al. recently gave an algorithm for computing the LZ78 parse of a string from a straight-line program for that string. A key idea in their algorithm is to build the LZ78 trie superimposed on the suffix tree for the string. To compute the LZ78 parse normally, we start at the left with an empty dictionary; at each step, we take as the next phrase the shortest prefix of the remainder of the string that is not yet in the dictionary; we add that phrase to the dictionary and delete it from the beginning of the remainder of the string. The trie of the phrases in the dictionary when we finish parsing is the LZ78 trie. If we delete only the first character of the remainder of the string at each step, instead, then the trie of the phrases when we finish parsing is the position heap. In this section we use this idea to turn a suffix tree into a position heap in linear time independent of the alphabet size.

A simple way to build *Heap* is to build the suffix trie for S (i.e., the trie of all its suffixes); label each leaf with the starting position of the suffix which is its path label; label the root 0; for $1 \leq i \leq n$ in increasing order, move each leaf's label to its highest unlabelled ancestor (or, if there are no unlabelled ancestors, leave the label on the leaf); and finally, for $1 \leq i \leq n$, add a maximal-reach pointer from the node labelled i to the deepest labelled ancestor of the leaf originally labelled i. The correctness of this algorithm follows from the definition of the position heap; see Figure 1.

Suppose we already have built and preprocessed the suffix trie of S such that in $\mathcal{O}(1)$ time, first, we can mark nodes; second, given a node, we can find its lowest marked ancestor; and third, given a node and a depth, we can find that node's ancestor at that depth. Then we can perform the algorithm we have just described in linear time independent of the alphabet size, marking nodes whenever we move a label to them. Since the suffix trie has size $\Theta(n^2)$, however, building it explicitly and preprocessing it takes $\Omega(n^2)$ time.

Bannai et al. showed how we can use the suffix tree ST for S as a representation of the suffix trie. Suppose we have already built two copies ST_1 and ST_2 of ST with the same nodes. Westbrook [12] showed how we can preprocess ST_1 in linear time such that in $\mathcal{O}(1)$ amortized time, first, we can mark nodes; second, given a node, we can find its lowest marked ancestor; and third, we can insert a node in the middle of an edge. Berkman and Vishkin [4] showed how we can preprocess ST_2 in linear time such that, given a node and a depth, we can find that node's ancestor at that depth in $\mathcal{O}(1)$ time. We work in ST_1, which is dynamic; ST_2 remains static.

We start by labelling with 0 the root of ST_1 and marking it. For $1 \leq i \leq n$, we find the lowest marked ancestor u in ST_1 of the leaf w labelled i. (This is the difference between building a position heap and Bannai et al.'s algorithm for building the LZ78 trie: they consider only values of i that are the starting positions of phrases in the LZ78 parse.) If u is w itself, then we simply mark it; otherwise, we find the child v of u that is also an ancestor of w. If u has a constant number of children then finding v takes $\mathcal{O}(1)$ time even in ST_1. If u has more than a constant number of children then we use ST_2 to find v, as we explain next. If v's stringdepth (i.e., the length of its path label) is 1 more than u's, then we move the label i to v and mark it in ST_1. Otherwise, we insert a new node v' between u and v in ST_1; assign the first character of the edge label of the old edge (u, v) to the new edge (u, v') and assign the rest to the new edge (v', v); move the label i to v'; and mark v'. This all takes $\mathcal{O}(1)$ amortized time. Finally, for $1 \leq i \leq n$, we add a maximal-reach pointer from the node labelled i in ST_1 to the deepest marked ancestor of the leaf originally labelled i. Due to space constraints, we do not include a figure showing a position heap overlaid on a suffix tree; however, that figure for $S =$ abaababbabbab$ would look like Figure 1 but without the internal nodes of the trie that are not part of the heap. In this case, building *Heap* requires us to insert into ST_1 the nodes of the heap labelled 3, 7, 9 and 10.

Notice that, if u has more than one child then, first, u exists in both ST_1 and ST_2 and, second, we have not inserted any nodes in u's subtree in ST_1. Therefore, v also exists and is u's child in both ST_1 and ST_2. We can find v in $\mathcal{O}(1)$ time in ST_2 by finding the ancestor of w whose depth is 1 more than u's. Summing up, we have the following theorem.

Theorem 2. *Given the suffix tree for a string, we can build the position heap for that string in linear time independent of the size of the alphabet.*

Since Farach's construction of suffix trees takes linear time independent of the alphabet size, we have answered affirmatively Kucherov's question of whether there is an algorithm for building position heaps that takes linear time independent of the alphabet size.

5   Using a Position Heap as a Suffix Array

The order in which we see positions in a traversal of *Heap* may not be the order in which they appear from left to right on the leaves of the suffix tree for S, which is the same as their order in the suffix array $SA[1..n]$ for S. For example, if $S = $ abaababbabbab$ then $SA = [14, 3, 12, 1, 4, 9, 6, 13, 2, 11, 8, 5, 10, 7]$; since the node labelled 4 is the child of the node labelled 1 and the parent of the node labelled 12 in *Heap*, no traversal of *Heap* can produce the order 12, 1, 4. Nevertheless, by the definition of a position heap, if positions are labels of nodes at the same depth in *Heap*, then their left-to-right order is the same as the lexicographic order of the suffixes starting at those positions and, so, the same as their left-to-right order in the suffix tree or suffix array for S.

Let $D[1..n]$ be the array in which $D[i]$ is the depth in *Heap* of the value $SA[i]$. In other words, if $D[i]$ is the rth copy of d in D, then the label of the rth node from the left at depth d in *Heap* is $SA[i]$. For example, if $S = $ abaababbabbab$ then $D = [1, 2, 3, 1, 2, 4, 3, 2, 1, 4, 3, 2, 3, 2]$. Since $D[8]$ is the third copy of 2 in D, the label on the third node from the left at depth 2 in *Heap* is $SA[8] = 13$, as shown in Figure 1. It follows that, if we can answer access and partial rank queries on D and access nodes in *Heap* given their depths and their ranks from the left at those depths, then we can support access to SA.

We can store D in $nH_0(D) + o(n(H_0(D) + 1))$ bits, where $H(D) \leq \log h$ is the 0th-order empirical entropy of D and h is the height of *Heap*, such that access and partial rank queries take $\mathcal{O}(1)$ time [3]. Ehrenfeucht et al. showed that, although h can be as large as n in the worst case, it is typically $\mathcal{O}(\log n)$. There are $(2n + o(n))$-bit data structures that support access in $\mathcal{O}(1)$ time to any node in *Heap* given its rank in pre-order, in-order or post-order traversals; given a pointer to a node, they also return its rank in the appropriate traversal. Notice that any of these traversals visits the nodes at any particular depth in *Heap* in their left-to-right order. For the sake of simplicity, we now consider only pre-order traversal.

Let $E[1..n]$ be the array in which $E[i]$ is the depth of the $(i + 1)$st node (or ith if we ignore the root) visited in a pre-order traversal of *Heap*. In other words, if $E[i]$ is the rth copy of d in E, then the rth node from the left at depth d is the $(i+1)$st visited in a pre-order traversal. For example, if $S = $ abaababbabbab$ then $E = [1, 1, 2, 2, 3, 3, 4, 1, 2, 2, 3, 4, 2, 3]$. Since $E[9]$ is the third copy of 2 in E, the third node from the left at depth 2 is the 9th node visited in a pre-order traversal of *Heap*. It follows that, if we can answer select queries on E, then we can access nodes in *Heap* given their depths and their ranks from the left at those depths.

We can store E in $(1+\epsilon)nH_0(E)+o(n)$ bits such that select queries take $\mathcal{O}(1)$ time [2], where ϵ is any positive constant. Notice that E is a permutation of D so $H_0(E) = H_0(D) \leq \log h$. Therefore, we can add $(2 + \epsilon)nH_0(D) + o(n(H_0(D) + 1)) = \mathcal{O}(n \log h)$ bits to *Heap* and support access to SA in $\mathcal{O}(1)$ time.

The inverse suffix array $SA^{-1}[1..n]$ stores the lexicographic ranks of the suffixes in left-to-right order. For example, if $S = $ abaababbabbab$ then $SA^{-1} = [4, 9, 2, 5, 12, 7, 14, 11, 6, 13, 10, 3, 8, 1]$. Suppose we store data structures supporting access and partial rank queries on E and select queries on D, which take another $(2 + \epsilon)nH_0(D) + o(n(H_0(D) + 1)) = \mathcal{O}(n \log h)$ bits. If we want to access $SA^{-1}[i]$, then we follow the pointer to the node v labelled i in *Heap*; find v's rank t in the pre-order traversal of *Heap*; find the partial rank r of $E[t-1] = d$ in E; and use select to find the position of the rth copy of d in D. This takes a total of $\mathcal{O}(1)$ time. For example, to access $SA^{-1}[13]$, we find the node labelled 13 in *Heap*, which is the 10th visited in a pre-order traversal; find the partial rank 3 of $E[9] = 2$ in E; and return the position 8 of the third 2 in D.

Theorem 3. *We can add $\mathcal{O}(n \log h)$ bits to a position heap, where h is the height of the heap, such that it supports access to the corresponding suffix array and inverse suffix array in $\mathcal{O}(1)$ time.*

6   Using a Compressed Suffix Array as a Position Heap

Many compressed suffix arrays (see, e.g., [9] for a survey) support efficient access to both SA and SA^{-1}. Suppose we have access to SA and SA^{-1} and want to represent a position heap, including

- its structure as a tree;
- the nodes' labels;
- the edges' labels;
- the maximal-reach pointers;
- an array of pointers such that, given i, in $\mathcal{O}(1)$ time we can find the node labelled i;
- a data structure such that, given i and j, in $\mathcal{O}(1)$ time we can determine whether the node labelled i is an ancestor of the node labelled j.

We can represent the heap's structure as a tree using any of the $(2n + o(n))$-bit data structures mentioned in Section 5; assume we use the one based on pre-order traversal. Without increasing the size of the data structure by more than $o(n)$ bits, we can support queries to determine whether one node is the ancestor of another, given pointers to them. We now show that, with the data structures for access, partial rank and selection on D and E, we can represent the nodes' labels and the array of pointers.

To find the label of a given node v, we find v's rank t in the pre-order traversal of *Heap*; find the partial rank r of $E[t-1] = d$ in E; use select to find the position p of the rth copy of d in D; and return $SA[p]$. For example, if $S = $ abaababbabbab\$ and we are asked for the label of the 10th node visited in a pre-order traversal of *Heap*, then we find the partial rank 3 of $E[9] = 2$; find the position 8 of the third 2 in D; and return $SA[8] = 13$.

To find a node given its label i, we find the position $SA^{-1}[i] = p$ in SA of i; find the partial rank r of $D[i] = d$ in D; find the partial rank $t - 1$ of the rth copy of d in E; and return a pointer to the t node visited in a pre-order traversal of *Heap*. For example, if $S = $ abaababbabbab\$ and we are asked to find the node in *Heap* with label 13, then we find the position $SA^{-1}[13] = 8$ in SA of 13; find the partial rank 3 of $D[8] = 2$; find the partial rank 9 of the 3rd copy of 2 in E; and return a pointer to the 10th node visited in a pre-order traversal of *Heap*.

To be able to return edges' labels, we store a bitvector indicating, for each distinct character a, the interval of SA containing the positions of copies of a in S. Assuming the size σ of the alphabet is at most n, this bitvector takes $\sigma \log(n/\sigma) + o(n) = \mathcal{O}(n)$ bits, and lets us determine in $\mathcal{O}(1)$ time the first character $S[i]$ in suffix $S[i..n]$ given $S[i..n]$'s lexicographic rank among the suffixes of S. If we are using a compressed suffix array that already supports this functionality, then we do not need the bitvector.

To find an edge's label, we find the label i and depth d of the node at the bottom of that edge, find the position $SA^{-1}[i + d - 1]$ in SA of $i + d - 1$, then use the bitvector to determine the character $S[i + d - 1]$. For example, if $S = \text{abaababbabbab\$}$ and we are asked to find the label of the edge above the node labelled 13, which is at depth 2, then we find the position $SA^{-1}[14] = 1$ of 14 in SA and use the bitvector to determine $S[14] = \$$.

To be able to return nodes' maximal-reach pointers, we store the balanced-parentheses representation of the tree structure of $Heap$, with copies of a special symbol $*$ interleaved so that the ith copy of $*$ occurs after the jth copy of '(' if, in a pre-order traversal of the position heap overlaid on the suffix trie (see Figure 1), we visit the ith leaf of the trie after we visit the jth node of the heap; the ith copy of $*$ occurs before the jth copy of ')' if, in a post-order traversal of the position heap overlaid on the suffix trie, we visit the ith leaf of the trie before visiting the jth node of the heap. For example, if $S = \text{abaababbabbab\$}$ then we store

$$((*) ((*) ((*) * * ((*) *))) ((*) (* ((*) * *)) ((* *)))) .$$

To clarify this example, we now attach subscripts and superscripts showing the labels of the nodes of the heap to which parentheses correspond, and superscripts showing the labels of the leaves of the trie to which copies of $*$ correspond:

$$(_0 \ (_{14} \ *^{14} \ _{14}) \ (_1 \ (_3 \ *^3 \ _3) \ (_4 \ (_{12} \ *^{12} \ _{12}) \ *^1 \ *^4 \ (_6 \ (_9 \ *^9 \ _9) \ *^6 \ _6) \ _4) \ _1) \cdots$$

$$\cdots (_2 \ (_{13} \ *^{13} \ _{13}) \ (_5 \ *^2 \ (_8 \ (_{11} \ *^{11} \ _{11}) \ *^8 \ *^5 \ _8) \ _5) \ (_7 \ (_{10} \ *^{10} \ *^7 \ _{10}) \ _7) \ _2) \ _0) \ .$$

Recall from Section 4 that the maximal-reach pointer of the node labelled i in $Heap$ points to the deepest node of $Heap$ that, when $Heap$ is overlaid on the suffix trie, is an ancestor of the leaf labelled i in the suffix trie. For example, if $S = \text{abaababbabbab\$}$, then the node labelled 5 in $Heap$ points to the node labelled 8 (see Figure 1). It follows that the maximal-reach pointer of the node labelled i is to the node corresponding to the matching pair of parentheses most closely enclosing the $SA^{-1}[i]$th copy of $*$ in our augmented balanced-parentheses representation of $Heap$. For example, if $S = \text{abaababbabbab\$}$, then the $SA^{-1}[5] = 12$th copy of $*$ is most closely enclosed by the matching pair of parentheses corresponding to the 12th node visited in a pre-order traversal of $Heap$, which is labelled 8. We can store our augmented representation in $\mathcal{O}(n)$ bits such that we can find this matching pair of parentheses, and the corresponding node, in $\mathcal{O}(1)$ time. Carefully combining this with all the results in this section, we obtain the following theorem.

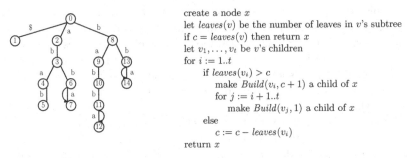

```
create a node x
let leaves(v) be the number of leaves in v's subtree
if c = leaves(v) then return x
let v₁, . . . , vₜ be v's children
for i := 1..t
    if leaves(vᵢ) > c
        make Build(vᵢ, c + 1) a child of x
        for j := i + 1..t
            make Build(vⱼ, 1) a child of x
    else
        c := c − leaves(vᵢ)
return x
```

Fig. 2. The suffix heap S-$Heap$ for S = abaababbabbab$ (left) and pseudocode for the recursive procedure $Build(v, c)$ (right)

Theorem 4. *Suppose we have a compressed suffix array that supports access to both the suffix array and the inverse suffix array in $\mathcal{O}(t)$ time, and the corresponding position heap has height h. Then we can add $\mathcal{O}(n \log h)$ bits to the compressed suffix array such that it simulates the position heap with an $\mathcal{O}(t)$-factor slowdown.*

7   Suffix Heaps

Suppose we modify the definition of a position heap so that, instead of the path label of the node labelled i being a prefix of $S[i..n]$, it is a prefix of $S[SA[i]..n]$. We call the resulting data structure the suffix heap S-$Heap$ for S. For example, if S = abaababbabbab$ then S-$Heap$ is as shown in Figure 2 (except that maximal-reach pointers are omitted there when they point back to the nodes themselves).

Searching in a suffix heap is similar to searching in a standard position heap but now, instead of reporting a node's label i, we report $SA[i]$; instead of computing $i + d$, we compute $SA^{-1}[SA[i] + d]$. Therefore, searching a suffix heap requires access to SA and SA^{-1}. For example, to search for P = aabab in S = abaababbabbab$, we start at the root and descend along the edge labelled $P[1]$ = a to the node v labelled 2 at depth 1. We return to the root and descend along the edges labelled $P[2]$ = a, $P[3]$ = b, $P[4]$ = a and $P[5]$ = b to the node v' labelled 5. Since $SA^{-1}[SA[2] + 1]$ = 5 is the label of v', we report position $SA[2]$ = 3. We will give more examples in the full version of this paper.

We can build a suffix heap using the linear-time algorithm described in Section 4 but first labelling the leaves of the suffix tree by their ranks from left to right. There is a simpler recursive algorithm, however, to build the suffix heap from the suffix trie; we can make it linear-time by simulating the suffix trie with the suffix tree, as before. We start by creating the root of S-$Heap$; for each child v of the root of the suffix trie, we call $Build(v, 1)$, where $Build(v, c)$ is the procedure given in Figure 2. We will prove this algorithm correct and analyze it in the full version of this paper. If we store trees using their balanced-parentheses representation, then this algorithm takes $\mathcal{O}(n)$ bits of work space.

Notice that nodes' labels are simply their ranks in a pre-order traversal of S-$Heap$; therefore, in a total of $2n + o(n)$ bits we can store

- S-$Heap$'s structure as a tree;
- the nodes' labels
- an array of pointers such that, given i, in $\mathcal{O}(1)$ time we can find the node labelled i;
- a data structure such that, given i and j, in $\mathcal{O}(1)$ time we can determine whether the node labelled i is an ancestor of the node labelled j.

Suppose we have stored the bitvector described in Section 6 that indicates, for each distinct character a, the interval of SA containing the positions of copies of a in S. Again, assuming the size σ of the alphabet is at most n, this bitvector takes $\mathcal{O}(n)$ bits and lets us determine in $\mathcal{O}(1)$ time the first character $S[i]$ in suffix $S[i..n]$ given $S[i..n]$'s lexicographic rank among the suffixes of S. If we are using a compressed suffix array that already supports this functionality, then we do not need the bitvector.

To find an edge's label, we find the label j and depth d of the node at the bottom of that edge; find the starting position $i = SA[j]$ in S of the lexicographically jth suffix; find the position $SA^{-1}[i + d - 1]$ in SA of $SA[j] + d - 1$; then use the bitvector to determine the character $S[i + d - 1]$. For example, if $S = $ abaababbabbab$ and we are asked to find the label of the edge above the node labelled 13, which is at depth 2, then we find the starting position $SA[13] = 10$ of the lexicographically 13th suffix; find the position $SA^{-1}[11] = 10$ of 11 in SA; and use the bitvector to determine $S[11] = $ b.

If the maximal-reach pointer of the node labelled i is to the node labelled j at depth d, then $S\left[SA[i]..SA[i] + d - 1\right] = \ldots = S\left[SA[j]..SA[j] + d - 1\right]$. It follows that, if the maximal reach pointer of the node labelled $i' > i$ is to the node labelled j', then $j' > j$. Therefore, we can store the nodes' maximal-reach pointers in S-$Heap$ as a balanced-parentheses representation of the tree structure with copies of a special symbol $*$ interleaved so that the ith copy of $*$ occurs after the jth copy of '(' if the maximal-reach pointer of the node labelled i is to the node labelled j. For example, if $S = $ abaababbabbab$ then we store

$$(_0 \ (_1 \ *^1 \ _1) \ (_2 \ *^2 \ (_3 \ *^3 \ (_4 \ *^4 \ (_5 \ *^5 \ _5)(_6 \ (_7 \ *^6 \ *^7 \ _7) \ _6) \ (_8 \ *^8 \ (_9 \ *^9 \ (_{10} \ *^{10} \ \ldots$$
$$\ldots (_{11} \ (_{12} \ *^{11} \ *^{12} \ _{12}) \ _{11}) \ _{10}) \ _9) \ (_{13} \ (_{14} \ *^{13} \ *^{14} \ _{14}) \ _{13}) \ _8) \ _0) \ ;$$

again, we have shown subscripts and superscripts only to clarify the example. Given a pointer to the node labelled i, we can find where its maximal-reach pointer points by using a select query to find the position of the ith copy of $*$, using a rank query to find the number of copies of '(' preceding it, and subtracting 1 for the root. For example, if $S = $ abaababbabbab$ and we want the maximal-reach pointer of the node labelled 6, then we compute $\mathrm{rank}_{(}(\mathrm{select}_*(6)) - 1 = 7$. We can store our augmented balanced-parentheses representation in $\mathcal{O}(n)$ bits such that rank and select queries take $\mathcal{O}(1)$ time. Carefully combining this with all the results in this section, we obtain the following theorem.

Theorem 5. *Suppose we have a compressed suffix array that supports access to both the suffix array and the inverse suffix array in $\mathcal{O}(t)$ time. Then we can add $\mathcal{O}(n)$ bits such that it simulates the corresponding suffix heap with an $\mathcal{O}(t)$-factor slowdown.*

Acknowledgments. Many thanks to Hideo Bannai, Gregory Kucherov and Gonzalo Navarro for their insightful comments. The first author also thanks his colleagues and students at Aalto and the University of Helsinki for helpful discussions.

References

1. Bannai, H., Inenaga, S., Takeda, M.: Efficient LZ78 factorization of grammar compressed text. In: Calderón-Benavides, L., González-Caro, C., Chávez, E., Ziviani, N. (eds.) SPIRE 2012. LNCS, vol. 7608, pp. 86–98. Springer, Heidelberg (2012)
2. Barbay, J., Claude, F., Gagie, T., Navarro, G., Nekrich, Y.: Efficient fully-compressed sequence representations. Algorithmica (to appear), doi:10.1007/s00453-012-9726-3
3. Belazzougui, D., Gagie, T., Navarro, G.: Better space bounds for parameterized range majority and minority. Technical Report 1210.1765 (2012), http://www.arxiv.org
4. Berkman, O., Vishkin, U.: Finding level-ancestors in trees. J. Comput. Syst. Sci. 48(2), 214–230 (1994)
5. Coffman Jr., E.G., Eve, J.: File structures using hashing functions. Commun. ACM 13(7), 427–432 (1970)
6. Ehrenfeucht, A., McConnell, R.M., Osheim, N., Woo, S.-W.: Position heaps: A simple and dynamic text indexing data structure. J. Discrete Algorithms 9(1), 100–121 (2011)
7. Kucherov, G.: On-line construction of position heaps. J. Discrete Algorithms (to appear), doi:10.1016/j.jda.2012.08.002
8. Nakashima, Y., Tomohiro, I., Inenaga, S., Bannai, H., Takeda, M.: The position heap of a trie. In: Calderón-Benavides, L., González-Caro, C., Chávez, E., Ziviani, N. (eds.) SPIRE 2012. LNCS, vol. 7608, pp. 360–371. Springer, Heidelberg (2012)
9. Navarro, G., Mäkinen, V.: Compressed full-text indexes. ACM Comput. Surv. 39(1) (2007)
10. Salson, M., Lecroq, T., Léonard, M., Mouchard, L.: A four-stage algorithm for updating a Burrows-Wheeler transform. Theor. Comput. Sci. 410(43), 4350–4359 (2009)
11. Salson, M., Lecroq, T., Léonard, M., Mouchard, L.: Dynamic extended suffix arrays. J. Discrete Algorithms 8(2), 241–257 (2010)
12. Westbrook, J.: Fast incremental planarity testing. In: Kuich, W. (ed.) ICALP 1992. LNCS, vol. 623, pp. 342–353. Springer, Heidelberg (1992)

Document Listing on Repetitive Collections

Travis Gagie[1], Kalle Karhu[2], Gonzalo Navarro[3],
Simon J. Puglisi[1], and Jouni Sirén[3]

[1] Helsinki Institute for Information Technology (Aalto),
Department of Computer Science, University of Helsinki
{travis.gagie, simon.j.puglisi}@gmail.com
[2] Department of Computer Science and Engineering, Aalto University
kalle.karhu@aalto.fi
[3] Department of Computer Science, University of Chile
{gnavarro, jsiren}@dcc.uchile.cl

Abstract. Many document collections consist largely of repeated material, and several indexes have been designed to take advantage of this. There has been only preliminary work, however, on document retrieval for repetitive collections. In this paper we show how one of those indexes, the run-length compressed suffix array (RLCSA), can be extended to support document listing. In our experiments, our additional structures on top of the RLCSA can reduce the query time for document listing by an order of magnitude while still using total space that is only a fraction of the raw collection size. As a byproduct, we develop a new document listing technique for general collections that is of independent interest.

1   Introduction

Document listing is a fundamental and well-studied problem in information retrieval. It is known how to store a collection of documents in entropy-compressed space such that, given a pattern, we can quickly list the distinct documents in which that pattern occurs [15,8]. If the collection is repetitive, however — e.g., genomes of individuals of the same or related species, software repositories, or versioned document collections — then its statistical entropy may not capture its true compressibility (the statistical entropy does not decrease if we concatenate the same text several times). Several indexes for exact pattern matching [9,4,2] take good advantage of repetitiveness, but to date there has been no work on document retrieval in this setting.

In this paper we show how Mäkinen et al.'s [9] run-length compressed suffix array (RLCSA) can be extended to support fast document listing. We present two different solutions. In Section 3, we show that interleaving the longest common

---

[1] This work was supported in part by Academy of Finland grants 250345 (CoECGR) and 134287; Fondecyt grant 1-110066, Chile; the Helsinki Doctoral Programme in Computer Science; the Jenny and Antti Wihuri Foundation, Finland; and the Millennium Nucleus for Information and Coordination in Networks (ICM/FIC P10-024F), Chile.

J. Fischer and P. Sanders (Eds.): CPM 2013, LNCS 7922, pp. 107–119, 2013.

prefix (LCP) arrays of the individual documents, in the order given by the global LCP of the collection, yields long runs of equal values on repetitive collections, which makes this so-called interleaved LCP (ILCP) array highly compressible. Further, we show that a classical document listing technique [11], designed for a completely different array, works almost verbatim over the ILCP, and this yields a new document listing technique of independent interest for generic document collections (not only repetitive). In Section 4 we explore the idea, dubbed PDL, of precomputing the answers of document listing queries for all suffix tree nodes with enough leaves, and exploiting repetitiveness by grammar-compressing the resulting sets of answers. In Section 5 we experimentally show that the ILCP takes very little extra space on top of the RLCSA, and can speed up the RLCSA when the pattern appears many times in the documents; PDL is an order of magnitude faster and still uses only a fraction of the original text size.

2   Related Work

The best current solutions for document listing are based on an idea by Muthukrishnan [11]. Let $T[1..n]$ be the concatenation of the collection of d documents separated by copies of a special character "$". Muthukrishnan's solution stores the suffix tree [18] of T, which in particular includes the suffix array [10] $\mathsf{SA}[1..n]$. The solution also stores a so-called document array $D[1..n]$ of T, in which each cell $D[i]$ stores the identifier of the document containing $T[\mathsf{SA}[i]]$; an array $C[1..n]$, in which each cell $C[i]$ stores the largest value $h < i$ such that $D[h] = D[i]$, or 0 if there is no such value h; and a data structure supporting range-minimum queries (RMQs) over C, $\mathrm{RMQ}_C(i,j) = \mathrm{argmin}_{i \leq k \leq j} C[k]$. These data structures take a total of $\mathcal{O}(n \lg n)$ bits. Given a pattern $P[1..m]$, the suffix tree is used to find the interval $\mathsf{SA}[\ell..r]$ that contains the starting positions of the suffixes prefixed by P. It follows that every value $C[i] < \ell$ in $C[\ell..r]$ corresponds to a distinct document in $D[i]$. Thus a recursive algorithm finding all those positions i starts with $k = \mathrm{RMQ}_C(\ell, r)$. If $C[k] \geq \ell$ it stops. Otherwise it reports document $D[k]$ and continues recursively with the ranges $C[\ell, k-1]$ and $C[k+1, r]$ (the condition $C[k] \geq \ell$ always uses the original ℓ value). In total, the algorithm uses $\mathcal{O}(m + \mathsf{ndoc})$ time, where ndoc is the number of documents returned.

Sadakane [15] gave a compressed version of Muthukrishnan's solution, which stores only a compressed suffix array CSA of T, a sparse bitvector $B[1..n]$ indicating where in T each document starts, an RMQ data structure for C that returns the position of the leftmost minimum in a range without accessing C, and a bitmap $V[1..d]$ to record which document identifiers we have already returned. Fischer [3] showed that such an RMQ data structure takes only $2n + o(n)$ bits and can answer queries in $\mathcal{O}(1)$ time. These data structures take a total of $|\mathsf{CSA}| + 2n + d \lg(n/d) + \mathcal{O}(d) + o(n)$ bits. Here $d \lg(n/d) + \mathcal{O}(d) + o(n)$ bits are for a sparse bitvector representation (e.g., [13]) of B, which has only d 1s. This representation answers in constant time query $\mathsf{rank}(B, i)$, which gives the number of 1s in $B[1..i]$. Now, given P, we use CSA to find ℓ and r, then emulate Muthukrishnan's algorithm: After each RMQ giving position k we use CSA and

B to compute $D[k] = \mathsf{rank}(B, \mathsf{CSA}[k])$, then check the bitmap V to see whether we have already returned that document. If $V[D[k]] = 1$, we stop that recursive branch, else we return $D[k]$, mark $V[D[k]] \leftarrow 1$, and continue recursing. In total we use $\mathcal{O}(\mathsf{search}(m) + \mathsf{ndoc} \cdot \mathsf{lookup}(n))$ time, where $\mathsf{search}(m)$ is the time to find ℓ and r and $\mathsf{lookup}(n)$ is the time to access a cell of SA, using CSA.

Hon et al. [8] push the space down further by sampling array C. The array is divided into blocks of length b, and an array $C'[1..n/b]$ stores the minima of the blocks. The recursive RMQs algorithm is run over C', so that each position $C'[k]$ found requires exploring the documents in one block of D, $D[(k-1)b+1..kb]$. By setting, say, $b = \lg^\epsilon n$ for a constant $\epsilon > 0$, the space becomes $|\mathsf{CSA}| + d \lg(n/d) + \mathcal{O}(d) + o(n)$ bits and the time raises to $\mathcal{O}(\mathsf{search}(m) + \mathsf{ndoc} \cdot \mathsf{lookup}(n) \lg^\epsilon n)$.

In a repetitive environment, one can use an RLCSA [9] as the CSA. However, those $2n + o(n)$ bits of Sadakane [15], and even the $o(n)$ bits of Hon et al. [8], are likely to dominate the space requirement.

Another trend to simulate Muthukrishnan's algorithm is to represent the document array $D[1..n]$ explicitly using a wavelet tree [7], which uses $n \lg d + o(n)$ bits and can access any $D[i]$, as well as compute $\mathsf{rank}_c(D, i)$ and $\mathsf{select}_c(D, j)$, in time $\mathcal{O}(\lg d)$. The first query counts the number of times c occurs in $D[1..i]$, whereas the second gives the position in D of the jth occurrence of c. The wavelet tree root divides values $\leq d/2$ and $> d/2$ in $D[1..n]$, storing only a bitmap $B[1..n]$ where $B[i] = 0$ iff $D[i] \leq d/2$. Then, recursively, the left child of the root represents the subsequence of D with values $\leq d/2$, and the right child the subsequence with values $> d/2$. The leaves represent runs of a single value in $[1..d]$, and the tree has height $\lg d$.

Mäkinen and Välimäki [17] showed that the wavelet tree of D can also emulate array C, as $C[i] = \mathsf{select}_{D[i]}(\mathsf{rank}_{D[i]}(D, i) - 1)$. Then, Gagie et al. [6] showed that just the CSA and the wavelet tree of D provided document listing in time $\mathcal{O}(\mathsf{search}(m) + \mathsf{ndoc} \lg(n/\mathsf{ndoc}))$, without using any RMQ structure. Navarro et al. [12] showed that this wavelet tree is grammar-compressible, as D contains repeated substrings at almost the same positions of the runs found in SA.

3 Interleaved LCP Array

The longest-common-prefix array $\mathsf{LCP}_S[1..|S|]$ of a string S is defined such that $\mathsf{LCP}_S[1] = 0$ and, for $2 \leq i \leq |S|$, $\mathsf{LCP}_S[i]$ is the length of the longest common prefix of the lexicographically $(i-1)$th and ith suffixes of S, that is, between $S[\mathsf{SA}_S[i-1]..|S|]$ and $S[\mathsf{SA}_S[i]..|S|]$, where SA_S is the suffix array of S. We define the interleaved LCP array of T, ILCP, to be the interleaving of the LCP arrays of the individual documents according to the document array.

Definition 1. *Let $T[1, n] = S_1 \cdot S_2 \cdots S_d$ be the concatenation of documents S_j, D the document array of T, and LCP_{S_j} the longest common prefix array of string S_j. Then the* interleaved LCP array *of T is defined, for all $1 \leq i \leq n$, as*

$$\mathsf{ILCP}[i] \;\; = \;\; \mathsf{LCP}_{S_{D[i]}} \left[\mathsf{rank}_{D[i]}(D, i) \right].$$

The following property of ILCP makes it suitable for document retrieval.

Lemma 1. *Let $T[1, n] = S_1 \cdot S_2 \cdots S_d$ be the concatenation of documents S_j, SA its suffix array and D its document array. Let $\mathsf{SA}[\ell..r]$ be the interval that contains the starting positions of suffixes prefixed by a pattern $P[1..m]$. Then the values strictly less than m in $\mathsf{ILCP}[\ell..r]$ are in the same positions as the leftmost occurrences in $D[\ell..r]$ of the distinct document identifiers in that range.*

Proof. Let $\mathsf{SA}_{S_j}[\ell_j..r_j]$ be the interval of all the suffixes of S_j starting with $P[1..m]$. Then it must hold that $\mathsf{LCP}_{S_j}[\ell_j] < m$, as otherwise $S_j[\mathsf{SA}[\ell_j-1]..\mathsf{SA}[\ell_j - 1] + m - 1] = S_j[\mathsf{SA}[\ell_j]..\mathsf{SA}[\ell_j] + m - 1] = P$ as well, contradicting the definition of ℓ_j. For the same reason, it holds that $\mathsf{LCP}_{S_j}[\ell_j + k] \geq m$ for all $1 \leq k \leq r_j - \ell_j$. Now, let S_j start at position $p_j + 1$ in T, where $p_j = |S_1 \cdots S_{j-1}|$. Because each S_j is terminated by the special symbol "$\$$", the lexicographic ordering between the suffixes $S_j[k..]$ in SA_{S_j} is the same as of the corresponding suffixes $T[p_j + k..]$ in SA. That is, it holds that $\langle \mathsf{SA}[i], D[i] = j, 1 \leq i \leq n \rangle = \langle p_j + \mathsf{SA}_{S_j}[i], 1 \leq i \leq |S_j| \rangle$. Or, put another way, $\mathsf{SA}[i] = p_j + \mathsf{SA}_{S_j}[\mathrm{rank}_j(D, i)]$ whenever $D[i] = j$. Now, let f_j be the leftmost occurrence of j in $D[\ell..r]$. This means that $\mathsf{SA}[f_j]$ is the lexicographically first suffix of S_j that starts with P. By definition of ℓ_j, it holds that $\ell_j = \mathrm{rank}_j(D, f_j)$. Thus, by definition of ILCP, it holds that $\mathsf{ILCP}[f_j] = \mathsf{LCP}_{S_j}[\mathrm{rank}_j(D, f_j)] = \mathsf{LCP}_{S_j}[\ell_j] < m$, whereas all the other $\mathsf{ILCP}[k]$ values, for $\ell \leq k \leq r$, where $D[k] = j$, must be $\geq m$. □

Therefore, for the purposes of document listing, we can replace the C array by ILCP in Muthukrishnan's algorithm: instead of recursing until listing all the positions k such that $C[k] < \ell$, we recurse until listing all the positions k such that $\mathsf{ILCP}[k] < m$.

3.1 Document Listing in General Collections

Under Szpankowski's very general A2 probabilistic model [16] (which includes Bernoulli and Markov chains of fixed memory), the maximum LCP value in a string S is almost surely (a very strong kind of convergence[2], which we abbreviate a.s.) $\mathcal{O}(\lg |S|)$ [16]. This means that storing ILCP explicitly requires a.s. at most $n \lg \lg(n/d) + \mathcal{O}(n)$ bits, usually far less than the $n \lg d$ bits required by C.

The fact that we are interested in the values 0 to $m - 1$ in ILCP gives a new relevant index for document listing in general collections. Grossi et al. [7] proved that, if we give the wavelet tree of a sequence S any shape (i.e., not necessarily balanced) and represent the wavelet tree bitmaps using a compressed representation (e.g., [13]), then the total space is the zero-order entropy of the represented sequence, $H_0(S)$, plus $o(nh)$ bits, where h is the wavelet tree's height. The $o(nh)$ bits can become $\mathcal{O}(nh/\lg n)$ if we use the bitmap representation of Pătraşcu [14] instead. Now consider a representation where the leftmost leaf is at depth 1, the next 2 leaves are at depth 3, the next 4 leaves are at depth 5, and in general the 2^{d-1}th to $(2^d - 1)$th leftmost leaves are at depth $2d - 1$. Then the ith

---

[2] A sequence X_n tends to a value β almost surely if, for every $\epsilon > 0$, the probability that $|X_N/\beta - 1| > \epsilon$ for some $N > n$ tends to zero as n tends to infinity, $\lim_{n \to \infty} \sup_{N > n} \Pr(|X_N/\beta - 1| > \epsilon) = 0$.

leftmost leaf is at depth $\mathcal{O}(\lg i)$. If we build this wavelet tree on sequence ILCP, the total space is $H_0(\text{ILCP}) + \mathcal{O}(n \lg d / \lg n)$, which is a.s. $n \lg \lg(n/d) + \mathcal{O}(n)$. What is interesting about this shape is that, using the traversal of Gagie et al. [6] to reach the leaves with values 0 to $m - 1$, we need only reach m leaves at depth $\mathcal{O}(\lg m)$ (i.e., the leftmost m in the wavelet tree), and thus we need to traverse only $\mathcal{O}(m)$ wavelet tree nodes. Array D can be stored in plain form, but permuted so that it is aligned to the wavelet tree leaves, which allows determining each distinct document identifier in $\mathcal{O}(1)$ time.

Theorem 1. *Let $T[1..n] = S_1 \cdot S_2 \cdots S_d$ be the concatenation of d documents S_j and let l be the maximum length of a repeated string in any S_j. Let CSA be a compressed suffix array on T that searches for any pattern $P[1..m]$ in time $\text{search}(m) \geq m$. Then we can store T in $|\text{CSA}| + n(\lg d + \lg l + \mathcal{O}(1))$ bits such that the ndoc documents where $P[1..m]$ occurs can be listed in time $\mathcal{O}(\text{search}(m) + \text{ndoc})$. If T is generated under Szpankowski's A2 model [16], then the space is $|\text{CSA}| + n(\lg d + \lg \lg(n/d) + \mathcal{O}(1))$ bits.*

In particular, if we use the CSA of Belazzougui and Navarro [1], we recover the optimal time of Muthukrishnan's solution, using (in most cases) less space.

Corollary 1. *Under the conditions of Theorem 1, we can obtain $nH_k(T) + o(nH_k(T)) + n(\lg d + \lg l + \mathcal{O}(1))$ bits and $\mathcal{O}(m + \text{ndoc})$ time, where $H_k(T)$ is the k-th order empirical entropy of T, for any $k \leq \alpha \lg_\sigma n$, σ the alphabet size of T, and $0 < \alpha < 1$ any constant.*

3.2   Document Listing in Repetitive Collections

Array ILCP has yet another property, which also makes it attractive for repetitive collections.

Lemma 2. *Let S be a string generated under Szpankowski's A2 model. Let T be formed by concatenating d copies of S, each terminated with the special symbol "$\$$", and then carrying out s edits (symbol insertions, deletions, or substitutions) at arbitrary positions in T (excluding the '$\$$'s). Then, a.s., the ILCP array of T is formed by $\rho \leq r + \mathcal{O}(s \lg(r + s))$ runs of equal values, where $r = |S|$.*

Proof. Before applying the edit operations, we have $T = S_1 \cdots S_d$ and $S_j = S\$$ for all j. At this point, ILCP is formed by at most $r + 1$ runs of equal values, since the d equal suffixes $S_j[\text{SA}_{S_j}[i]..r+1]$ must be contiguous in the suffix array SA of T, in the area $\text{SA}[(i-1)d + 1..id]$. Since the values $l = \text{LCP}_{S_j}[i]$ are also equal, and ILCP values are the LCP_{S_j} values listed in the order of SA, it follows that $\text{ILCP}[(i-1)d + 1..id] = l$ forms a run, and thus there are $r + 1 = n/d$ runs in ILCP. Now, if we carry out s edit operations on T, any S_j will be of length at most $r + s + 1$. Consider an arbitrary edit operation at $T[k]$. It changes all the suffixes $T[k - h..n]$ for all $0 \leq h < k$. However, since a.s. the string depth of a leaf in the suffix tree of S is $\mathcal{O}(\lg(r + s))$ [16], the suffix will possibly be moved in SA only for $h = \mathcal{O}(\lg(r + s))$. Thus, a.s., only $\mathcal{O}(\lg(r + s))$ suffixes are moved in SA, and possibly the corresponding runs in ILCP are broken. Hence $\rho \leq r + \mathcal{O}(s \lg(r + s))$ a.s. $\square$

This proof generalizes Mäkinen et al.'s [9] arguments, which hold for uniformly distributed strings S. There is also experimental evidence [9] that, in real-life text collections, a small change to a string usually causes only a small change to its LCP array. Next we design a document listing data structure whose size is bound in terms of ρ.

Let LILCP$[1..\rho]$ be the array containing the partial sums of the lengths of the ρ runs in ILCP, and let VILCP$[1..\rho]$ be the array containing the values in those runs. We can store LILCP as a bitvector $L[1..n]$ with ρ 1s, so that LILCP$[i] = \text{select}(L, i)$. Bitmap L can be stored using a structure by Okanohara and Sadakane [13] that requires $\rho \lg(n/\rho) + \mathcal{O}(\rho)$ bits and answers select queries in $\mathcal{O}(1)$ time[3]. For rank it requires $\mathcal{O}(\lg(n/\rho))$ time, but we can reduce it to $\mathcal{O}(\lg \lg n)$ by building a y-fast trie [19] on every $(\lg n)$th value of LILCP and completing the query with a binary search using select, adding $\mathcal{O}(\rho)$ bits.

With this representation, it holds that ILCP$[i] = \text{VILCP}[\text{rank}(L, i)]$. We can map from any position i to its run $i' = \text{rank}(L, i)$ in time $\mathcal{O}(\lg \lg n)$, and from any run i' to its starting position in ILCP, $i = \text{select}(L, i')$, in constant time.

This is sufficient to emulate Sadakane's algorithm [15] on a repetitive collection. We will use RLCSA as the CSA. The sparse bitvector $B[1..n]$ marking the document beginnings in T will be represented just like L, so that it requires $d \lg(n/d) + \mathcal{O}(d)$ bits and lets us compute any value $D[i] = \text{rank}(B, \text{SA}[i])$ in time $\mathcal{O}(\lg \lg n + \text{lookup}(n))$. Finally, we build an RMQ data structure on VILCP, requiring $2\rho + o(\rho)$ bits and without needing access to VILCP [3].

Assume we have already used RLCSA to find ℓ and r in $\mathcal{O}(\text{search}(m))$ time. Now we compute $\ell' = \text{rank}(L, \ell)$ and $r' = \text{rank}(L, r)$, which are the endpoints of the interval VILCP$[\ell'..r']$ containing the values in the runs in ILCP$[\ell..r]$. Now we run the recursive RMQs algorithm on VILCP$[\ell'..r']$. Each time we find a minimum at VILCP$[i']$, we remap it to the run ILCP$[i..j]$, where $i = \max(\ell, \text{select}(L, i))$ and $j = \min(r, \text{select}(L, i+1) - 1)$. For each $i \leq k \leq j$, we compute $D[k]$ using B and RLCSA as explained, mark it in $V[D[k]] \leftarrow 1$, and report it. Since we do not have access to the values in ILCP nor in VILCP, the condition to stop the recursion at some value i' is that $V[D[i]] = 1$ is already marked. We show next that this is correct as long as RMQ returns the leftmost minimum in the range and that we recurse first to the left and then to the right of each minimum VILCP$[i']$ found.

Lemma 3. *Using the procedure described, we correctly find all the positions $\ell \leq k \leq r$ such that* ILCP$[k] < m$.

Proof. Let $j = D[k]$ be the leftmost occurrence of document j in $D[\ell..r]$. By Lemma 1, among all the positions where $D[k'] = j$ in $D[\ell..r]$, k is the only one where ILCP$[k] < m$. Since we find a minimum ILCP value in the range, and then explore the left subrange before the right subrange, it is not possible to find first another occurrence $D[k'] = j$, since it has a larger ILCP value and is to the right of k. Therefore, when $V[D[k]] = 0$, that is, the first time we find a $D[k] = j$, it must hold ILCP$[k] < m$, and the same is true for all the other ILCP values in the run. Hence it is correct to list all those documents and mark them in V.

---

[3] Using a constant-time rank/select data structure for their internal array H.

Conversely, whenever we find a $V[D[k']] = 1$, the document has already been reported, thus this is not its leftmost occurrence and then $\mathsf{ILCP}[k'] \geq m$ holds, as well as for the whole run. Hence it is correct to avoid reporting the whole run and to stop the recursion in the range, as the minimum value is already $\geq m$.   □

We have thus obtained our first result for repetitive collections:

Theorem 2. *Let* $T = S_1 \cdot S_2 \cdots S_d$ *be the concatenation of d documents S_j, and* RLCSA *be a suffix array on* T, *searching for any pattern $P[1..m]$ in time* search(m) *and accessing* SA$[i]$ *in time* lookup(n). *Let ρ be the number of runs in the* ILCP *array of T. We can store T in* $|\mathsf{RLCSA}| + \rho \lg(n/\rho) + \mathcal{O}(\rho) + d \lg(n/d) + \mathcal{O}(d)$ *bits such that document listing takes* $\mathcal{O}(\mathsf{search}(m) + \mathsf{ndoc} \cdot (\lg \lg n + \mathsf{lookup}(n)))$ *time.*

3.3   Document Counting

Finally, array ILCP allows us to efficiently count the number of distinct documents where P appears, without listing them all. Sadakane [15] showed how to compute it in constant time adding just $2n + o(n)$ bits of space. With ILCP we can obtain a variant that is suitable for repetitive collections.

We represent VILCP using a skewed wavelet tree as in Section 3.1. We can visit the first m leaves in time $\mathcal{O}(m)$. Moreover, the traversal algorithm [6] tells us how many times each value $0 \leq l < m$ occurs in VILCP$[\ell'..r']$. More precisely, we arrive at each leaf l with an interval $[\ell'_l, r'_l]$ such that VILCP$[\ell'..r']$ contains from the ℓ'_lth to the r'_lth occurrences of value l in VILCP$[\ell'..r']$. We store a reordering of the run lengths so that the runs corresponding to each value l are collected left to right in ILCP and stored aligned to the wavelet tree leaf l. Those are concatenated into another bitmap $L'[1..n]$ with ρ 1s, similar to L, which allows us, using select$(L', \cdot)$, to count the total length spanned by the ℓ'_lth to r'_lth runs in leaf l. By adding the areas spanned over the m leaves, we count the total number of documents where P occurs. Note that we need to correct the lengths of runs ℓ' and r', as they may overlap the original interval ILCP$[\ell..r]$.

Theorem 3. *Let* $T = S_1 \cdot S_2 \cdots S_d$ *be the concatenation of d documents S_j, and* RLCSA *a compressed suffix array on T that searches for any pattern $P[1..m]$ in time* search$(m) \geq m$. *Let ρ be the number of runs in the* ILCP *array of T and l be the maximum length of a repeated substring inside any S_j. Then we can store T in* $|\mathsf{RLCSA}| + \rho(\lg l + 2 \lg(n/\rho) + \mathcal{O}(1))$ *bits such that the number of documents where a pattern $P[1..m]$ occurs can be computed in time* $\mathcal{O}(\mathsf{search}(m))$.

4   Precomputed Document Listing

When the document collection is repetitive, the document array is also repetitive. Let SA$[i..j]$ be a run in the suffix array, so that there is another area SA$[i'..j']$, where SA$[i + k] = $ SA$[i' + k] - 1$ for all $k \leq j - i$. Then $D[i + k] = D[i' + k]$ for all $k \leq j - i$, except for at most d cells in the entire array D [5]. Navarro et al. [12] used this repetitiveness in grammar-based compression of the wavelet tree of D. We can also use it to compress the precomputed answers to document listing queries covering long intervals of suffixes.

Let v be a suffix tree node. We write SA_v to denote the interval of the suffix array covered by node v, and D_v to denote the set of distinct document identifiers occurring in the same interval of the document array. Given block size b and a constant $\beta \geq 1$, we build a sparse suffix tree that allows us to answer document listing queries efficiently. For any suffix tree node v, it holds that

1. $|\mathsf{SA}_v| < b$, and thus documents can be listed in time $\mathcal{O}(b \cdot \mathsf{lookup}(n))$ by using CSA and bitvector B; or
2. we can compute the set D_v as a union of some sets $D_{u_1}, \ldots, D_{u_k}$ of total size at most $\beta \cdot |D_v|$, where nodes $u_1, \ldots, u_k$ are in the sparse suffix tree.

We start by selecting suffix tree nodes $v_1, \ldots, v_L$, so that no selected node is an ancestor of another, and the intervals SA_{v_i} of the selected nodes cover the entire suffix array. Given node v and its parent w, we select v if $|\mathsf{SA}_v| \leq b$ and $|\mathsf{SA}_w| > b$, and store D_v with the node. These nodes become the leaves of the sparse suffix tree, and we assume that they are numbered from left to right. Next we proceed upward in the suffix tree. Let v be an internal node, $u_1, \ldots, u_k$ its children, and w its parent. If the total size of sets $D_{u_1}, \ldots, D_{u_k}$ is at most $\beta \cdot |D_v|$, we remove node v from the tree, and add nodes $u_1, \ldots, u_k$ to the children of node w. Otherwise we keep node v in the sparse suffix tree, and store D_v there.

Let $v_1, \ldots, v_L$ be the leaf nodes and $v_{L+1}, \ldots, v_{L+I}$ the internal nodes of the sparse suffix tree. We use grammar-based compression to replace frequent subsets in sets $D_{v_1}, \ldots, D_{v_{L+I}}$ with grammar rules expanding to those subsets. Given a set Z and a grammar rule $X \to Y$, where $Y \subseteq \{1, \ldots, d\}$, we replace Z with $(Z \cup \{X\}) \setminus Y$, if $Y \subseteq Z$. As long as $|Y| \geq 2$ for all grammar rules $X \to Y$, each set D_{v_i} can be decompressed in $\mathcal{O}(|D_{v_i}|)$ time.

When all rules have been applied, we store the reduced sets $D_{v_1}, \ldots, D_{v_{L+I}}$ as an array A of document and rule identifiers. The array takes $|A| \lg(d + n_R)$ bits of space, where n_R is the total number of rules. We mark the first cell in the encoding of each set with a 1 in a bitvector $B_A[1..|A|]$, so that set D_{v_i} can be retrieved by decompressing $A[\mathsf{select}(B_A, i), \mathsf{select}(B_A, i+1) - 1]$. The bitvector takes $|A|(1 + o(1))$ bits of space and answers select queries in $\mathcal{O}(1)$ time [13]. The grammar rules are stored similarly, in an array G taking $|G| \lg d$ bits and a bitvector $B_G[1..|G|]$ of $|G|(1 + o(1))$ bits separating the array into rules (note that right hand sides of rules are formed only by terminals).

In addition to the sets and the grammar, we also have to store the sparse suffix tree. Bitvector $B_L[1..n]$ marks the first cell of interval SA_{v_i} for all leaf nodes v_i, allowing us to convert interval $\mathsf{SA}[\ell, r]$ into a range of nodes $[ln, rn] = [\mathsf{rank}(B_L, \ell), \mathsf{rank}(B_L, r+1) - 1]$. By using the same bitvector as for LILCP in Section 3.2, we can store B_L in $L \lg(n/L) + \mathcal{O}(L)$ bits and answer rank queries in $\mathcal{O}(\lg \lg n)$ time and select queries in constant time. Another bitvector $B_F[1..L+I]$ of $(L + I)(1 + o(1))$ bits marks the nodes that are the first children of their respective parents, supporting rank queries in constant time [13]. Array F of $I \lg I$ bits stores pointers to parent nodes, so that if node v_i is a first child, its

```
function listDocuments(ℓ, r)                        function parent(i)
    (res, ln) ← (∅, rank(B_L, ℓ))                       par ← F[rank(B_F, i)]
    if select(B_L, ln) < ℓ:                             return (par + L, N[par])
        r' ← min(select(B_L, ln + 1) − 1, r)
        (res, ln) ← (list(ℓ, r'), ln + 1)          function set(i)
        if r' = r: return res                           res ← ∅
    rn ← rank(B_L, r + 1) − 1                           ℓ ← select(B_A, i)
    if select(B_L, rn + 1) ≤ r:                         r ← select(B_A, i + 1) − 1
        ℓ' ← select(B_L, rn + 1)                        for j ← ℓ to r:
        res ← res ∪ list(ℓ', r)                             if A[j] ≤ d: res ← res ∪ {A[j]}
    return res ∪ decompress(ln, rn)                         else: res ← res ∪ rule(A[j] − d)
                                                        return res
function decompress(ℓ, r)
    (res, i) ← (∅, ℓ)                               function rule(i)
    while i ≤ r:                                         ℓ ← select(B_G, i)
        next ← i + 1                                     r ← select(B_G, i + 1) − 1
        while B_F[i] = 1:                                return G[ℓ . . . r]
            (i', next') ← parent(i)
            if next' > r + 1: break                 function list(ℓ, r)
            (i, next) ← (i', next')                     res ← ∅
        res ← res ∪ set(i)                              for i ← ℓ to r:
        i ← next                                            res ← res ∪ {rank(B, SA[i])}
    return res                                          return res
```

Fig. 1. Pseudocode for document listing using precomputed answers. Function listDocuments(ℓ, r) lists the documents from interval SA$[\ell, r]$; decompress(ℓ, r) decompresses the sets stored in nodes $v_\ell, \ldots, v_r$; parent(i) returns the parent node and the leaf node following it for a first child v_i; set(i) decompresses the set stored in v_i; rule(i) expands the ith grammar rule; and list(ℓ, r) lists the documents from interval SA$[\ell, r]$ by using CSA and bitvector B.

parent node is v_j, where $j = L + F[\text{rank}(B_F, i)]$. Finally, array N of $I \lg L$ bits stores a pointer to the leaf node following each internal node.

Figure 1 contains pseudocode for document listing using the precomputed answers. Function list(ℓ, r) takes $\mathcal{O}((r + 1 - \ell)(\lg \lg n + \text{lookup}(n)))$ time, set(i) takes $\mathcal{O}(|D_{v_i}|)$ time, and parent(i) takes $\mathcal{O}(1)$ time. Function decompress(ℓ, r) requires $\mathcal{O}(|res|)$ time to decompress the sets. Traversing the tree takes additional $\mathcal{O}(h)$ time per decompressed set, where h is the height of the sparse suffix tree. As each set contains at least one document, and we may have to list each document up to β times, this sums to $\mathcal{O}(\beta h \cdot |res|)$ time in the worst case. Hence the total time for listDocuments(ℓ, r) is $\mathcal{O}(\text{ndoc} \cdot \beta h + \lg \lg n)$, if the answer has been precomputed, and $\mathcal{O}(b \cdot (\lg \lg n + \text{lookup}(n)))$ otherwise.

5   Experiments

We implemented the document listing approaches described in preceding sections, and measured their performance on two datasets. All experiments were

Table 1. Means and standard deviations (SD) of ndoc and the ratio $\frac{occ}{ndoc}$ for the pattern sets

	High		Medium		Low	
	Mean	SD	Mean	SD	Mean	SD
FIWIKI, ndoc	1810.8	1369.8	602.7	654.9	327.0	556.7
FIWIKI, ratio	32.04	378.62	4.26	22.72	1.75	2.46
INFLUENZA, ndoc	111021.3	29379.4	69666.5	19056.8	46304.3	17082.8
INFLUENZA, ratio	1.55	0.26	1.23	0.08	1.11	0.06

run on an Intel i7 860 2.8 GHz (8192 KB cache), with 16 GB RAM, running Ubuntu 12.04 and compiling with gcc-4.6.3 -O3.

Test Data. We used two repetitive text collections. FIWIKI is a 400 MB prefix of Finnish Wikipedia version history. Each version of each Wikipedia article is considered a separate document, giving 20,433 documents. INFLUENZA is composed of genomes of the influenza virus, totaling 321.2 MB, and 227,356 documents.

Test Patterns. Let occ be the number of times a pattern occurs in the whole collection, and recall ndoc is the number of documents containing the pattern. Document listing queries for patterns with similar occ and ndoc are easily handled by just enumerating all the positions of pattern occurrences (with the RLCSA) and mapping them to document identifiers. This approach however becomes less feasible as the separation between occ and ndoc grows, and at some point specialized document listing approaches become necessary. With this in mind, for each collection we constructed three sets of patterns as follows. First, we listed all patterns of length k present, and then ordered the patterns in descending order by value $occ - ndoc$, picking specific intervals of this list for testing.

For FIWIKI, the pattern length is 8, and each pattern set contains 20,000 patterns, starting at ranks 1,001, 40,001 and 100,001 of the full list of patterns. For INFLUENZA, the pattern length is 6, the set size 1000, and starting ranks are 1, 1,001 and 2,001. We call these three sets in both collections the *high*, *medium* and *low* pattern sets, respectively. Table 1 gives pattern statistics.

Results. Figure 2 shows the space-time tradeoff achieved by our document listing methods. The interleaved LCP array approach (Section 3) is called ilcp, and values following underscores represent the RLCSA sample rate. The precomputed document listing approach (Section 4) is called pdl, and values following underscores represent block size and the β value.

As a baseline we measured the time for a brute force (brute) approach, which simply enumerates pattern occurrences with the RLCSA, collecting distinct documents. This approach adds no space to the index. Like ilcp, brute's tradeoff comes from the sample period of the RLCSA.

Our first observation is that the new approaches achieve small space overhead, particularly on the FIWIKI set. Specifically, the RLCSA with sample period 128

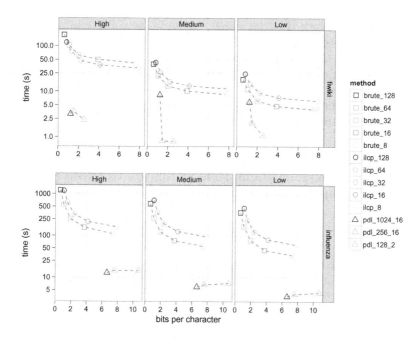

Fig. 2. Document listing times and memory required by different document listing approaches. Bits per character are shown on the x-axis, and time taken to list the documents on the y-axis (note the logarithmic scale). The time taken to find suffix array intervals corresponding to each pattern is not included in times shown here.

takes 29 MB and 27 MB for the FIWIKI and INFLUENZA collections, respectively (about 7% and 8% of the uncompressed collection sizes). Including such RLCSA, ilcp took 40 MB and 45 MB (about 10% and 14%). With block size $b = 1024$ and $\beta = 16$, pdl took 61 MB and 267 MB (about 15% and 83%).

With respect to query time, pdl significantly outperforms ilcp and brute on both data sets and is around an order of magnitude faster than the others when memory is equated. On the other hand ilcp is beaten by brute, except when the separation between occ and ndoc becomes large (the *high* FIWIKI pattern set).

Our most important experimental result is that, on the FIWIKI collection, pdl speeds up document listing by around an order of magnitude over brute while still using total space that is only a fraction of the uncompressed collection size. We were unable to compare to more sophisticated document listing techniques [12] designed for non-highly-repetitive collections because we could not construct them on our data sets. We leave an extensive comparison for the full paper.

6   Conclusions

We have described two approaches to document listing in highly repetitive collections — using an interleaved LCP array (ilcp) and precomputed document listing (pdl) — and shown that, on some representative collections, pdl significantly reduces the query time of a brute-force solution, while still using only a fraction of the space of the uncompressed collection.

Aside from further experimental analysis, there are many directions for future work. Probably the most interesting one is to apply the ilcp approach over faster document listing indices, such as the wavelet tree of Theorem 3, which would yield an interesting space/time tradeoff.

Acknowledgements. We thank Giovanni Manzini for suggesting this line of research, Veli Mäkinen and Jorma Tarhio for helpful discussions, Cecilia Hernández for her grammar compressor, and Meg Gagie for righting our own grammar.

References

1. Belazzougui, D., Navarro, G.: Alphabet-independent compressed text indexing. In: Demetrescu, C., Halldórsson, M.M. (eds.) ESA 2011. LNCS, vol. 6942, pp. 748–759. Springer, Heidelberg (2011)
2. Claude, F., Navarro, G.: Improved grammar-based compressed indexes. In: Calderón-Benavides, L., González-Caro, C., Chávez, E., Ziviani, N. (eds.) SPIRE 2012. LNCS, vol. 7608, pp. 180–192. Springer, Heidelberg (2012)
3. Fischer, J.: Optimal succinctness for range minimum queries. In: López-Ortiz, A. (ed.) LATIN 2010. LNCS, vol. 6034, pp. 158–169. Springer, Heidelberg (2010)
4. Gagie, T., Gawrychowski, P., Kärkkäinen, J., Nekrich, Y., Puglisi, S.J.: A faster grammar-based self-index. In: Dediu, A.-H., Martín-Vide, C. (eds.) LATA 2012. LNCS, vol. 7183, pp. 240–251. Springer, Heidelberg (2012)
5. Gagie, T., Navarro, G., Puglisi, S.J.: Colored range queries and document retrieval. In: Chavez, E., Lonardi, S. (eds.) SPIRE 2010. LNCS, vol. 6393, pp. 67–81. Springer, Heidelberg (2010)
6. Gagie, T., Navarro, G., Puglisi, S.J.: New algorithms on wavelet trees and applications to information retrieval. Theor. Comp. Sci. 426-427, 25–41 (2012)
7. Grossi, R., Gupta, A., Vitter, J.S.: High-order entropy-compressed text indexes. In: Proc. SODA, pp. 636–645 (2003)
8. Hon, W.-K., Shah, R., Vitter, J.: Space-efficient framework for top-k string retrieval problems. In: Proc. FOCS, pp. 713–722 (2009)
9. Mäkinen, V., Navarro, G., Sirén, J., Valimäki, N.: Storage and retrieval of highly repetitive sequence collections. J. Computational Biology 17(3), 281–308 (2010)
10. Manber, U., Myers, G.: Suffix arrays: a new method for on-line string searches. SIAM J. Comput. 22(5), 935–948 (1993)
11. Muthukrishnan, S.: Efficient algorithms for document retrieval problems. In: Proc. SODA, pp. 657–666 (2002)
12. Navarro, G., Puglisi, S.J., Valenzuela, D.: Practical compressed document retrieval. In: Pardalos, P.M., Rebennack, S. (eds.) SEA 2011. LNCS, vol. 6630, pp. 193–205. Springer, Heidelberg (2011)
13. Okanohara, D., Sadakane, K.: Practical entropy-compressed rank/select dictionary. In: Proc. ALENEX (2007)
14. Pătraşcu, M.: Succincter. In: Proc. FOCS, pp. 305–313 (2008)

15. Sadakane, K.: Succinct data structures for flexible text retrieval systems. J. Disc. Alg. 5(1), 12–22 (2007)
16. Szpankowski, W.: A generalized suffix tree and its (un)expected asymptotic behaviors. SIAM J. Comput. 22(6), 1176–1198 (1993)
17. Välimäki, N., Mäkinen, V.: Space-efficient algorithms for document retrieval. In: Ma, B., Zhang, K. (eds.) CPM 2007. LNCS, vol. 4580, pp. 205–215. Springer, Heidelberg (2007)
18. Weiner, P.: Linear pattern matching algorithm. In: Proc. SAT, pp. 1–11 (1973)
19. Willard, D.: Log-logarithmic worst-case range queries are possible in space $\theta(n)$. Inf. Pr. Lett. 17(2), 81–84 (1983)

Approximating Shortest Superstring Problem
Using de Bruijn Graphs

Alexander Golovnev[1], Alexander S. Kulikov[2,3], and Ivan Mihajlin[4]

[1] New York University
[2] St. Petersburg Department of Steklov Institute of Mathematics
[3] Algorithmic Biology Laboratory, St. Petersburg Academic University
[4] St. Petersburg Academic University

Abstract. The best known approximation ratio for the shortest super-string problem is $2\frac{11}{23}$ (Mucha, 2012). In this note, we improve this bound for the case when the length of all input strings is equal to r, for $r \leq 7$. E.g., for strings of length 3 we get a $1\frac{1}{3}$-approximation. An advantage of the algorithm is that it is extremely simple both to implement and to analyze. Another advantage is that it is based on de Bruijn graphs. Such graphs are widely used in genome assembly (one of the most important practical applications of the shortest common superstring problem). At the same time these graphs have only a few applications in theoretical investigations of the shortest superstring problem.

1 Introduction

1.1 Problem Statement

The *superstring problem* (also known as shortest common superstring problem, SCS, or shortest superstring problem, SSP) is: given n strings $s_1, \ldots, s_n$ to find a shortest string containing each s_i as a substring. By *r-superstring problem* (or just *r*-SCS) we denote the SCS problem for the special case when all input strings have length exactly r. Gallant et al. [10] showed that both SCS over the binary alphabet and 3-SCS are NP-hard, while 2-SCS can be solved in linear time. Crochemore et al. [7] proved that 2-SCS with multiplicities can be solved in quadratic time. (Note however that when both parameters, the length of input strings and the size of the alphabet, are bounded by constants then the problem degenerates since then the number of possible input strings is bounded by a constant.) The problem has received a lot of attention as it is interesting as a purely theoretical problem and has many practical applications including genome assembly and data compression.

   In this note, we present a simple polynomial time algorithm that finds an $(r^2 + r - 4)/(4r - 6)$-approximation to r-SCS. This is better than the best known approximation ratio $2\frac{11}{23}$ by Mucha [16] for $r = 3, \ldots, 7$. The algorithm first finds an approximate longest traveling salesman path in the overlap graph. It then finds an approximate shortest rural postman path in the de Bruijn graph. We show that if a permutation of the input strings given by one of these two paths does not give a good enough superstring then the other permutation does.

J. Fischer and P. Sanders (Eds.): CPM 2013, LNCS 7922, pp. 120–129, 2013.

1.2 General Setting

For strings s and t by $overlap(s,t)$ we denote the longest suffix of s that is also a prefix of t. By $prefix(s,t)$ we denote the first $|s| - |overlap(s,t)|$ symbols of s. Similarly, $suffix(s,t)$ is the last $|t| - |overlap(s,t)|$ symbols of t. Clearly, for any strings s and t,

$$\text{prefix}(s,t) \circ \text{overlap}(s,t) = s, \text{overlap}(s,t) \circ \text{suffix}(s,t) = t.$$

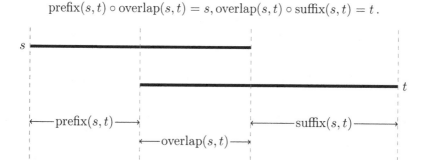

E.g.,

$$\text{overlap}(\texttt{ABACBA}, \texttt{BABCA}) = \texttt{BA}, \text{prefix}(\texttt{ABACBA}, \texttt{BABCA}) = \texttt{ABAC}.$$

For a non-empty string s, by $prefix(s)$ and $suffix(s)$ we denote the string resulting from s by removing the last and the first symbol, respectively.

Now let $\mathcal{S} = \{s_1, \ldots, s_n\}$ be a set of strings over an alphabet Σ and s be a superstring of $\mathcal{S}$. By $\text{OPT}(\mathcal{S})$ we denote the length of a shortest possible superstring for $\mathcal{S}$. A *compression* of s (w.r.t. $\mathcal{S}$) is

$$|s_1| + |s_2| + \cdots + |s_n| - |s|.$$

Clearly minimizing the length of a superstring corresponds to maximizing the compression.

2 Known Results for SCS and Related Graph Problems

2.1 Superstring Problem

Improving the approximation ratio for the SCS problem is interesting both from practical and theoretical points of view. In this subsection, we review known results in this direction. Table 1 shows the sequence of known approximation algorithms and inapproximability results (under the P$\neq$NP assumption) both for minimizing the length and maximizing the compression of superstrings. The well-known Greedy Conjecture [4] says that repeatedly combining two strings with maximal overlap gives a 2-approximation for SCS. Blum et al. [4] proved that this simple algorithm has ratio 4, and Kaplan and Shafrir [13] improved the ratio to 3.5.

Vassilevska [25] showed that an α-approximation of SCS over the binary alphabet implies an α-approximation over any alphabet. Hence, approximating SCS over the binary alphabet cannot be easier than for an arbitrary alphabet.

Table 1. Known approximation ratios and inapproximability results for length and compression of superstrings

ratio	authors	year
	approximating SCS	
3	Blum, Jiang, Li, Tromp and Yannakakis [4]	1991
$2\frac{8}{9}$	Teng, Yao [23]	1993
$2\frac{5}{6}$	Czumaj, Gasieniec, Piotrow, Rytter [8]	1994
$2\frac{50}{63}$	Kosaraju, Park, Stein [15]	1994
$2\frac{3}{4}$	Armen, Stein [1]	1994
$2\frac{50}{69}$	Armen, Stein [2]	1995
$2\frac{2}{3}$	Armen, Stein [3]	1996
$2\frac{25}{42}$	Breslauer, Jiang, Jiang [5]	1997
$2\frac{1}{2}$	Sweedyk [21]	1999
$2\frac{1}{2}$	Kaplan, Lewenstein, Shafrir, Sviridenko [12]	2005
$2\frac{1}{2}$	Paluch, Elbassioni, van Zuylen [18]	2012
$2\frac{11}{23}$	Mucha [16]	2013
	approximating compression	
$\frac{1}{2}$	Tarhio, Ukkonen [22]	1988
$\frac{1}{2}$	Turner [24]	1989
$\frac{2}{3}$	Kaplan, Lewenstein, Shafrir, Sviridenko [12]	2005
$\frac{2}{3}$	Paluch, Elbassioni, van Zuylen [18]	2012
	inapproximability for SCS	
$1\frac{1}{17245}$	Ott [17]	1999
$1\frac{1}{1216}$	Vassilevska [25]	2005
$1\frac{1}{332}$	Karpinski, Schmied [14]	2012
	inapproximability for compression	
$1\frac{1}{11216}$	Ott [17]	1999
$1\frac{1}{1071}$	Vassilevska [25]	2005
$1\frac{1}{203}$	Karpinski, Schmied [14]	2012

Note that SCS is a typical *permutation problem*: if we know the order of the input strings in a shortest superstring then we can recover this superstring by overlapping the strings in this given order. For this reason, it will be convenient for us to identify a superstring with the order of input strings in it. Below we describe several related graph permutation problems.

2.2   Prefix/Overlap Graphs and Traveling Salesman Problem

Many known approximation algorithms for SCS work with the so-called overlap graph. The *overlap graph* $OG(\mathcal{S})$ of the set of strings $\mathcal{S} = \{s_1, \ldots, s_n\}$ is a complete weighted directed graph on a set of vertices $V = \{1, \ldots, n\}$. The weight of an edge from i to j equals $|\text{overlap}(s_i, s_j)|$. It is easy to see that solving SCS corresponds to solving the *asymmetric maximum traveling salesman path* (MAX-ATSP) problem in $OG(\mathcal{S})$ where one is asked to find a longest path visiting each vertex of the graph exactly once (such a path is called *Hamiltonian*). Note that the length of any Hamiltonian path in this graph equals the compression of the corresponding superstring. The best known approximation ratio 2/3 for MAX-ATSP is due to Kaplan et al. [12]. This immediately gives a 2/3-approximation for the compression. Also, Breslauer et al. [5] showed that an α-approximation for MAX-ATSP implies a $3.5 - 1.5\alpha$ approximation for SCS. Plugging in the result by Kaplan et al. [12] gives a 2.5-approximation for SCS.

An alternative way is to find a *minimum traveling salesman path* (MIN-ATSP) in the *prefix graph* $PG(\mathcal{S})$ where vertices i and j are joined by an edge of weight $|\text{prefix}(s_i, s_j)|$. However MIN-ATSP cannot be approximated within any polynomial time computable function unless P=NP [20].

2.3   De Bruijn Graphs and Rural Path Problem

Another important concept is the de Bruijn graph $DG(\mathcal{S})$. In this graph each input string $s_i \in \mathcal{S}$ is represented as a directed (unweighted) edge from $\text{prefix}(s_i)$ to $\text{suffix}(s_i)$. De Bruijn graphs are widely used in genome assembly, one of the practical applications of the SCS problem [19]. A useful property of de Bruijn graphs is the following: if $\mathcal{S}$ is the set of all substrings of length k of some unknown string s (this is called a k-*spectrum of* s) then we can solve SCS for $\mathcal{S}$ in polynomial time. Indeed, in this case there is an Eulerian path in the de Bruijn graph $DG(\mathcal{S})$ spelling the string s. The advantage is that an Eulerian path in a graph can be found in linear time (as opposed to Hamiltonian path that is NP-hard to find). The found Eulerian path in $DG(\mathcal{S})$ does not necessarily need to spell the initial string s (as a graph may contain many Eulerian paths) but it spells a shortest superstring. See Figure 1 for an illustration. A more detailed description of this algorithm can be found, e.g., in [19].

In general, solving the r-SCS problem corresponds to finding a shortest rural postman path in the following *extended de Bruijn graph* $EDG(\mathcal{S})$: the set of vertices is Σ^{r-1}, and every two vertices s and t are joined by a directed edge of weight $|\text{suffix}(s,t)|$. A path $t_1, \ldots, t_k$ spells a string of length

$$|t_1| + |\text{suffix}(t_1, t_2)| + |\text{suffix}(t_2, t_3)| + \cdots + |\text{suffix}(t_{k-1}, t_k)|.$$

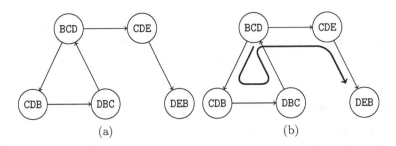

Fig. 1. Solving SCS for a k-spectrum of an unknown string s is easy. (a) De Bruijn graph of a set of strings {CDEB, CDBC, DBCD, BCDB, BCDE}. (b) An Eulerian path in this graph spells a shortest superstring BCDBCDEB.

Thus, the extended de Bruijn graph may be viewed as a weighted analogue of de Bruijn graph. The *shortest directed rural postman path* (DRPP) problem is: given a graph and a subset of edges to find a shortest path going through all these edges. DRPP has many practical applications (see, e.g., [9], [11]), and many papers study heuristic algorithms for it ([6], [9], [11]). At the same time almost no non-trivial theoretical bounds are known for DRPP.

This approach is particularly useful for solving 2-SCS. For this, we first construct the de Bruijn graph of the given set of 2-strings, then for each weakly connected component we add edges between imbalanced vertices (i.e., vertices with non-zero difference of in-degree and out-degree) so that the resulting component contains an Eulerian path. Finally, we add edges between components so that the graph contains an Eulerian path. Figure 2 gives an example. For a more detailed explanation of this algorithm see [10]. Crochemore et al. [7] used a similar technique to solve 2-SCS with multiplicities.

Note that the algorithm described above works for 2-SCS, but not for general r-SCS for the following reason: in case of 2-SCS, strings from different weakly connected components do not share letters (and hence have empty overlap) so the components can be traversed in any order.

3 Algorithm

In this section, we present a simple $(r^2+r-4)/(4r-6)$-approximation algorithm for the r-SCS problem. This ratio is better than the best known ratio $2\frac{11}{23}$ [16] for $r \leq 7$. Before presenting the algorithm we explain its main idea for the case of 3-strings.

3.1 Informally

Let $\mathcal{S} \subseteq \Sigma^3$ be a set of n strings of length 3. Note that $n + 2 \leq \mathrm{OPT}(\mathcal{S}) \leq 3n$ (the former inequality corresponds to the case when in a shortest superstring all input strings have overlaps of size 2, the latter one corresponds to the case when

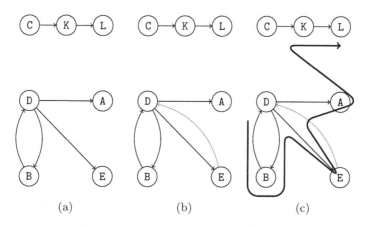

Fig. 2. 2-SCS can be solved in polynomial time. (a) de Bruijn graph of a set of strings {KL, DB, DE, CK, BD, DA}. (b) After adding an edge ED each weakly connected component contains an Eulerian path. (c) The string DBDEDACKL spelled by a path going through all the edges is a shortest superstring.

the input strings do not overlap at all in a shortest superstring). Note that in the two extreme cases when $OPT(\mathcal{S}) = n+2$ or $OPT(\mathcal{S}) = 3n$ a shortest superstring can be easily found. Indeed, if $OPT(\mathcal{S}) = n+2$ then $\mathcal{S}$ is the set of *all* substrings of length 3 of an unknown superstring of length $n + 2$. Such a superstring can be found by traversing an Eulerian path in the de Bruijn graph of $\mathcal{S}$, $DG(\mathcal{S})$. On the other hand, if $OPT(\mathcal{S}) = 3n$ then the input strings just do not overlap with each other and any concatenation of them is a shortest superstring. The case $OPT(\mathcal{S}) = n + 2$ corresponds to the maximal possible compression while the case $OPT(\mathcal{S}) = 3n$ corresponds to the minimal (i.e., zero) compression.

The algorithm proceeds as follows. We first construct the overlap graph $OG(\mathcal{S})$ and find a 2/3-approximation of the maximum TSP path in it. Such a path provides a good approximation to $OPT(\mathcal{S})$ in case $OPT(\mathcal{S})$ is large.

We then construct the de Bruijn graph $DG(\mathcal{S})$. Note that this graph can be viewed as the de Bruijn graph of a set of strings of length 2 over the alphabet Σ^2. Namely, in the original de Bruijn graph a string ABC is represented as an edge from AB to BC. This edge can be viewed as corresponding to the string (AB)(BC) of length 2 over the new alphabet. We then find a shortest superstring to this new set of 2-strings (recall that 2-SCS can be solved exactly in polynomial time) and translate the found solution back to the original problem. This gives a good approximation in case $OPT(\mathcal{S})$ is small. The crucial fact is that if two input strings overlap a lot, then the corresponding 2-strings also overlap a lot and hence many overlaps are found by an algorithm for 2-SCS.

3.2   Formally

We are now ready to give all the details, see Algorithm 3.1. Note that the algorithm is quite easy to implement. Its only black-box part is a 2/3-approximation of MAX-ATSP. A recent algorithm achieving this ratio is due to Paluch et al. [18] and it is essentially based on finding a maximum weight matching. Thus, the running time of the presented algorithm is $O(n^3 \cdot \sum_{i=1}^{n} |s_i|) = O(n^4)$.

Algorithm 3.1. $(r^2 + r - 4)/(4r - 6)$-approximation algorithm r-SCS

Input: $S = \{s_1, \ldots, s_n\} \subseteq \Sigma^r$.
Output: A superstring of S that is at most $(r^2 + r - 4)/(4r - 6)$ times longer than a shortest superstring.

// first, find a long traveling salesman path in the overlap graph
1: let π be a 2/3-approximate maximum traveling salesman path in $\mathrm{OG}(S)$

// then, find a short rural postman path in the de Bruijn graph
2: let $S' = \{s_1', \ldots, s_n'\} \subseteq \Sigma_1^2$ be a set of 2-strings over the alphabet $\Sigma_1 = \Sigma^{r-1}$; s_i' is the 2-string consisting of prefix of s_i of length $r - 1$ and suffix of s_i of length $r - 1$
3: let π_1 be a shortest superstring for the set of 2-strings S'
4: **return** the better one among π and π_1

Theorem 1. *Algorithm 3.1 finds an $\alpha(r)$-approximation for r-SCS where*

$$\alpha(r) = \frac{r^2 + r - 4}{4r - 6}.$$

Proof. Let H be a shortest Hamiltonian path in $\mathrm{OG}(S)$. Then clearly

$$\mathrm{OPT}(S) = rn - w(H),$$

where $w(H)$ is the weight of H. A 2/3-approximate maximum traveling salesman path has weight at least $2w(H)/3$. Thus, the permutation π gives a superstring of length at most $rn - 2w(H)/3$ (formally, to get a superstring from a permutation one just overlaps all the strings in this given order). The corresponding approximation ratio is

$$\frac{rn - 2w(H)/3}{rn - w(H)}. \tag{1}$$

Now let u denote the number of edges of weight at most $(r - 2)$ in H. Then the number of edges of weight exactly $(r - 1)$ in H is $(n - 1 - u)$. Then $w(H) \leq (r - 1)(n - 1 - u) + (r - 2)u$ and hence

$$u \leq (r - 1)(n - 1) - w(H). \tag{2}$$

Note that

$$\mathrm{overlap}(s_i', s_j') = \begin{cases} 1 & \text{if } \mathrm{overlap}(s_i, s_j) = r - 1, \\ 0 & \text{otherwise.} \end{cases}$$

Since S' is a 2-SCS instance, a shortest superstring for S' has the maximal possible number of overlaps of size 1. This number is in turn equal to the maximal possible number of overlaps of size $r - 1$ for S. Since the number of overlaps of

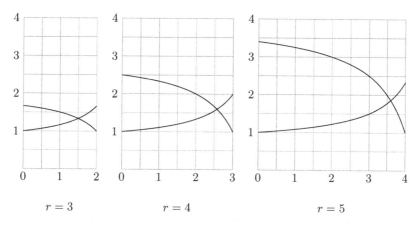

$r = 3$ $r = 4$ $r = 5$

Fig. 3. Plots of $\frac{r-2x/3}{r-x}$ and $\frac{(r^2-2r+2)-(r-1)x}{r-x}$ for $r = 3, 4, 5$ and $0 \le x \le r - 1$

size $r-1$ in H is $(n-1-u)$ the length of a shortest superstring for $\mathcal{S}'$ has length at most $2n - (n - 1 - u) = n + u + 1$. Hence π_1 gives a superstring of length at most $rn - (r - 1)(n - 1 - u)$ in $\mathcal{S}$. Because of (2), this is at most

$$rn - (r - 1)(n - 1 - (r - 1)(n - 1) + w(H)) < (r^2 - 2r + 2)n - (r - 1)w(H).$$

The corresponding approximation ratio is

$$\frac{(r^2 - 2r + 2)n - (r - 1)w(H)}{rn - w(H)}. \tag{3}$$

From (1) and (3) and a simple observation that $0 \le w(H)/n \le (r - 1)$ we conclude that the approximation ratio of the constructed algorithm is

$$\alpha(r) = \max_{0 \le x \le r-1} \left\{ \min \left\{ \frac{r - 2x/3}{r - x}, \frac{(r^2 - 2r + 2) - (r - 1)x}{r - x} \right\} \right\}.$$

Fig. 3 shows plots of the considered functions for $r = 3, 4, 5$.

By taking the derivatives it is easy to see that the former function increases while the latter one decreases on $[0, r - 1]$. This means that the maximum of their minimum is attained at x where they meet, namely

$$x = \frac{r^2 - 3r + 2}{r - 5/3}.$$

Plugging in this x gives

$$\alpha(r) = \frac{r^2 + r - 4}{4r - 6}. \qquad \square$$

Acknowledgements. Research is partially supported by Russian Foundation for Basic Research (11-01-00760-a, 12-01-31057), RAS Program for Fundamental Research, Grant of the President of Russian Federation (NSh-3229.2012.1), the Ministry of Education and Science of the Russian Federation (8216), the Government of the Russian Federation (11.G34.31.0018), and Computer Science Club scholarship.

Also, we would like to thank the anonymous referees for many valuable comments.

References

1. Armen, C., Stein, C.: A $2\frac{3}{4}$-Approximation Algorithm for the Shortest Superstring Problem. Tech. rep., Dartmouth College, Hanover, NH, USA (1994)
2. Armen, C., Stein, C.: Improved length bounds for the shortest superstring problem. In: Sack, J.-R., Akl, S.G., Dehne, F., Santoro, N. (eds.) WADS 1995. LNCS, vol. 955, pp. 494–505. Springer, Heidelberg (1995)
3. Armen, C., Stein, C.: A $2\frac{2}{3}$-Approximation for the Shortest Superstring Problem. In: Hirschberg, D.S., Meyers, G. (eds.) CPM 1996. LNCS, vol. 1075, pp. 87–101. Springer, Heidelberg (1996)
4. Blum, A., Jiang, T., Li, M., Tromp, J., Yannakakis, M.: Linear approximation of shortest superstrings. In: Proceedings of the Twenty-Third Annual ACM Symposium on Theory of Computing, STOC 1991, pp. 328–336. ACM, New York (1991)
5. Breslauer, D., Jiang, T., Jiang, Z.: Rotations of periodic strings and short superstrings. J. Algorithms 24(2), 340–353 (1997)
6. Christofides, N., Campos, V., Corberan, A., Mota, E.: An algorithm for the Rural Postman problem on a directed graph. In: Netflow at Pisa, Mathematical Programming Studies, vol. 26, pp. 155–166. Springer, Heidelberg (1986)
7. Crochemore, M., Cygan, M., Iliopoulos, C., Kubica, M., Radoszewski, J., Rytter, W., Waleń, T.: Algorithms for three versions of the shortest common superstring problem. In: Amir, A., Parida, L. (eds.) CPM 2010. LNCS, vol. 6129, pp. 299–309. Springer, Heidelberg (2010)
8. Czumaj, A., Gasieniec, L., Piotrow, M., Rytter, W.: Parallel and sequential approximation of shortest superstrings. In: Schmidt, E.M., Skyum, S. (eds.) SWAT 1994. LNCS, vol. 824, pp. 95–106. Springer, Heidelberg (1994)
9. Eiselt, H.A., Gendreau, M., Laporte, G.: Arc Routing Problems, Part II: The Rural Postman Problem. Operations Research 43(3), 399–414 (1995)
10. Gallant, J., Maier, D., Storer, J.A.: On finding minimal length superstrings. Journal of Computer and System Sciences 20(1), 50–58 (1980)
11. Groves, G., van Vuuren, J.: Efficient heuristics for the Rural Postman Problem. ORiON 21(1), 33–51 (2005)
12. Kaplan, H., Lewenstein, M., Shafrir, N., Sviridenko, M.: Approximation Algorithms for Asymmetric TSP by Decomposing Directed Regular Multigraphs. J. ACM 52, 602–626 (2005)
13. Kaplan, H., Shafrir, N.: The greedy algorithm for shortest superstrings. Inf. Process. Lett. 93(1), 13–17 (2005)
14. Karpinski, M., Schmied, R.: Improved Lower Bounds for the Shortest Superstring and Related Problems. CoRR abs/1111.5442v3 (2012)
15. Kosaraju, S.R., Park, J.K., Stein, C.: Long tours and short superstrings. In: Proceedings of the 35th Annual Symposium on Foundations of Computer Science, SFCS 1994, pp. 166–177. IEEE Computer Society, Washington, DC (1994)
16. Mucha, M.: Lyndon Words and Short Superstrings. In: Proceedings of the Twenty-Fourth Annual ACM-SIAM Symposium on Discrete Algorithms, SODA 2013. Society for Industrial and Applied Mathematics (2013)

17. Ott, S.: Lower bounds for approximating shortest superstrings over an alphabet of size 2. In: Widmayer, P., Neyer, G., Eidenbenz, S. (eds.) WG 1999. LNCS, vol. 1665, p. 55. Springer, Heidelberg (1999)
18. Paluch, K., Elbassioni, K., van Zuylen, A.: Simpler Approximation of the Maximum Asymmetric Traveling Salesman Problem. In: STACS 2012. LIPIcs, vol. 14, pp. 501–506 (2012)
19. Pevzner, P.A., Tang, H., Waterman, M.S.: An Eulerian path approach to DNA fragment assembly. Proc. Natl. Acad. Sci. 98(17), 9748–9753 (2001)
20. Sahni, S., Gonzalez, T.: P-Complete Approximation Problems. J. ACM 23, 555–565 (1976)
21. Sweedyk, Z.: $2\frac{1}{2}$-Approximation Algorithm for Shortest Superstring. SIAM J. Comput. 29(3), 954–986 (1999)
22. Tarhio, J., Ukkonen, E.: A greedy approximation algorithm for constructing shortest common superstrings. Theoretical Computer Science 57(1), 131–145 (1988)
23. Teng, S.H., Yao, F.: Approximating shortest superstrings. In: Proceedings of the 1993 IEEE 34th Annual Foundations of Computer Science, SFCS 1993, pp. 158–165. IEEE Computer Society, Washington, DC (1993)
24. Turner, J.S.: Approximation algorithms for the shortest common superstring problem. Information and Computation 83(1), 1–20 (1989)
25. Vassilevska, V.: Explicit Inapproximability Bounds for the Shortest Superstring Problem. In: Jedrzejowicz, J., Szepietowski, A. (eds.) MFCS 2005. LNCS, vol. 3618, pp. 793–800. Springer, Heidelberg (2005)

Local Search for String Problems: Brute Force Is Essentially Optimal

Jiong Guo[1], Danny Hermelin[2], and Christian Komusiewicz[3]

[1] Universität des Saarlandes,
Campus E 1.7, D-66123 Saarbrücken, Germany
jguo@mmci.uni-saarland.de
[2] Department of Industrial Management and Engineering,
Ben-Gurion University, Beer Sheva, Israel
hermelin@bgu.ac.il
[3] Institut für Softwaretechnik und Theoretische Informatik,
Technische Universität Berlin, D-10587 Berlin, Germany
christian.komusiewicz@tu-berlin.de

Abstract. We address the problem of whether the brute-force procedure for the local improvement step in a local search algorithm can be substantially improved when applied to classical NP-hard string problems. We examine four problems in this domain: CLOSEST STRING, LONGEST COMMON SUBSEQUENCE, SHORTEST COMMON SUPERSEQUENCE, and SHORTEST COMMON SUPERSTRING. Herein, we consider arguably the most fundamental string distance measure, namely the Hamming distance, which has been applied in practical local search implementations for string problems. Our results indicate that for all four problems, the brute-force algorithm is essentially optimal.

1 Introduction

Local search is a universal algorithmic approach for coping with computationally hard optimization problems. It is typically applied on problems which can be formulated as finding a solution maximizing or minimizing a criterion among a number of feasible solutions. The main idea is to start with some solution, then search inside the *local neighborhood* of this solution for a better solution until a locally optimal solution has been found. The hope is then that the locally optimal solution is almost as good as a globally optimal one. See the book by Aarts and Lenstra [1] for further background and results concerning local search.

There are two main theoretical approaches to study local search: PLS-completeness [11] and parameterized local search [13,6]. PLS-completeness can be used to show that finding a locally optimal solution is computationally hard since a lot of improvement steps might be needed until it has been found. In contrast, parameterized local search is concerned with the parameterized complexity of the problem of searching the local neighborhood of a solution in order to find a better solution. Usually the size of the neighborhood is $n^{O(k)}$, where n is the total input length, and k is a parameter measuring the "radius" of the neighborhood; that is, the maximum distance to the current solution. It is therefore

J. Fischer and P. Sanders (Eds.): CPM 2013, LNCS 7922, pp. 130–141, 2013.

natural to ask whether $n^{O(k)}$ time is required for searching this neighborhood, or whether $f(k) \cdot \mathrm{poly}(n)$ time can be achieved. This is precisely the main question underlying the theory of *parameterized complexity* [5].

There is substantial work in parameterized local search. For example, concerning the TRAVELING SALESMAN problem, Balas [2] showed that one can find, if it exists, a better tour with "shift" distance at most k to the old one in $4^k \cdot \mathrm{poly}(n)$ time. Marx [13] proved the non-existence of such an algorithm for the edge-exchange neighborhood. Subsequently, the complexity of local search for further neighborhood measures of EUCLIDEAN TRAVELING SALESMAN was examined [10]. Notably, the fixed-parameter tractability of EUCLIDEAN TRAV-ELLING SALESMAN and the edge-exchange neighborhood remains open [13,10]. Fellows et al. [6] provided fixed-parameter algorithms for local search variants of diverse graph problems such as VERTEX COVER, ODD CYCLE TRANSVERSAL, MAX CUT, and MIN-BISECTION on planar graphs and proved W[1]-hardness for the general case. Fomin et al. [8] considered the FEEDBACK ARC SET IN TOUR-NAMENTS problems and presented a subexponential-time algorithm for its edge-exchange local search version. Further results concerning parameterized local search have been achieved for clustering problems [4], BOOLEAN CONSTRAINT SATISFACTION [12], STABLE MARRIAGE variants [14], and SATISFIABILITY [16].

In this paper, we add a new realm to the study of parameterized local search by considering string problems. Stringology is one of the most widely studied areas in computer science, particularly motivated by direct applications in text mining and computational biology. Here, we consider four of the most prominent NP-hard string problems: CLOSEST STRING, LONGEST COMMON SUB-SEQUENCE, SHORTEST COMMON SUPERSEQUENCE, and SHORTEST COMMON SUPERSTRING. Local search seems to be a natural approach for dealing with string problems. For instance, a local search heuristic using the Hamming distance neighborhood has been implemented and evaluated with real-world data for problems closely related to CLOSEST STRING [7,15].

We examine all four string problems above in the framework of parameterized local search. Herein, we consider the Hamming distance neighborhood of a temporary solution and prove that the local search version of all these problems are W[1]-hard even on alphabets of constant size, with the Hamming distance k between the old and new solutions as parameter. Since the Hamming distance seems to be the most simple distance between strings, our results could serve as the basis for proving the hardness for other distance neighborhoods. Moreover, for all problems except SHORTEST COMMON SUPERSEQUENCE, we can exclude the existence of algorithms with running-times $n^{o(k)}$. Thus, for these three problems, the $n^{O(k)}$-time brute-force cannot be substantially improved. We remark that these results do not exclude the existence of all parameterized local search algorithms for these problems, but rather motivate the study of further parameterizations, for instance by considering the combined parameter "number m of strings and neighborhood radius k".

2   Preliminaries

For a string S we write $|S|$ to denote the *length* of S. We use $S[i]$, $1 \leq i \leq |S|$, to denote the letter at position i in S and use $S[i, j]$, $1 \leq i < j \leq |S|$, to denote the *substring* $S[i] \cdots S[j]$ of S from position i to position j. A substring of the form $S[i, n]$ is called a *suffix* of S, and a substring $S[1, j]$ is called a *prefix*. For a given suffix T of S, we write $S - T$ to denote the string $S[1, |S| - |T|]$. We use $S-$ as a shorthand for $S - S[|S|]$. A string T is a *subsequence* of S if T can be obtained from S by deleting some letters; that is, if there exists a sequence of positions $i_1 < \cdots < i_{|T|}$ with $S[i_j] = T[j]$ for all $j \in \{1, \ldots, |T|\}$. If T is a subsequence of S, then S is called a *supersequence* of T. The *Hamming distance* $d_H(S, T) := |\{i : S[i] \neq T[i]\}|$ between two string S and T of equal length is defined as the number of positions in which the two strings differ. We define the Hamming distance of a string S to a set $\mathcal{T}$ of strings as $d_H(S, \mathcal{T}) := \max_{T \in \mathcal{T}} d_H(S, T)$.

We analyze our local search string problems in the framework of parameterized complexity [5]. A *parameterized reduction* from a parameterized problem L to another parameterized problem L' is an algorithm with running time $f(k) \cdot \text{poly}(|x|)$ for some computable $f()$, that maps an instance $(x, k) \in \{0, 1\}^* \times \mathbb{N}$ to an instance $(x', k') \in \{0, 1\}^* \times \mathbb{N}$ such that:

(i) $k' \leq g(k)$ for some computable $g()$, and
(ii) $(x, k) \in L \iff (x', k') \in L'$.

If $g()$ is linearly bounded, *i.e.* $g(k) \leq ck$ for some constant c, then we say that the reduction is a *linear parameterized reduction*. Two basic classes of parameterized intractability are W[1] and W[2]; if there is a parameterized reduction from a W[1]-hard (W[2]-hard) problem to a parameterized problem L, then L is W[1]-hard (W[2]-hard).

The hardness results in this paper are obtained by parameterized reductions from the following three problems which all have the solution size k as parameter: In the W[2]-hard MULTICOLORED HITTING SET(k) (MHS(k)), the input is a hypergraph $(V, \mathcal{E})$ and a coloring function $c : V \to \{c_1, \ldots, c_k\}$. The goal is to determine whether there exists a size-k subset $H \subseteq V$ with $H \cap E \neq \emptyset$ for all $E \in \mathcal{E}$, such that H is *multicolored*, that is, $|\{v \in H : c(v) = c_i\}| = 1$ for all $c_i \in \{c_1, \ldots, c_k\}$. In the W[1]-hard MULTICOLORED INDEPENDENT SET(k) (MIS(k)), the input is a graph (V, E) and a coloring function $c : V \to \{c_1, \ldots, c_k\}$, and the goal is to determine if (V, E) has a multicolored independent set $I \subseteq V$. The W[1]-hard MULTICOLORED CLIQUE(k) (MC(k)) is defined similarly, except the goal is to determine the existence of a multicolored clique instead of a multicolored independent set. We make use of the following result [3].[1]

Lemma 1. *Let L be a parameterized problem with parameter k, and assume that there is a linear parameterized reduction from either MHS(k), MIS(k), or MC(k) to L. Then unless all problems in SNP can be solved in subexponential time, size-n instances of L cannot be solved in $n^{o(k)}$ time.*

---

[1] For HITTING SET, Chen et al. [3] do not explicitly make this statement, but it can be inferred via a simple reduction from DOMINATING SET.

3   Closest String

The first local search string problem we consider is a local search variant of the CLOSEST STRING problem. Let Σ denote some arbitrary alphabet, and n be a positive integer. In CLOSEST STRING, the input is a set $\mathcal{T} \subseteq \Sigma^n$ of strings and an integer d, and the goal is to determine whether there is a string $S \in \Sigma^n$ such that $d_H(S, \mathcal{T}) \leq d$. The local search variant of this problem that we consider is defined as follows:

LOCAL SEARCH CLOSEST STRING (LSCS):
Input: A set $\mathcal{T} := \{T_1, \ldots, T_m\} \subseteq \Sigma^n$ of input strings, a temporary solution string $S \in \Sigma^n$ with $d := d_H(S, \mathcal{T})$, and a nonnegative integer k.
Question: Is there a string $\widetilde{S}$ of length n such that $d_H(\widetilde{S}, \mathcal{T}) < d$ and $d_H(\widetilde{S}, S) \leq k$?

Thus, we are given a temporary solution string S, and we want to find a better solution $\widetilde{S}$ in the k-neighborhood of S, where this neighborhood is defined w.r.t. Hamming distance.

We denote the different parameterizations of this problem by appending the parameters to the problem name in parenthesis. Thus, $\mathrm{LSCS}(k)$ for instance, is the LSCS problem parameterized by k. Observe that LSCS can be solved by a brute-force algorithm in $O(n^{k+1} \cdot m)$ time. It is also not difficult to devise a $d^k \cdot poly(n, m)$ algorithm for this problem based on the following observation: as long as S differs from some input string at least d positions, then one of these positions in S has to be changed. Achieving an $f(m) \cdot \mathrm{poly}(n)$-time algorithm by modifying the Integer Linear Programming-based algorithm of Gramm et al. [9] is also possible. Below, we show that for the parameter k, one cannot substantially improve on the brute-force algorithm in general, even when the strings are binary. We begin with the easier case of parameterized-size alphabets.

Proposition 1. *There is a linear parameterized reduction from* $\mathrm{MHS}(k)$ *to* $\mathrm{LSCS}(k + |\Sigma|)$.

We next consider the binary case. Let $(V := \{1, \ldots, |V|\}, \mathcal{E}, c)$ be an instance of $\mathrm{MHS}(k)$, and assume, w.l.o.g., that $|E| \leq |V| - k$ for each $E \in \mathcal{E}$. Set the individual input string length to $n := |V| + |V| \cdot |\mathcal{E}| + 2^k \cdot |V|$, and set the temporary solution S to 0^n. For each $E \in \mathcal{E}$ create a string T_E of length n. For each $v \in \{1, \ldots, |V|\}$, set $T_E[v] := 1$ if $v \in E$ and $T_E[v] := 0$ otherwise. Note that the Hamming distance between $T_E[1, |V|]$ and $S[1, |V|]$ is exactly $|E| \leq |V| - k$. The remaining positions are used to "pad" the distance between T_E and S to $|V| - k$. To this end, assign a unique number $i \in \{1, \ldots, |\mathcal{E}|\}$, and use the substring $T_E[i \cdot |V| + 1, (i+1) \cdot |V|]$ to pad the distance between T_E and S; that is, set the first $|V| - k - |E|$ positions in this substring to 1 and all other positions in $T_E[|V| + 1, n]$ to 0.

Next, add an additional set of strings which enforce that for each proper subset of colors $C \subset \{c_1, \ldots, c_k\}$, the set of colors used by a solution string $C(\widetilde{S}) := \{c(v) : \widetilde{S}[v] = 1\}$ is not C. Since we enforce this for each proper subset, it will follow that $C(\widetilde{S}) = \{c_1, \ldots, c_k\}$. For each proper $C \subset \{c_1, \ldots, c_k\}$, construct a string T_C such that, for each $v \in \{1, \ldots, |V|\}$, we have $T_C[v] = 0$ if $c(v) \in C$ and $T_C[v] = 1$ otherwise. Note that the distance between $S[1, |V|]$ and $T_C[1, |V|]$ equals the number of vertices in V not colored by a color in C. Pad the distance between T_C and S to $|V| - |C|$ by assigning T_C a unique number $i \in \{1, \ldots, 2^k - 1\}$, and let x denote the number of positions v in $T_C[1, |V|]$ with $T_C[v] = 0$. Note that $x \geq |C|$ since for each color $c \in C$ there is at least one vertex colored c. Consequently, set the first $x - |C|$ positions in $T_C[|V| \cdot (|\mathcal{E}| + i) + 1, |V| \cdot (|\mathcal{E}| + i + 1)]$ to 1, and all remaining unspecified positions to 0. Observe that in this way $T_{\emptyset} = 1^{|V|} 0^{n - |V|}$.

This concludes the construction of the set $\mathcal{T}$ of input strings, and the instance $(\mathcal{T}, S, k)$ of LSCS(k). Clearly this construction can be performed in $2^k \cdot \text{poly}(n, m)$ time, and therefore it is a parameterized reduction. Furthermore, observe that $d_H(S, \mathcal{T}) = |V|$, and that this distance is obtained by the distance between S and $T_{\emptyset}$.

Theorem 1. *There is a linear parameterized reduction from* MHS(k) *to* LSCS(k) *for binary strings.*

Corollary 1. LSCS(k) *for binary strings is* W[2]*-hard, it cannot be solved* $n^{o(k)}$ *time unless all problems in* SNP *can be solved in subexponential time.*

4   Longest Common Subsequence

The LONGEST COMMON SUBSEQUENCE (LCS) problem asks to determine whether an input set $\mathcal{T}$ of strings has a string S of some specified length ℓ such that S is a subsequence of each string $T \in \mathcal{T}$. In this section we consider the following local search variant of LCS:

LOCAL SEARCH LONGEST COMMON SUBSEQUENCE (LSLCS):
Input: A set $\mathcal{T} := \{T_1, \ldots, T_m\}$ of input strings over an alphabet Σ, a temporary solution string S such that S is a subsequence of each string in $\mathcal{T}$, and a nonnegative integer k.
Question: Is there a letter $\sigma \in \Sigma$ and a string $\widetilde{S}$ of length $|S|$ such that $\widetilde{S}\sigma$ is a subsequence of each string in $\mathcal{T}$ and $d_H(\widetilde{S}, S) \leq k$?

Observe that LSLCS can be solved in $\binom{|S|}{k} \cdot |\Sigma|^k \cdot \text{poly}(n) = n^{O(k)}$ time by brute-force (n denotes the overall instance size). We show that it is unlikely to substantially improve on this algorithm, even in the case of constant-size alphabets. As a warm-up, we begin with the very easy case of unbounded alphabets.

Lemma 2. *There is a linear parameterized reduction from the* W[2]*-hard* LCS(ℓ) *problem to* LSLCS(k) *with unbounded alphabets.*

We next proceed to the more involved case where $|\Sigma|$ is part of the parameter. We present a reduction from MIS(k) to LSLCS$(k + |\Sigma|)$. Let $(G = (V, E), c)$ denote an instance of MIS(k), where G is a graph and c is a coloring function

$c : V \to \{c_1 \ldots, c_k\}$. By padding (G, c), we can assume, w.l.o.g, that each color class in G has precisely n vertices, that is, $|\{v : c(v) = c_i\}| = n$ for each $i \in \{1, \ldots, k\}$.

We begin by describing the solution string S. The string S consists of a suffix $S^* := (\$\pounds^{k+1}\$)^{k+1}$, where $\$$ and $\pounds$ are two letters of the alphabet that do not appear elsewhere in S. The prefix of S consists of k substrings, or blocks, one for each color class. The substring $S(c_i)$ corresponding to c_i is defined as the string $S(c_i) := \overrightarrow{c_i}(0\#)^n \overleftarrow{c_i}$ where $\overrightarrow{c_i}$ and $\overleftarrow{c_i}$ are letters corresponding to color class c_i. The whole string S is thus constructed as

$$S := S(c_1) \cdots S(c_k) S^*.$$

Next we construct the two enforcement strings $T_1, T_2 \in \mathcal{T}$. The string T_1 contains the string S as its suffix. Its prefix contains k blocks, one for each color class of G, where the i'th block $T_1(c_i)$ is defined as $T_1(c_i) := \overrightarrow{c_i}(0\#1\#)^{n-1} \overleftarrow{c_i}$. The prefix of T_1 is separated from its suffix with the string S^* to form the string

$$T_1 := T_1(c_1) \cdots T_1(c_k) S^* S.$$

The string T_2 also contains k blocks, each corresponding to a color of G, where the block corresponding to c_i is constructed as $T_2(c_i) := \overrightarrow{c_i}(01\#)^n \overleftarrow{c_i}$. We concatenate all these blocks with the suffix $S^*\$$ to obtain the string

$$T_2 := T_2(c_1) \cdots T_2(c_k) S^*\$.$$

Finally, for each edge $e \in E$, we construct an input string T_e as follows. Assume that the vertices in each color class are ordered. Let e be an edge between the x'th vertex of color c_i and the y'th vertex of color c_j, where $i < j$. The string T_e consists of two blocks for each color class of G, defined by

- $T_e^1(c_i) := \overrightarrow{c_i}(01\#)^{x-1} 0\#(01\#)^{n-x} \overleftarrow{c_i}$,
- $T_e^2(c_j) := \overrightarrow{c_j}(01\#)^{y-1} 0\#(01\#)^{n-y} \overleftarrow{c_j}$,
- $T_e^2(c_i) := \overrightarrow{c_i}(01\#)^n \overleftarrow{c_i}$,
- $T_e^1(c_j) := \overrightarrow{c_j}(01\#)^n \overleftarrow{c_j}$,
- $T_e^1(c_\ell) := T_e^2(c_\ell) := \overrightarrow{c_\ell}(01\#)^n \overleftarrow{c_\ell}$, for all $\ell \neq i, j$.

We then construct T_e by concatenating all these blocks, along with the suffix $S^*\$$ to form
$$T_e := T_e^1(c_1) \cdots T_e^1(c_k) T_e^2(c_1) \cdots T_e^2(c_k) S^*\$.$$

Setting $\mathcal{T} := \{T_1, T_2\} \cup \{T_e : e \in E\}$ completes the construction of our LSLCS$(k + |\Sigma|)$ instance $(\mathcal{T}, S, k)$. Observe that S is indeed a subsequence of all strings in $\mathcal{T}$, and that Σ is an alphabet of size $2k + 5$ consisting of the letters $\overrightarrow{c_1}, \overleftarrow{c_1}, \ldots, \overrightarrow{c_k}, \overleftarrow{c_k}, 0, 1, \#, \$$, and $\pounds$. We now make two observations that lead to the soundness and completeness of our reduction.

Lemma 3. *Suppose that $\widetilde{S}\sigma$ is a solution string for the constructed instance $(\mathcal{T}, S, k)$. Then $\widetilde{S}\sigma = \widetilde{S}(c_1) \cdots \widetilde{S}(c_k) S^* \sigma$, where for each $i \in \{1, \ldots, k\}$, the substring $\widetilde{S}(c_i)$ is obtained from $S(c_i)$ by replacing exactly one occurrence of the letter 0 with the letter 1.*

According to Lemma 3 above, we can think of the positions in which $\widetilde{S}(c_1)\cdots\widetilde{S}(c_k)$ differs from $S(c_1)\cdots S(c_k)$ as an encoding the selection of k vertices, one for each color class of G. We refer to these vertices as the *set of vertices selected by $\widetilde{S}$*.

Lemma 4. *The set $I \subseteq V(G)$ of vertices selected by $\widetilde{S}$ is a multicolored independent set in G.*

Theorem 2. *There is a linear parameterized reduction from* MIS(k) *to* LSLCS$(k + |\Sigma|)$.

We next sketch how to reduce the alphabet in our construction to constant size. For each $i \in \{1, \ldots, k\}$, replace the letters $\overrightarrow{c_i}$ and $\overleftarrow{c_i}$ with the substrings $p^{\alpha(i)}$ and $q^{\alpha(i)}$ respectively, where $\alpha(k) := 1$ and $\alpha(i) := \alpha(k) + \cdots + \alpha(i + 1) + 1$ for $i < k$. The new alphabet is of size 7. It is not difficult to verify that Lemma 3 still holds under this modification. The rest of the proof remains unchanged.

Corollary 2. LSLCS(k) *restricted to strings over a constant-size alphabet is* W[1]*-hard. Moreover, the problem has no $n^{o(k)}$ algorithm unless all problems in* SNP *can be solved in subexponential time.*

5   Shortest Common Supersequence

In this section, we consider a local search version of SHORTEST COMMON SUPERSEQUENCE (SCSEQ). In SCSEQ, the input is a set of strings $\mathcal{T}$ and an integer ℓ, and the question is whether there exists a string S of length ℓ which is a supersequence of all strings in $\mathcal{T}$. The local search variant of this problem that we consider is given by:

> LOCAL SEARCH SHORTEST COMMON SUPERSEQUENCE (LSSCSEQ):
> **Input:** A set $\mathcal{T} = \{T_1, \ldots, T_m\}$ of strings over an alphabet Σ, a string S which is a supersequence of all T_i's, and a positive integer k.
> **Question:** Is there a string $\widetilde{S}$ of length $|S| - 1$ which is a supersequence of all T_i's such that $d_H(S-, \widetilde{S}) \leq k$?

In other words, the new solution supersequence $\widetilde{S}$ is created from S by removing the last position of S and modifying at most k remaining positions. The main result of this section is the theorem below.

Theorem 3. *There is a linear parameterized reduction from* MIS(k) *to* LSSCSEQ(k) *restricted to strings over an alphabet of constant size.*

Let $(G = (V, E), c)$ denote an arbitrary input of MIS(k) with $c : V \to \{c_1, \ldots, c_k\}$. We assume, w.l.o.g., that there are n vertices colored c_i, for each color $c_i \in \{c_1, \ldots, c_k\}$, and that any pair of vertices with equal color are adjacent in G. Furthermore, to ease our presentation, we assume that the edges in G are directed; that is, E contains the two ordered pairs (u, v) and (v, u) for every pair of adjacent vertices u and v in G.

We begin by constructing the temporary solution S. First we create three substrings for each color $c_i \in \{c_1, \ldots, c_k\}$, which we refer to as *selection blocks*:

$$S^1(c_i) := \overrightarrow{c_i}(01\#)^n\overleftarrow{c_i}, \ S^2(c_i) := \overrightarrow{c_i}(00\#)^n\overleftarrow{c_i}, \text{ and } S^3(c_i) := \overrightarrow{c_i}(01\#)^n\overleftarrow{c_i}.$$

We construct S by concatenating the selection blocks, using the letter $\&$ to separate the three sets of selection blocks. We then add a suffix to S: The string $S^* := (\$\pounds^n\$)^{k+1}$ concatenated to the input string $T_1 \in \mathcal{T}$ which will be specified later. The string S is then given by

$$S := S^1(c_1) \cdots S^1(c_k) \& S^2(c_1) \cdots S^2(c_k) \& S^3(c_1) \cdots S^3(c_k) S^* T_1.$$

Next, we construct the input string T_1 which is the first of two input strings that will act as enforcement strings, enforcing the changes in S to occur in its selection blocks in a controlled fashion. For $c_i \in \{c_1, \ldots, c_k\}$, define

$$T_1^1(c_i) := \overrightarrow{c_i}\,0^{n+1}\overleftarrow{c_i}, \ T_1^2(c_i) := \overrightarrow{c_i}\,1\,\overleftarrow{c_i}, \text{ and } T_1^3(c_i) := \overrightarrow{c_i}\,0^{n+1}\overleftarrow{c_i}.$$

We construct T_1 using these substrings, the separation letter $\&$, and the suffix S^*:

$$T_1 := T_1^1(c_1) \cdots T_1^1(c_k) \& T_1^2(c_1) \cdots T_1^2(c_k) \& T_1^3(c_1) \cdots T_1^3(c_k) S^*.$$

The second enforcement string T_2 is constructed using the following substrings corresponding to a color $c_i \in \{c_1, \ldots, c_k\}$:

$$T_2^1(c_i) := \overrightarrow{c_i}\,(0\#)^n\overleftarrow{c_i}, \ T_2^2(c_i) := \overrightarrow{c_i}\,(0\#)^n\overleftarrow{c_i}, \text{ and } T_2^3(c_i) := \overrightarrow{c_i}\,(0\#)^n\overleftarrow{c_i}.$$

The string T_2 is then constructed as

$$T_2 := T_2^1(c_1) \cdots T_2^1(c_k) \& T_2^2(c_1) \cdots T_2^2(c_k) \& T_2^3(c_1) \cdots T_2^3(c_k) S^* T_1 - .$$

To complete the construction of $\mathcal{T}$, we construct a string T_e for each $e \in E$. These strings are composed of substrings that correspond to vertices of G. Let $v \in V$ with $c(v) := c_i$, and assume v is the x'th vertex of color c_i. The string $T(v)$ is defined by

$$T(v) := \overrightarrow{c_i}\,(0\#)^{x-1}\,01\,(0\#)^{n-x}\overleftarrow{c_i}.$$

The string T_e is constructed as $T_e := T(u) \& T(v)$ if $e := (u, v)$ (recall that we assume that the edges are directed, and that any pair of vertices with the same color are adjacent).

To finalize our construction, we set the parameter k' of the LSSCSEQ instance to $3k$. Clearly the instance $(\mathcal{T}, S, k')$ can be constructed in polynomial time. We proceed to show that this instance is equivalent to the MIS(k) instance. The first crucial step is given by the following lemma.

Lemma 5. *Let $(\mathcal{T}, S, k')$ be an LSSCSEQ instance constructed as described above. If $(\mathcal{T}, S, k') \in$ LSSCSEQ, then there exists a solution string $\widetilde{S}$ for $(\mathcal{T}, S, k')$ where $\widetilde{S}$ can be written as $\widetilde{S} := S' S^* T_1-$ with*

$$S' := \widetilde{S}^1(c_1) \cdots \widetilde{S}^1(c_k) \& \widetilde{S}^2(c_1) \cdots \widetilde{S}^2(c_k) \& \widetilde{S}^3(c_1) \cdots \widetilde{S}^3(c_k),$$

such that for each $i \in \{1, \ldots, k\}$ we have:

- $\widetilde{S}^1(c_i)$ *is obtained from* $S^1(c_i)$ *by replacing exactly one occurrence of* 01 *by* 00,
- $\widetilde{S}^2(c_i)$ *is obtained from* $S^2(c_i)$ *by replacing exactly one occurrence of* 00 *by* 01,
- $\widetilde{S}^3(c_i)$ *is obtained from* $S^3(c_i)$ *by replacing exactly one occurrence of* 01 *by* 00.

Let $\widetilde{S}$ be a solution string for $(\mathcal{T}, S, k')$ as in Lemma 5. We interpret the positions in S' that differ from $S-$ as a set of *selected vertices* $\{v_1^1, v_1^2, v_1^3, \ldots, v_k^1, v_k^2, v_k^3\}$ of G, where for each $i \in \{1, \ldots, k\}$, the vertex v_i^1 (resp. v_i^2, v_i^3) is the x-th vertex in c_i if the x-th substring 01 (resp. 00, 01) in $S(c_i)$ is modified in $\widetilde{S}(c_i)$. The next lemma shows that the set of selected vertices includes in fact only k vertices.

Lemma 6. *For each* $i \in \{1, \ldots, k\}$ $v_i^1 = v_i^2 = v_i^3$.

According to Lemma 6, we let v_i be the single vertex corresponding to $v_i^1 = v_i^2 = v_i^3$, giving us a multicolored set $\{v_1, \ldots, v_k\}$ of vertices in G. The next lemma shows that this set is independent in G.

Lemma 7. *The set of vertices* $I := \{v_1, \ldots, v_k\}$ *forms an independent set in* G.

Corollary 3. LSSCSEQ(k) *restricted to strings over a constant-size alphabet is* W[1]*-hard, and has no* $n^{o(k)}$ *algorithm unless all problems in* SNP *can be solved in subexponential time.*

6   Shortest Common Superstring

In this section we deal with a local search variant of SHORTEST COMMON SUPERSTRING. In this problem, the input is a set of strings $\mathcal{T}$ and an integer ℓ, and the question is whether there is a string S of length at most ℓ which is a superstring of all strings in $\mathcal{T}$. The local search version of SHORTEST COMMON SUPERSTRING is defined as follows:

> LOCAL SEARCH SHORTEST COMMON SUPERSTRING (LSSCStr):
> **Input**: A set $\mathcal{T} = \{T_1, \ldots, T_m\}$ of strings over an alphabet Σ, a string S which is a superstring of all T_i's, and a positive integer k.
> **Question**: Is there a string $\widetilde{S}$ of length $|S| - 1$ which is a superstring of all T_i's such that $d_H(\widetilde{S}, S-) \leq k$?

Theorem 4. LSSCStr(k) *is* W[1]*-hard, even with an alphabet of constant size.*

For ease of presentation, we describe here only the case that the alphabet size $|\Sigma|$ is part of the parameter. The case with constant-size alphabets can be coped with the method introduced in Section 4. The reduction is from the W[1]-complete MULTICOLORED CLIQUE problem, where, given a graph $G = (V, E)$ and a coloring function $c : V \to \{c_1, \ldots, c_k\}$, we ask for a multicolored clique of size k. We assume, w.l.o.g., that c is a *proper* coloring, that is, there is no edge $\{u, v\}$ between vertices u and v with $c(u) = c(v)$ (such edges can be removed in linear time), and that each color class contains exactly $|V|/k$ vertices.

The alphabet Σ consists of $k^* \cdot k(k-1) + 4k + 4$ letters with $k^* := 2^k(k^2 + k)$. The letters \$ and # are separating letters, where \$ does not occur in the input strings.

The letters 0 and 1 are encoding letters. The other letters correspond to colors and color pairs. For each color c_i, we have 4 letters: a_i, b_i, c_i, and d_i. For each ordered color pair (c_i, c_j) with $i \neq j$, there are k^* letters, namely, $c_{i,j}^1, \ldots, c_{i,j}^{k^*}$. Assume that each color class in G contains n vertices. The LSSCSTR(k)-instance consists of the superstring S and a set $\mathcal{T}$ of $1 + (k-1)k \cdot n + (k-1) \cdot n$ input strings: one special input string T_0, k input strings for each vertex from color classes c_1 to c_{k-1}, and $k-1$ input strings for each vertex from the color class c_k.

To construct these strings, we first introduce some strings, which are used as "building blocks" in the construction. First, we describe the "separating blocks". For each color c_i with $2 \leq i \leq k$, we introduce two such blocks: $A_i := a_i^{g(i)}$ and $B_i := b_i^{g(i)}$, where $g(i) := 2^{k-i} \cdot (k^2 + k)$. For each ordered pair of colors c_i and c_j with $i \neq j$, we construct one separating block: $C_{i,j} := (c_{i,j}^1 \#)^n \cdots (c_{i,j}^{k^*} \#)^n$. Moreover, we construct two "color-pair matching" blocks for each color c_i:

- $M_i^1 := 0C_{i,1} \cdots 0C_{i,i-1}\, C_{i,i+1} \cdots C_{i,k}$, and
- $M_i^2 := C_{i,1} \cdots C_{i,i-1}\, C_{i,i+1}0 \cdots C_{i,k}0$.

Finally, for every vertex v we construct an "identifying block". Let $c_i := c(v)$. Here we distinguish $i = 1$ and $i > 1$. Assume v is the x'th vertex colored c_i. The identifying block for v is constructed as

- $I(v) := d_1\, 0^{x-1}\, 1\, 0^{n-x}\, d_1$ for $i = 1$, and
- $I(v) := d_i\, (0A_i)^{x-1} 1\, A_i (0A_i)^{n-x}\, d_i d_{i-1}$, for $i > 1$.

We are now ready to describe the set of input strings $\mathcal{T}$ in our LSSCSTR instance. First, for each vertex v colored c_i with $1 \leq i < k$, we construct one "triggering" input string. If v is the x'th vertex colored c_i, its triggering input string $T(v)$ is constructed as:

$$T(v) := c_{i+1}\, M_{i+1}^1\, d_{i+1}\, (0A_{i+1})^{x-1} 0\, B_{i+1}\, (0A_{i+1})^{n-x}\, d_{i+1} d_i.$$

Then, for each vertex v colored c_i with $1 \leq i \leq k$, we add $k - 1$ "pairing" input strings, each corresponding to a color class c_j with $j \neq i$. Here, we distinguish $i < j$ and $i > j$:

- $T(v, c_j) := C_{i,j+1} \cdots C_{i,k}\, I(v)\, C_{i,1} \cdots C_{i,i-1} C_{i,i+1} 0 \cdots C_{i,j} 0$ $(i < j)$,
- $T(v, c_j) := 0C_{i,j} \cdots 0C_{i,i-1} C_{i,i+1} \cdots C_{i,k}\, I(v)\, C_{i,1} \cdots C_{i,j-1}$ $(i > j)$.

To finalize our construction of $\mathcal{T}$, we set the special input string T_0:

$$T_0 := c_1\, M_1^1\, d_1\, 0^n\, d_1.$$

Now, it remains to describe the temporary solution S. To this end, we introduce some further building blocks. For each edge $e = \{u, v\} \in E$, where u is colored c_i, and v is colored c_j with $i < j$, we construct one "edge block" $S(e)$ for S as:

$$S(e) := T(u, c_j) - 1 - T(v, c_i),$$

where $T(u, c_j)-$ as usual denotes the prefix of the pairing input string $T(u, c_j)$ without the last 0, and $-T(v, c_i)$ denotes the suffix of $T(v, c_i)$ without the first 0. Furthermore, for each vertex $v \in V$ colored c_i in G we construct the selection block $S(v)$ of v by:

- $S(v) := T(v) M_i^1 I(v) M_i^2$ for $i < k$, and
- $S(v) := M_k^1 I(v) M_k^2$ for $i = k$.

The solution S then consists of three parts, $S := S(V)S(E)T_0$, where the first part $S(V)$ is the concatenation of the selection blocks $S(v)$ separated by \$'s in any arbitrary order, the second part $S(E)$ is the concatenation of edge blocks $S(e)$ separated by \$'s, and T_0 is the special input string described above.

Finally, we set the parameter for the LSSCSTR-instance to $k' := 2k + k(k - 1)/2 + (2^{k-1} - 1)(k^2 + k)$. It is easy to verify that S is a superstring of all input strings: The string T_0 occurs at the end of S. Furthermore, for each vertex $v \in V$, the triggering input string $T(v)$ is a prefix of $S(v)$, while the pairing strings $T(v, c_j)$ are clearly substrings of $M_i^1 I(v) M_i^2$. We next turn to showing the equivalence of the two instances.

Lemma 8. *If G has a multicolored clique K then $(S, \mathcal{T}, k')$ has a solution string $\widetilde{S}$.*

We next consider the reversed direction. Suppose that a solution $\widetilde{S}$ exists for $(\mathcal{T}, S, k')$. We use $\widetilde{S}(v)$ to denote substring of $\widetilde{S}$ corresponding to the selection block $S(v)$ of S.

Lemma 9. *If there is a solution $\widetilde{S}$ for $(\mathcal{T}, S, k)$ constructed above, then the input string T_0 is a substring of some $\widetilde{S}(v_1)$ for some $v_1 \in V$ with $c(v_1) = c_1$.*

Let v_1 be the vertex in Lemma 9. By construction, we have to match M_1^1 of T_0 to the M_1^1-substring of $\widetilde{S}(v_1)$. This implies that the letter 1 in the corresponding identifying block has to be changed to 0. Moreover, the last letter d_1 of the corresponding triggering block must be changed to c_1. These two changes cause that the pairing input strings and the triggering string for v_1 are matched to somewhere else in $\widetilde{S}$ than in S. We consider first the triggering string $T(v_1)$.

Lemma 10. *The triggering input string $T(v_1)$ can only be matched to a substring of $\widetilde{S}(v_2)$ for some vertex v_2 with $c(v_2) = c_2$.*

It can be shown that the matching of $T(v_1)$ to some $\widetilde{S}(v_2)$ causes $2 + |A_2|$ modifications after which the triggering input string $T(v_2)$ and $k-1$ pairing input strings are unmatched. Furthermore, $T(v_2)$ has to be matched to some $\widetilde{S}(v_3)$ with $c(v_3) = c_3$: after performing the $2 + |A_2| = 2 + 2^{k-2}(k^2 + k)$ changes to match $T(v_1)$, one cannot afford to perform $2|B_3| = 2 \cdot 2^{k-3}(k^2 + k) = 2^{k-2}(k^2 + k)$ changes which are necessary for matching $T(v_2)$ to the selection block of another vertex colored c_2. The same argument applies inductively for all $i > 2$. In this way, the string $\widetilde{S}$ differs from S in exactly k selection blocks corresponding to a multicolored set of vertices $\{v_1, \ldots, v_k\}$ in G, and the Hamming distance between the remaining suffixes of $S-$ and $\widetilde{S}$ is at most $k(k - 1)/2$.

Lemma 11. *The set of vertices $\{v_1, \ldots, v_k\}$ specified above forms a clique in G.*

Combining all lemmas above completes the proof of Theorem 4 when $|\Sigma|$ is part of the parameter. Using the method from Section 4 gives the proof for constant-size alphabets.

References

1. Aarts, E.H.L., Lenstra, J.K.: Local Search in Combinatorial Optimization. Wiley-Interscience (1997)
2. Balas, E.: New classes of efficiently solvable generalized traveling salesman problems. Ann. Oper. Res. 86, 529–558 (1999)
3. Chen, J., Chor, B., Fellows, M., Huang, X., Juedes, D.W., Kanj, I.A., Xia, G.: Tight lower bounds for certain parameterized NP-hard problems. Inform. Comput. 201(2), 216–231 (2005)
4. Dörnfelder, M., Guo, J., Komusiewicz, C., Weller, M.: On the parameterized complexity of consensus clustering. In: Asano, T., Nakano, S.-I., Okamoto, Y., Watanabe, O. (eds.) ISAAC 2011. LNCS, vol. 7074, pp. 624–633. Springer, Heidelberg (2011)
5. Downey, R.G., Fellows, M.R.: Parameterized Complexity. Springer (1999)
6. Fellows, M.R., Rosamond, F.A., Fomin, F.V., Lokshtanov, D., Saurabh, S., Villanger, Y.: Local search: Is brute-force avoidable? J. Comput. Syst. Sci. 78(3), 707–719 (2012)
7. Festa, P., Pardalos, P.M.: Efficient solutions for the far from most string problem. Ann. Oper. Res. 196(1), 663–682 (2012)
8. Fomin, F.V., Lokshtanov, D., Raman, V., Saurabh, S.: Fast local search algorithm for weighted feedback arc set in tournaments. In: Proc. 24th AAAI. AAAI Press (2010)
9. Gramm, J., Niedermeier, R., Rossmanith, P.: Fixed-parameter algorithms for closest string and related problems. Algorithmica 37(1), 25–42 (2003)
10. Guo, J., Hartung, S., Niedermeier, R., Suchý, O.: The parameterized complexity of local search for TSP, more refined. In: Asano, T., Nakano, S.-I., Okamoto, Y., Watanabe, O. (eds.) ISAAC 2011. LNCS, vol. 7074, pp. 614–623. Springer, Heidelberg (2011)
11. Johnson, D.S., Papadimitriou, C.H., Yannakakis, M.: How easy is local search? J. Comput. Syst. Sci. 37(1), 79–100 (1988)
12. Krokhin, A., Marx, D.: On the hardness of losing weight. ACM T. Alg. 8(2), 19 (2012)
13. Marx, D.: Searching the k-change neighborhood for TSP is W[1]-hard. Oper. Res. Lett. 36(1), 31–36 (2008)
14. Marx, D., Schlotter, I.: Stable assignment with couples: Parameterized complexity and local search. Discr. Optim. 8(1), 25–40 (2011)
15. Meneses, C., Oliveira, C.A.S., Pardalos, P.M.: Optimization techniques for string selection and comparison problems in genomics. IEEE Eng. Med. Bio. Mag. 24(3), 81–87 (2005)
16. Szeider, S.: The parameterized complexity of k-flip local search for SAT and MAX SAT. Discr. Optim. 8(1), 139–145 (2011)

Space-Efficient Construction Algorithm for the Circular Suffix Tree

Wing-Kai Hon[1], Tsung-Han Ku[1], Rahul Shah[2], and Sharma V. Thankachan[2]

[1] National Tsing Hua University, Taiwan
{wkhon, thku}@cs.nthu.edu.tw
[2] Louisiana State University, USA
{rahul, thanks}@csc.lsu.edu

Abstract. Hon et al. (2011) proposed a variant of the suffix tree, called *circular suffix tree*, and showed that it can be stored succinctly and can be used to solve the circular dictionary matching problem efficiently. In this paper, we give the first construction algorithm for the circular suffix tree, which takes $O(n \log n)$ time and requires $O(n \log \sigma + d \log n)$ bits of working space, where n is the total length of the patterns in $\mathcal{D}$, d is the number of patterns in $\mathcal{D}$, and σ is the alphabet size.

1 Introduction

Given a set $\mathcal{D}$ of d patterns, the dictionary matching problem is to index $\mathcal{D}$ such that for any online query text T, we can quickly locate the occurrences of any pattern of $\mathcal{D}$ within T. This problem has been well-studied in the literature [1,3], and an index taking optimal space and simultaneously supporting optimal-time query is achieved [2,6]. In some practical bioinformatics and computational geometry applications [10], such as indexing a collection of viruses, we are interested in searching for, not only the original patterns in $\mathcal{D}$, but also all of their cyclic shifts. We call this the *circular dictionary matching problem*. Hon et al. [9] recently proposed a variant of suffix tree [14,18], called *circular suffix tree* and showed that it can be compressed into succinct space. With a tree structure augmented to a circular pattern matching index called *circular suffix array*, the circular suffix tree can be used to solve the circular dictionary matching problem efficiently. Although there are several efficient construction algorithms for the suffix tree in the literature, none of them can be applied directly to construct circular suffix tree due to the different nature of the patterns being indexed. In this paper, we give the first construction algorithm for the circular suffix tree, which takes $O(n \log n)$ time and requires $O(n \log \sigma + d \log n)$ bits of working space, where n denotes the total length of the patterns in $\mathcal{D}$ and σ denotes the alphabet size.

Briefly speaking, the framework of the construction is as follows. First, we directly apply the result of [7] to obtain the *circular suffix array* of $\mathcal{D}$ in the succinct form, and use this as a succinct representation of the leaf labels in the circular suffix tree. Next comes to the major technical challenge, where we

J. Fischer and P. Sanders (Eds.): CPM 2013, LNCS 7922, pp. 142–152, 2013.
© Springer-Verlag Berlin Heidelberg 2013

construct the tree structure of the circular suffix tree. Our approach is based on Kasai et al.'s algorithm [12], which is originally used for contructing the tree structure of a (non-circular) suffix tree. However, a non-trivial adaptation (yet with simple modifications) to the algorithm is proposed to handle the circular case, and new observation about the auxiliary data structure is given so as to control the working space of the algorithm. Finally, we mark the nodes in the circular suffix tree as specified in the definition, which can be done efficiently with standard techniques, even though the tree structure and the marking are both represented in the succinct forms. As we shall see, each part of the construction algorithm runs in $O(n \log n)$ time, and all the data structures involved (including the final output and the intermediate auxiliary strucutres) require $O(n \log \sigma + d \log n)$ bits of storage, the result of the paper thus follows.

The remainder of the paper is organised as follows. Section 2 provides basic notation, the definitions of the circular suffix array, and the circular suffix tree. Section 3 reviews Kasai et al.'s algorithm and explains how we can adapt it to construct the desired tree structure. Finally, in Section 4, we show how the marking of nodes is performed.

2 Preliminaries

Let $P = P[1..|P|]$ be a pattern of length $|P|$. We use P^∞ to denote the string formed by repeating P an infinite number of times. For any $i \in [1, |P|]$, the string $Q = P[i..|P|]P[1..i-1]$ is called the ith *cyclic shift* of P, and the string Q^∞ is called the ith circular suffix of P^∞. We have the following definition.

Definition 1. *Let P and Q be two patterns. We say P is circularly larger than Q, or simply P is larger than Q, if and only if P^∞ is lexicographically larger than Q^∞. The notions of "smaller than", or "equal to", are defined analogously.*

Lemma 1 ([7]). *Let P and Q be two strings, and let m denote the maximum of $|P|$ and $|Q|$. Then, P is circularly smaller than Q if and only if $P^\infty[1..2m]$ is lexicographically smaller than $Q^\infty[1..2m]$. In other words, to compare the "circular" lexicographical order of P and Q, we only need to compare P^∞ and Q^∞ directly up to length $2m$.[1]*

Let $\mathcal{D} = \{P_1, P_2, \ldots, P_d\}$ be a set of d patterns to be indexed for the circular dictionary matching problem. Let n be the total length of the d patterns, and σ be the size of the alphabet. Without loss of generality, we assume that the lengths of the patterns in $\mathcal{D}$ are monotonically decreasing (i.e., $|P_i| \geq |P_j|$ if $i < j$). For ease of discussion, we assume that each pattern P_j in $\mathcal{D}$ cannot be written as P^k for some string P and some integer $k > 1$, and that no pattern is a cyclic shift of another one. These two assumptions ensure that all cyclic shifts of all patterns are distinct in the circular sense.

---

[1] In fact, Mantaci et al. [13] have shown a better bound of $|P| + |Q| - gcd(|P|, |Q|)$, based on the Fine and Wilf theorem.

2.1 Circular Suffix Array

The *circular suffix array* $SA_{circ}[1..n]$ for $\mathcal{D}$ is an arrangement of all circular suffixes of all P_j^∞ according to the lexicographical order. Precisely, $SA_{circ}[i] = (j, k)$ if the ith lexicographically smallest circular suffix is the kth circular suffix of P_j^∞. For example, $\mathcal{D} = \{\mathsf{bbba}, \mathsf{ba}, \mathsf{b}\}$, and

$$SA_{circ}[1..7] = [(2,2), (1,4), (2,1), (1,3), (1,2), (1,1), (3,1)].$$

See Figure 2 for the lexiographical arrangement of the respective circular suffixes. We also define $SA_{circ}^{-1}(j, k)$ to be i, if and only if $SA_{circ}[i] = (j, k)$.

Lemma 2 ([7]). *The circular suffix array can be stored succinctly in $O(n \log \sigma)$ bits, and can be constructed in $O(n \log n)$ time using $O(n \log \sigma)$ bits working space. The functions SA_{circ} and SA_{circ}^{-1} can both be evaluated in $O(\log n)$ time.*

For ease of discussion, we use S_i to denote the ith lexicographically smallest circular suffix, and $\|S_i\|$ to denote the length of its corresponding cyclic shift, which is $|P_j|$ if $SA_{circ}[i] = (j, k)$. Note that the value $\|S_i\|$ can be computed in $O(\log n)$ time by first finding $SA_{circ}[i]$, and then reporting the length of P_j.

2.2 Circular Suffix Tree

Let $\mathcal{C}$ be a compact (i.e., path-compressed) trie. For each node v in $\mathcal{C}$, we use $path(v)$ to denote the concatenation of edge labels along the path from the root to v. The *circular suffix tree* ST_{circ} for $\mathcal{D}$ is a compact trie storing all circular suffixes of all P_j^∞. In addition, as in a (non-circular) suffix tree, the children of a node are arranged according to the lexicographical order of the labels in their incident edges. Consequently, the ith leftmost leaf in the circular suffix tree will correspond to the ith lexicographically smallest circular suffix, which is indexed by $SA_{circ}[i]$.

To facilitate efficient reporting of pattern occurrences in answering a query, certain nodes in the circular suffix tree are marked. Precisely, a node u is marked if there exists a circular suffix S such that **(i)** $\|S\|$ is less than $|path(u)|$, **(ii)** $parent(u)$ is an ancestor of the leaf for S, but **(iii)** u is not an ancestor of the leaf for S; here, $parent(u)$ denotes the parent node of u. Intuitively, the marked node u is used when the searching path of a query text ends in the subtree rooted at u, so that it reveals *at least* one cyclic shift of some pattern appears as a prefix in the query text. (See [9] for the details of the query algorithm.) Moreover, each node in ST_{circ} maintains a pointer to its lowest marked ancestor.

3 Construction of Circular Suffix Tree

To save the working space, we shall directly construct the circular suffix tree in its succinct form, as introduced in [9], which contains three key components: **(1)** the circular suffix array, for representing the leaf labels and the edge labels;

(2) an *Hgt* array (to be defined), for representing the lengths of the edge labels; and **(3)** a string of balanced parentheses, for encoding the tree structure. As we can already construct the circular suffix array efficiently by Lemma 2, the remaining of this section is devoted to the efficient construction of the other two components.

3.1   Construction of the *Hgt* Array

Kasai et al.'s Algorithm. The *Hgt* array (called the *height array*) is introduced by Kasai et al. [12], which was originally defined for the (non-circular) suffix tree of a pattern P. Precisely, it is an array $Hgt[1..|P|]$, such that $Hgt[i]$ stores the length of the longest common prefix between S_i and its *predecessor* suffix S_{i-1} (the lexicographically ith and $i-1$th smallest suffixes). See Figure 1 for an example. It is shown [12] that given the SA^{-1}, the *Hgt* array can be constructed in linear time, using $O(n \log n)$ bits of working space. See Algorithm 1 for details.

index	suffix	SA	Hgt
1	abc	5	-1
2	abcdabc	1	3
3	bc	6	0
4	bcdabc	2	2
5	c	7	0
6	cdabc	3	1
7	dabc	4	0

index	suffix	SA_{circ}	Hgt
1	$(ab)^\infty$	$(2,2)$	-1
2	$(abbb)^\infty$	$(1,4)$	2
3	$(ba)^\infty$	$(2,1)$	0
4	$(babb)^\infty$	$(1,3)$	3
5	$(bbab)^\infty$	$(1,2)$	1
6	$(bbba)^\infty$	$(1,1)$	2
7	$(b)^\infty$	$(3,1)$	3

Fig. 1. *Hgt* array for $P = abcdabc$ **Fig. 2.** *Hgt* array for $\mathcal{D} = \{\text{bbba}, \text{ba}, \text{b}\}$

Kasai et al.'s algorithm consists of $|P|$ rounds. In round k, it examines the suffix of $P[k..|P|]$ and sets up the corresponding value $Hgt[i]$, where i is the rank of $P[k..|P|]$. Briefly speaking, the efficiency of the algorithm comes from maintaining the following invariant: Before each round starts, the length of the LCP to be computed is at least h, where the value is inferred from the previous round. This allows us to avoid redundant matching, and start matching the $h+1$th character of the corresponding suffixes (Line 6) in the current round to compute the LCP; consequently the overall running time is bounded by $O(|P|)$.

The *Hgt* array can naturally be defined for the circular suffix tree for $\mathcal{D}$. See Figure 2 for an example. We can also extend Kasai et al.'s algorithm to construct the *Hgt* array in this case, as shown in Algorithm 2. Nevertheless, when we process the circular suffixes of P_j^∞ in round j, the total time required will become $O(|P_j| + L_j)$, where L_j denotes the maximum length of LCP between a circular suffix of P_j^∞ and its predecessor circular suffix. In the worst case, $L_j = O(n)$ for each j, and the overall running time will become $O(dn)$.

Algorithm 1. Kasai et al.'s Algorithm

1: $h \leftarrow 0$;
2: **for** $k \leftarrow 1$ to $|P|$ **do**
3:     $i \leftarrow SA^{-1}[k]$;   { Finding i such that $S_i = P[k..|P|]$ }
4:     **if** $i > 1$ **then**
5:         $m \leftarrow SA[i-1]$;   { Finding m such that $S_{i-1} = P[m..|P|]$ }
6:         **while** $P[k+h] = P[m+h]$ **do**
7:             $h \leftarrow h + 1$;
8:         **end while**
9:         Set $Hgt[i]$ to be h;
10:        **if** $h > 0$ **then**
11:            $h \leftarrow h - 1$;
12:        **end if**
13:    **else**
14:        Set $Hgt[i]$ to be -1;
15:    **end if**
16: **end for**

Algorithm 2. Constructing the Hgt array for ST_{circ}

1: **for** $j \leftarrow 1$ to d **do**
2:     Apply Kasai et al.'s algorithm to compute the Hgt entries corresponding to the circular suffixes of P_j^∞;
        { those $Hgt[i]$ entries with $SA_{circ}[i] = (j, k)$, for $k = 1, 2, \ldots, |P_j|$ }
3: **end for**

The $O(L_j)$ bound becomes problematic when it is $\omega(|P_j|)$. This occurs when the length of the LCP of a certain suffix of P_j^∞ and its predecessor suffix is not bounded by $O(|P_j|)$, so that the computation of the corresponding Hgt value will take too much time. By Lemma 1, this happens only if the predecessor suffix is a circular suffix of P_i^∞ for some $i < j$. To overcome the above problem, we make use of a very simple, yet powerful idea: to construct an analogous array Hgt' *in phase* with Hgt, where $Hgt'[i]$ stores the length of the LCP between S_i and its *successor* S_{i+1}. Precisely, each round will update both $Hgt[i]$ and $Hgt'[i]$ that correspond to the same circular suffix S_i. The benefit is that, if $Hgt'[i-1]$ is already computed, we can immediately set $Hgt[i] = h = Hgt'[i-1]$ without performing any character comparison.[2] Consequently, the time for round j can be bounded by $O(|P_j| + L'_j)$, where L'_j is defined similarly as L_j, but with an extra condition that *the predecessor suffix is not a circular suffix of P_i^∞, with $i < j$.* (We defer the details of the analysis to the full paper.) Since L'_j is bounded by $O(|P_j|)$, the overall running time is thus $O(n)$.

So far, we have implicitly assumed that SA_{circ} and SA_{circ}^{-1} can be evaluated in constant time. If we use the succinct form of the circular suffix array instead, the evaluation takes $O(\log n)$ time, so that the overall running time is increased

---

[2] Similarly, if $Hgt'[i+1]$ is already computed, we set $Hgt'[i] = h' = Hgt[i+1]$, where h' is used by the invariant corresponding to the construction of Hgt'.

to $O(n \log n)$. For the working space, apart from storing the circular suffix array in $O(n \log \sigma)$ bits, we need $O(n \log n)$ bits for the storage of the Hgt array. In the following, we use an alternative representation for Hgt that is introduced in [16,9], namely the H array, and show that its storage can be bounded by $O(n + d \log n)$ bits.[3]

Succinct Representation for *Hgt* Array. The H array is a two-dimensional array which stores that Hgt values in the order they are computed. Precisely, we define $H[j, k] = Hgt[i]$, if the kth circular suffix of P_j^∞ is S_i. Then, when we process the circular suffixes of P_j^∞ in round j of Algorithm 2, the values $H[j, 1], H[j, 2], \ldots, H[j, |P_j|]$ will be computed sequentially by Kasai et al.'s algorithm (more precisely, the modified version that constructs Hgt and Hgt' in phase). We also define an analogous H' array where $H'[j, k] = Hgt'[i]$. In [9], it is observed that for each j,

$$H[j, 1] \leq H[j, 2] + 1 \leq H[j, 3] + 2 \leq \cdots.$$

Since P_j^∞ is circular, its $|P_j| + 1$th circular suffix is the same as itself. By the above observation, we obtain the following relationship:

$$H[j, 1] \leq H[j, |P_j|] + |P_j| - 1 \leq H[j, |P_j| + 1] + |P_j| = H[j, 1] + |P_j|.$$

Thus, the sequence of values $H[j, k] + k - 1$ is monotonically increasing, and we shall encode them with difference encoding, using $\log n + O(|P_i|)$ bits: **(1)** Store the value $H[j, 1]$ explicitly. **(2)** Initialize an empty bit-vector z. **(3)** For $k = 1$ to $|P_j| - 1$, compute the difference between the kth and $k + 1$th values. If the difference is x, append x 0s to z, followed by a 1.

In the end, the bit-vector z contains exactly $|P_j| - 1$ 1s and at most $|P_j|$ 0s, so that its length is bounded by $O(|P_j|)$. For example, in Figure 2, the sequence of values of $H[1, 1]$, $H[1, 2] + 1$, $H[1, 3] + 2$, and $H[1, 4] + 3$ are 2, 2, 5, and 5, respectively. We shall store $H[1, 1] = 2$ explicity, and represent the other values by the bit-vector $z = 100011$. Note that z can be created on the fly when we process P_j^∞ in Algorithm 2, as the values of $H[j, k]$ are computed in increasing order of k.

To compute a particular value $H[j, k]$, we first compute $H[j, k] + k - 1$ as follows: **(1)** Find the $k - 1$th 1 in z, and count the number of 0s in z before it, and **(2)** add $H[j, 1]$ to the count, and report the sum. Once $H[j, k] + k - 1$ is obtained, we just subtract $k - 1$ from it to obtain the desired $H[j, k]$. The above process can be performed in constant time, by maintaining Jacobson's rank and select structure [11]. This structure takes $o(|P_j|)$ bits of storage, and can be constructed in $o(|P_j|)$ time using $o(|P_j|)$ of working space once the bit-vector z is ready.

---

[3] We remark that this bound is not observed in [9]. Consequently, we can slightly improve the result of [9] by removing an assumption about the lengths of the patterns in $\mathcal{D}$. Details are deferred to the full paper.

As soon as the H array entries corresponding to the circular suffixes of P_j^∞ are stored, then for any i such that S_i is a circular suffix of P_j^∞ (say, the kth one), the value $Hgt[i]$ can be computed by

$$Hgt[i] \equiv H[j,k] \equiv H[SA_{circ}^{-1}[i]],$$

using $O(\log n)$ time. With similar arguments, we can derive the same bounds for the storage of the H' array and the computation of Hgt' value. In conclusion, we have the following theorem.

Theorem 1. *Suppose the succinct form of the circular suffix array is given. Using the H array as a succinct representation, the Hgt array can be constructed in $O(n \log n)$ time using $O(n \log \sigma + d \log n)$ bits of working space. When the construction process is over, each entry in the Hgt array can be computed in $O(\log n)$ time.*

Proof (sketch). In the modified version of Algorithm 2, the values of Hgt and Hgt' are accessed at most $O(n)$ times (where they are only accessed when the corresponding H array entries are stored during the previous rounds). These accesses create an overhead of $O(n \log n)$ time in total, which does not affect the overall time complexity. For the space, the number of bits for storing H is bounded by $\sum_{j=1}^{d} O(\log n + |P_j|)$, which is $O(n + d \log n)$. Together with the space for storing the succinct form of the circular suffix array, the space bound follows.

3.2 Construction of the Parentheses Encoding of ST_{circ}

The structure of the circular suffix tree is encoded by a string of balanced parentheses in the following way [15]: **(1)** Perform a pre-order traversal from the root. **(2)** When visiting a node in the first time, output an open parenthesis symbol '('. **(3)** When visiting a node in the last time, output a close parenthesis symbol ')'. Kasai et al. [12] showed that a left-to-right bottom-up traversal of the suffix tree can be performed in linear time based on the Hgt array. Here, we simulate Kasai et al.'s algorithm to compute the topology of the circular suffix tree, using $O(n)$ bits of additional working space, apart from those used by the circular suffix array and the Hgt array.[4] Let Q_1 be a list of $2n$ bits. During the traversal, we append '()' to the end of Q_1 if we visit a leaf, and we append ')' to the end of Q_1 if we last visit an internal node. See Algorithm 3 for details. Essentially, after the traversal, Q_1 is very close to the desired parentheses encoding, only with all the ('s corresponding to the internal nodes removed. Similarly, by performing a symmetric *right-to-left* traversal starting from the rightmost leaf, we can obtain another list of parentheses Q_2 which is equal to the parentheses encoding of the circular suffix tree, but with all the)'s corresponding to the internal node removed. Finally, we compute the desired parentheses encoding by merging Q_1 and Q_2, as shown in Algorithm 4.

---

[4] In fact, the same simulation has been used in Hon [5] (Chapter 5.1) to construct the topology of a suffix tree with succinct working space. We include the details here for the sake of completeness.

Algorithm 3. Computing Q_1

1: Initialize U as an empty stack;
2: **for** $k \leftarrow 1$ to n **do**
3:   Output '()' that represents the leaf of S_k;
4:   $\ell \leftarrow H[SA_{circ}[k]]$;
     { ℓ is equal to $Hgt[k]$, the length of LCP between S_k and S_{k+1} }
5:   **while** $\text{Top}(U) > \ell$ **do**
6:     $\text{Pop}(U)$ and output ')';
7:   **end while**
8:   **if** $\text{Top}(U) < \ell$ **then**
9:     $\text{Push}(U, \ell)$;
10: **end if**
11: **end for**

Algorithm 4. Computing the parentheses encoding from Q_1 and Q_2

1: Scan Q_1 and Q_2 from left to right;
2: **while** Q_1 and Q_2 are not both empty **do**
3:   **if** Q_1 begins with ')' **then**
4:     Output ')' and remove the symbol from the beginning of Q_1;
5:   **else if** Q_2 begins with '(' but not with '()' **then**
6:     Output '(' and remove the symbol from the beginning of Q_2;
7:   **else**
8:     {Both Q_1 and Q_2 begin with '()'}
9:     Output '()' and remove the symbols from the beginning of Q_1 and Q_2;
10: **end if**
11: **end while**

Theorem 2. *Given the circular suffix array as represented in Lemma 2 and the H array as represented in Theorem 1, the parentheses encoding of the circular suffix tree ST_{circ} can be constructed in $O(n \log n)$ time, using $O(n)$ bits of working space.*

Proof (sketch). The most time-consuming part of the algorithm is the computation of $H[SA_{circ}[k]]$, where each such computation can be performed in $O(\log n)$ time. The total time complexity is thus bounded by $O(n \log n)$. As for the space, apart from the stack, we just need $O(n)$ bits for storing each of the Q_1, Q_2, and the final parentheses encoding. For the stack, it is easy to check that the entries will always form a straightly increasing integer sequence (when reading the stack from bottom to top), so that we can represent each entry by its difference with the previous entry. Furthermore, the difference will be stored by using Elias's γ code or δ code [4], so that each entry can be encoded/decoded in $O(1)$ time during the push/pop operation, while the total number of bits required for representing all entries in the stack can always be bounded by $O(n)$. We defer the details to the full paper.

4   Marking the Nodes in ST_{circ}

We describe two preprocessing steps for speeding up our marking algorithm.

- We obtain the parentheses encoding of ST_{circ} from the previous section, where briefly speaking, such an encoding represents each node by its pre-order rank. Then, using $O(n)$ time, we construct an auxiliary data structure of $O(n)$ bits, so that given any integer k, and any nodes u and v, we can perform each of the following operations in constant time [17]: **(1)** return the kth leftmost leaf; **(2)** return the parent of u; **(3)** return the lowest common ancestor (LCA) of u and v; **(4)** return the *circular suffix range* $[i, j]$ of a node u, where $S_i, S_{i+1}, \ldots, S_j$ are exactly those circular suffixes with leaves in the subtree rooted at u.
- We define an array $L[1..n]$, where $L[i] = ||S_i||$ stores the length of the cyclic shift that corresponds to the circular suffix S_i. The L array will not be stored explicitly, but each entry can be retrieved in $O(\log n)$ time. Then, we construct the $o(n)$-bit range-minimum-query (RMQ) data structure of [16] on top of the array L; this structure can, on given any query range $[i..j]$, return the position $k \in [i, j]$ such that $L[k]$ is minimized among $L[i..j]$, in constant time. Immediately, this implies that the minimum of $L[i..j]$ can be reported in $O(\log n)$ time. The RMQ structure can be constructed in $o(n)$ time, in additional to $O(n)$ accesses to the L array, making the total construction time $O(n \log n)$.

After the above steps, we can apply Algorithm 5 to mark the desired nodes.

Algorithm 5. Marking nodes in ST_{circ}

1: **for** $k \leftarrow 1$ to $n - 1$ **do**
2:    $u \leftarrow$ the LCA of the kth and $k + 1$th leaves;
3:    Compute the circular suffix range $[\ell_u, r_u]$ of u;
4:    $v \leftarrow$ parent of u;
5:    Compute the circular suffix range $[\ell_v, r_v]$ of v;
6:    $m_1 \leftarrow$ minimum in $L[\ell_v..\ell_u - 1]$; $m_2 \leftarrow$ minimum in $L[r_u + 1..r_v]$;
7:    **if** either m_1 or m_2 is less than $H[SA_{circ}[k]]$ **then**
8:        Mark u;
9:    **end if**
10: **end for**

Theorem 3. *Given the circular suffix array as represented in Lemma 2, the H array as represented in Theorem 1, and the auxiliary data structures as constructed in the preprocessing steps, the marked nodes can be determined in $O(n \log n)$ time and can be represented in $O(n)$ bits. The working space is $O(n)$ bits.*

.

Proof (sketch). The construction time follows directly from the bounds of the individual data strucutres that are used. To mark a node u, we adopt the standard technique of marking directly the parentheses that represent u in the parentheses encoding, which can be done by using an additional copy of the original parentheses encoding. In the end, all the marked parentheses will form a parentheses encoding for the marked nodes, and by constructing the auxiliary data structure of [17], we can support finding the lowest marked ancestor for each node in constant time. We defer the details to the full paper.

References

1. Aho, A., Corasick, M.: Efficient String Matching: An Aid to Bibligoraphic Search. Communications of the ACM 18(6), 333–340 (1975)
2. Belazzougui, D.: Succinct Dictionary Matching with No Slowdown. In: Amir, A., Parida, L. (eds.) CPM 2010. LNCS, vol. 6129, pp. 88–100. Springer, Heidelberg (2010)
3. Chan, H.L., Hon, W.K., Lam, T.W., Sadakane, K.: Compressed Indexes for Dynamic Text Collections. ACM Transactions on Algorithms 3(2) (2007)
4. Elias, P.: Universal Codeword Sets and Representations of the Integers. IEEE Transactions on Information Theory 21(2), 194–203 (1975)
5. Hon, W.K.: On the Construction and Application of Compressed Text Indexes. PhD Thesis, Department of Computer Science, University of Hong Kong (2004)
6. Hon, W.-K., Ku, T.-H., Shah, R., Thankachan, S.V., Vitter, J.S.: Faster Compressed Dictionary Matching. In: Chavez, E., Lonardi, S. (eds.) SPIRE 2010. LNCS, vol. 6393, pp. 191–200. Springer, Heidelberg (2010)
7. Hon, W.-K., Ku, T.-H., Lu, C.-H., Shah, R., Thankachan, S.V.: Efficient Algorithm for Circular Burrows-Wheeler Transform. In: Kärkkäinen, J., Stoye, J. (eds.) CPM 2012. LNCS, vol. 7354, pp. 257–268. Springer, Heidelberg (2012)
8. Hon, W.K., Lam, T.W., Sadakane, K., Sung, W.K., Yiu, S.M.: A Space and Time Efficient Algorithm for Constructing Compressed Suffix Arrays. Algorithmica 48(1), 22–36 (2007)
9. Hon, W.-K., Lu, C.-H., Shah, R., Thankachan, S.V.: Succinct Indexes for Circular Patterns. In: Asano, T., Nakano, S.-I., Okamoto, Y., Watanabe, O. (eds.) ISAAC 2011. LNCS, vol. 7074, pp. 673–682. Springer, Heidelberg (2011)
10. Iliopoulos, C.S., Rahman, M.S.: Indexing circular patterns. In: Nakano, S.-I., Rahman, M.S. (eds.) WALCOM 2008. LNCS, vol. 4921, pp. 46–57. Springer, Heidelberg (2008)
11. Jacobson, G.: Space-efficient Static Trees and Graphs. In: Proc. FOCS, pp. 549–554 (1989)
12. Kasai, T., Lee, G., Arimura, H., Arikawa, S., Park, K.: Linear-Time Longest-Common-Prefix Computation in Suffix Arrays and Its Applications. In: Amir, A., Landau, G.M. (eds.) CPM 2001. LNCS, vol. 2089, pp. 181–192. Springer, Heidelberg (2001)
13. Mantaci, S., Restivo, A., Rosone, G., Sciortino, M.: An Extension of the Burrows Wheeler Transform. Theoretical Computer Science 387(3), 298–312 (2007)
14. McCreight, E.M.: A Space-economical Suffix Tree Construction Algorithm. Journal of the ACM 23(2), 262–272 (1976)

15. Munro, J.I., Raman, V.: Succinct Representation of Balanced Parentheses and Static Trees. SIAM Journal on Computing 31(3), 762–776 (2001)
16. Sadakane, K.: Compressed Suffix Trees with Full Functionality. Theory of Computing System 41(4), 589–607 (2007)
17. Sadakane, K., Navarro, G.: Fully-Functional Succinct Trees. In: Proc. SODA, pp. 134–149 (2010)
18. Weiner, P.: Linear Pattern Matching Algorithms. In: Proc. Switching and Automata Theory, pp. 1–11 (1973)

Efficient Lyndon Factorization of Grammar Compressed Text

Tomohiro I[1,2], Yuto Nakashima[1], Shunsuke Inenaga[1],
Hideo Bannai[1], and Masayuki Takeda[1]

[1] Department of Informatics, Kyushu University, Japan
{tomohiro.i,yuto.nakashima,inenaga,bannai,takeda}@inf.kyushu-u.ac.jp
[2] Japan Society for the Promotion of Science (JSPS)

Abstract. We present an algorithm for computing the Lyndon factorization of a string that is given in grammar compressed form, namely, a Straight Line Program (SLP). The algorithm runs in $O(n^4 + mn^3h)$ time and $O(n^2)$ space, where m is the size of the Lyndon factorization, n is the size of the SLP, and h is the height of the derivation tree of the SLP. Since the length of the decompressed string can be exponentially large w.r.t. n, m and h, our result is the first polynomial time solution when the string is given as SLP.

1 Introduction

Compressed string processing (*CSP*) is a task of processing compressed string data without explicit decompression. As any method that first decompresses the data requires time and space dependent on the decompressed size of the data, CSP without explicit decompression has been gaining importance due to the ever increasing amount of data produced and stored. A number of efficient CSP algorithms have been proposed, e.g., see [16,25,15,12,11,13]. In this paper, we present new CSP algorithms that compute the *Lyndon factorization* of strings.

A string ℓ is said to be a *Lyndon word* if ℓ is lexicographically smallest among its circular permutations of characters of ℓ. For example, aab is a Lyndon word, but its circular permutations aba and baa are not. Lyndon words have various and important applications in, e.g., musicology [4], bioinformatics [8], approximation algorithm [22], string matching [6,2,23], word combinatorics [10,24], and free Lie algebras [20].

The *Lyndon factorization* (a.k.a. *standard factorization*) of a string w, denoted $LF(w)$, is a unique sequence of Lyndon words such that the concatenation of the Lyndon words gives w and the Lyndon words in the sequence are lexicographically non-increasing [5]. Lyndon factorizations are used in a bijective variant of Burrows-Wheeler transform [17,14] and a digital geometry algorithm [3]. Duval [9] proposed an elegant on-line algorithm to compute $LF(w)$ of a given string w of length N in $O(N)$ time. Efficient parallel algorithms to compute the Lyndon factorization are also known [1,7].

We present a new CSP algorithm which computes the Lyndon factorization $LF(w)$ of a string w, when w is given in a *grammar-compressed form*. Let m

J. Fischer and P. Sanders (Eds.): CPM 2013, LNCS 7922, pp. 153–164, 2013.

be the number of factors in $LF(w)$. Our first algorithm computes $LF(w)$ in $O(n^4 + mn^3h)$ time and $O(n^2)$ space, where n is the size of a given *straight-line program (SLP)*, which is a context-free grammar in Chomsky normal form that derives only w, and h is the height of the derivation tree of the SLP. Since the decompressed string length $|w| = N$ can be exponentially large w.r.t. n, m and h, our $O(n^4 + mn^3h)$ solution can be efficient for highly compressive strings.

2   Preliminaries

2.1   Strings and Model of Computation

Let Σ be a finite *alphabet*. An element of Σ^* is called a *string*. The length of a string w is denoted by $|w|$. The empty string ε is a string of length 0, namely, $|\varepsilon| = 0$. Let Σ^+ be the set of non-empty strings, i.e., $\Sigma^+ = \Sigma^* - \{\varepsilon\}$. For a string $w = xyz$, x, y and z are called a *prefix*, *substring*, and *suffix* of w, respectively. A prefix x of w is called a *proper prefix* of w if $x \neq w$, i.e., x is shorter than w. The set of suffixes of w is denoted by $Suffix(w)$. The i-th character of a string w is denoted by $w[i]$, where $1 \leq i \leq |w|$. For a string w and two integers $1 \leq i \leq j \leq |w|$, let $w[i..j]$ denote the substring of w that begins at position i and ends at position j. For convenience, let $w[i..j] = \varepsilon$ when $i > j$. For any string w let $w^1 = w$, and for any integer $k > 2$ let $w^k = ww^{k-1}$, i.e., w^k is a k-time repetition of w.

A positive integer p is said to be a *period* of a string w if $w[i] = w[i + p]$ for all $1 \leq i \leq |w| - p$. Let w be any string and q be its smallest period. If p is a period of a string w such that $p < |w|$, then the positive integer $|w| - p$ is said to be a *border* of w. If w has no borders, then w is said to be *border-free*.

If character $a \in \Sigma$ is lexicographically smaller than another character $b \in \Sigma$, then we write $a \prec b$. For any non-empty strings $x, y \in \Sigma^+$, let $lcp(x, y)$ be the length of the longest common prefix of x and y. We denote $x \prec y$, if either of the following conditions holds: $x[lcp(x, y) + 1] \prec y[lcp(x, y) + 1]$, or x is a proper prefix of y. For a set $S \subseteq \Sigma^+$ of non-empty strings, let $\min_\prec S$ denote the lexicographically smallest string in S.

Our model of computation is the word RAM: We shall assume that the computer word size is at least $\lceil \log_2 |w| \rceil$, and hence, standard operations on values representing lengths and positions of string w can be manipulated in constant time. Space complexities will be determined by the number of computer words (not bits).

2.2   Lyndon Words and Lyndon Factorization of Strings

Two strings x and y are said to be *conjugate*, if there exist strings u and v such that $x = uv$ and $y = vu$. A string w is said to be a *Lyndon word*, if w is lexicographically strictly smaller than all of its conjugates of w. Namely, w is a

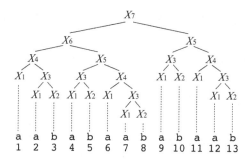

Fig. 1. The derivation tree of SLP $\mathcal{S} = \{X_1 \rightarrow \text{a}, X_2 \rightarrow \text{b}, X_3 \rightarrow X_1X_2, X_4 \rightarrow X_1X_3, X_5 \rightarrow X_3X_4, X_6 \rightarrow X_4X_5, X_7 \rightarrow X_6X_5\}$, representing string $S = val(X_7) = $ aababaababaab

Lyndon word, if for any factorization $w = uv$, it holds that $uv \prec vu$. It is known that any Lyndon word is border-free.

Definition 1 ([5]). *The* Lyndon factorization *of a string w, denoted $LF(w)$, is the factorization $\ell_1^{p_1} \cdots \ell_m^{p_m}$ of w, such that each $\ell_i \in \Sigma^+$ is a Lyndon word, $p_i \geq 1$, and $\ell_i \succ \ell_{i+1}$ for all $1 \leq i < m$.*

It is known that the Lyndon factorization is unique for each string w, and it was shown by Duval [9] that the Lyndon factorization can be computed in $O(N)$ time, where $N = |w|$.

$LF(w)$ can be represented by the sequence $(|\ell_1|, p_1), \ldots, (|\ell_m|, p_m)$ of integer pairs, where each pair $(|\ell_i|, p_i)$ represents the i-th Lyndon factor $\ell_i^{p_i}$ of w. Note that this representation requires $O(m)$ space.

2.3 Straight Line Programs

A *straight line program (SLP)* is a set of productions $\mathcal{S} = \{X_1 \rightarrow expr_1, X_2 \rightarrow expr_2, \ldots, X_n \rightarrow expr_n\}$, where each X_i is a variable and each $expr_i$ is an expression, where $expr_i = a$ $(a \in \Sigma)$, or $expr_i = X_{\ell(i)}X_{r(i)}$ $(i > \ell(i), r(i))$. It is essentially a context free grammar in Chomsky normal form, that derives a single string. Let $val(X_i)$ represent the string derived from variable X_i. To ease notation, we sometimes associate $val(X_i)$ with X_i and denote $|val(X_i)|$ as $|X_i|$, and $val(X_i)[u..v]$ as $X_i[u..v]$ for $1 \leq u \leq v \leq |X_i|$. An SLP $\mathcal{S}$ *represents* the string $w = val(X_n)$. The *size* of the program $\mathcal{S}$ is the number n of productions in $\mathcal{S}$. Let N be the length of the string represented by SLP $\mathcal{S}$, i.e., $N = |w|$. Then N can be as large as 2^{n-1}.

The derivation tree of SLP $\mathcal{S}$ is a labeled ordered binary tree where each internal node is labeled with a non-terminal variable in $\{X_1, \ldots, X_n\}$, and each leaf is labeled with a terminal character in Σ. The root node has label X_n. An example of the derivation tree of an SLP is shown in Fig. 1.

3   Computing Lyndon Factorization from SLP

In this section, we show how, given an SLP $\mathcal{S}$ of n productions representing string w, we can compute $LF(w)$ of size m in $O(n^4 + mn^3h)$ time. We will make use of the following known results:

Lemma 1 ([9]). *For any string w, let $LF(w) = \ell_1^{p_1}, \ldots, \ell_m^{p_m}$. Then, $\ell_m = \min_{\prec} Suffix(w)$, i.e., ℓ_m is the lexicographically smallest suffix of w.*

Lemma 2 ([18]). *Given an SLP $\mathcal{S}$ of size n representing a string w of length N, and two integers $1 \leq i \leq j \leq N$, we can compute in $O(n)$ time another SLP of size $O(n)$ representing the substring $w[i..j]$.*

Lemma 3 ([18]). *Given an SLP $\mathcal{S}$ of size n representing a string w of length N, we can compute the shortest period of w in $O(n^3 \log N)$ time and $O(n^2)$ space.*

For any non-empty string $w \in \Sigma^+$, let $LFCand(w) = \{x \mid x \in Suffix(w), \exists y \in \Sigma^+ \text{ s.t. } xy = \min_{\prec} Suffix(wy)\}$. Intuitively, $LFCand(w)$ is the set of suffixes of w which are a prefix of the lexicographically smallest suffix of string wy, for some non-empty string $y \in \Sigma^+$.

The following lemma may be almost trivial, but will play a central role in our algorithm.

Lemma 4. *For any two strings $u, v \in LFCand(w)$ with $|u| < |v|$, u is a prefix of v.*

Proof. If $v[1..|u|] \prec u$, then for any non-empty string y, $vy \prec uy$. However, this contradicts that $u \in LFCand(w)$. If $v[1..|u|] \succ u$, then for any non-empty string y, $vy \succ uy$. However, this contradicts that $v \in LFCand(w)$. Hence we have $v[1..|u|] = u$. □

Lemma 5. *For any string w, let $\ell = \min_{\prec} Suffix(w)$. Then, the shortest string of $LFCand(w)$ is ℓ^p, where $p \geq 1$ is the maximum integer such that ℓ^p is a suffix of w.*

Proof. For any string $x \in LFCand(w)$, and any non-empty string y, $xy = \min_{\prec} Suffix(wy)$ holds only if $y \succ \ell$.

Firstly, we compare ℓ^p with the suffixes s of w shorter than ℓ^p, and show that $\ell^p y \prec sy$ holds for *any* $y \succ \ell$. Such suffixes s are divided into two groups: (1) If s is of form ℓ^k for any integer $1 \leq k < p$, then $\ell^p y \prec \ell^k y = sy \prec y$ holds for any $y \succ \ell$; (2) If s is not of form ℓ^k, then since ℓ is border-free, ℓ is not a prefix of s, and s is not a prefix of ℓ, either. Thus $\ell^p \prec s$ holds, implying that $\ell^p y \prec sy$ for any $y \succ \ell$.

Secondly, we compare ℓ^p with the suffixes t of w longer than ℓ^p, and show that $\ell^p y \prec ty$ holds for *some* $y \succ \ell$. By Lemma 4, $t = \ell^q u$ holds, where $q \geq p$ is the maximum integer such that ℓ^q is a prefix of t, and $u \in \Sigma^+$. By definition, $\ell \prec u$ and ℓ is not a prefix of u. Choosing $y = \ell^{q-p}u'$ with $u' \prec u$, we have $\ell^p y = \ell^q u' \prec \ell^q u = t \prec ty$. Hence, $\ell^p \in LFCand(w)$ and no shorter strings exist in $LFCand(w)$. □

By Lemma 1 and Lemma 5, computing the last Lyndon factor $\ell_m^{p_m}$ of $w = val(X_n)$ reduces to computing $LFCand(X_n)$ for the last variable X_n. In what follows, we propose a dynamic programming algorithm to compute $LFCand(X_i)$ for each variable. Firstly we show the number of strings in $LFCand(X_i)$ is $O(\log N)$, where $N = |val(X_n)| = |w|$.

Lemma 6. *For any string w, let s_j be the jth shortest string of $LFCand(w)$. Then, $|s_{j+1}| > 2|s_j|$ for any $1 \leq j < |LFCand(w)|$.*

Proof. Let $\ell = \min_{\prec} Suffix(w)$, and y any string such that $y \succ \ell$. It follows from Lemma 4 that ℓ is a prefix of any string $s_j \in LFCand(w)$, and hence $s_j \prec y$ holds.

Assume on the contrary that $|s_{j+1}| \leq 2|s_j|$. If $|s_{j+1}| = 2|s_j|$, i.e., $s_{j+1} = s_j s_j$, then $s_{j+1}y = s_j s_j y \prec s_j y$ holds, but this contradicts that $s_j \in LFCand(w)$. Hence $s_{j+1} \neq s_j s_j$. If $|s_{j+1}| < 2|s_j|$, by Lemma 4, s_j is a prefix of s_{j+1}, and therefore s_j has a period q such that $s_{j+1} = u^k v$ and $s_j = u^{k-1}v$, where $u = s_j[1..q]$, $k \geq 1$ is an integer, and v is a proper prefix of u. There are two cases to consider: (1) If $uvy \prec vy$, then $u^k vy \prec u^{k-1}vy = s_j y$. (2) If $vy \prec uvy$, then $vy \prec uvy \prec u^2 vy \prec \cdots \prec u^{k-1}vy = s_j y$. It means that $\min_{\prec}\{u^k vy, vy\} \prec s_j y$ for any $y \succ \ell$, however, this contradicts that $s_j \in LFCand(w)$. Hence $|s_{j+1}| > 2|s_j|$ holds. □

Since s_j is a suffix of s_{j+1}, it follows from Lemma 4 and Lemma 6 that $s_{j+1} = s_j t s_j$ with some non-empty string $t \in \Sigma^+$. This also implies that the number of strings in $LFCand(w)$ is $O(\log N)$, where N is the length of w. By identifying each suffix of $LFCand(X_i)$ with its length, and using Lemma 6, $LFCand(X_i)$ for all variables can be stored in a total of $O(n \log N)$ space.

For any two variables X_i, X_j of an SLP S and a positive integer k satisfying $|X_i| \geq k + |X_j| - 1$, consider the FM function such that $FM(X_i, X_j, k) = lcp(val(X_i)[k..|X_i|], val(X_j))$, i.e., it returns the length of the lcp of the suffix of $val(X_i)$ starting at position k and X_j.

Lemma 7 ([21,19]). *We can preprocess a given SLP S of size n in $O(n^3)$ time and $O(n^2)$ space so that $FM(X_i, X_j, k)$ can be answered in $O(n^2)$ time.*

For each variable X_i we store the length $|X_i|$ of the string derived by X_i. It requires a total of $O(n)$ space for all $1 \leq i \leq n$, and can be computed in a total of $O(n)$ time by a simple dynamic programming algorithm. Given a position j of the uncompressed string w of length N, i.e., $1 \leq j \leq N$, we can retrieve the jth character $w[j]$ in $O(n)$ time by a simple binary search on the derivation tree of X_n using the lengths stored in the variables. Hence, we can lexicographically compare $val(X_i)[k..|X_i|]$ and $val(X_j)$ in $O(n^2)$ time, after $O(n^3)$-time preprocessing.

The following lemma shows a dynamic programming approach to compute $LFCand(X_i)$ for each variable X_i. We will mean by a sorted list of $LFCand(X_i)$ the list of the elements of $LFCand(X_i)$ sorted in increasing order of length.

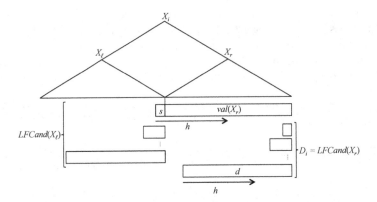

Fig. 2. Lemma 8: Initially $D_i = LFCand(X_r)$ and $h = s \cdot val(X_\ell)$ with s being the shortest string of $LFCand(X_\ell)$.

Lemma 8. *Let $X_i = X_\ell X_r$ be any production of a given SLP S of size n. Provided that sorted lists for $LFCand(X_\ell)$ and $LFCand(X_r)$ are already computed, a sorted list for $LFCand(X_i)$ can be computed in $O(n^3)$ time and $O(n^2)$ space.*

Proof. Let D_i be a sorted list of the suffixes of X_i that are candidates of elements of $LFCand(X_i)$. We initially set $D_i \leftarrow LFCand(X_r)$.

We process the elements of $LFCand(X_\ell)$ in increasing order of length. Let s be any string in $LFCand(X_\ell)$, and d the longest string in D_i. Since any string of $LFCand(X_r)$ is a prefix of d by Lemma 4, in order to compute $LFCand(X_i)$ it suffices to lexicographically compare $s \cdot val(X_r)$ and d. Let $h = lcp(s \cdot val(X_r), d))$. See also Fig. 2.

- If $(s \cdot val(X_r))[h + 1] \prec d[h + 1]$, then $s \cdot val(X_r) \prec d$. Since any string in D_i is a prefix of d by Lemma 4, we observe that any element in D_i that is longer than h cannot be an element of $LFCand(X_i)$. Hence we delete any element of D_i that is longer than h from D_i, then add $s \cdot val(X_r)$ to D_i, and update $d \leftarrow s \cdot val(X_r)$. See also Fig. 3.
- If $(s \cdot val(X_r))[h + 1] \succ d[h + 1]$, then $s \cdot val(X_r) \succ d$. Since $s \cdot val(X_r)$ cannot be an element of $LFCand(X_i)$, in this case neither D_i nor d is updated. See also Fig. 4.
- If $h = |d|$, i.e., d is a prefix of $s \cdot val(X_r)$, then there are two sub-cases:
  - If $|s \cdot val(X_r)| \leq 2|d|$, d has a period q such that $s \cdot val(X_r) = u^k v$ and $d = u^{k-1} v$, where $u = d[1..q]$, $k \geq 1$ is an integer, and v is a proper prefix of u. By similar arguments to Lemma 6, we observe that d cannot be a member of $LFCand(X_i)$ while $s \cdot val(X_r)$ may be a member of $LFCand(X_i)$. Thus we add $s \cdot val(X_r)$ to D_i, delete d from D_i, and update $d \leftarrow s \cdot val(X_r)$. See also Fig. 5.
  - If $|s \cdot val(X_r)| > 2|d|$, then both d and $s \cdot val(X_r)$ may be a member of $LFCand(X_i)$. Thus we add $s \cdot val(X_r)$ to D_i, and update $d \leftarrow s \cdot val(X_r)$. See also Fig. 6.

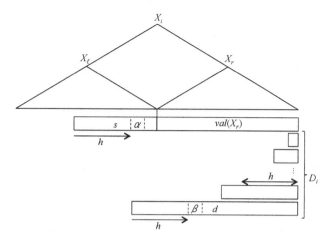

Fig. 3. Lemma 8: Case where $(s \cdot val(X_r))[h + 1] = \alpha \prec d[h + 1] = \beta$. d and any string in D_i that is longer than h are deleted from D_i. Then $s \cdot val(X_r)$ becomes the longest candidate in D_i.

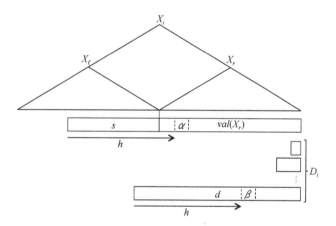

Fig. 4. Lemma 8: Case where $(s \cdot val(X_r))[h + 1] = \alpha \succ d[h + 1] = \beta$. There are no updates on D_i.

We represent the strings in $LFCand(X_\ell)$, $LFCand(X_r)$, $LFCand(X_i)$, and D_i by their lengths. Given sorted lists of $LFCand(X_\ell)$ and $LFCand(X_r)$, the above algorithm computes a sorted list for D_i, and it follows from Lemma 6 that the number of elements in D_i is always $O(\log N)$. Thus all the above operations on D_i can be conducted in $O(\log N)$ time in each step.

We now show how to efficiently compute $h = lcp(s \cdot val(X_r), d)$, for any $s \in LFCand(X_\ell)$. Let z be the longest string in $LFCand(X_\ell)$, and consider to process any string $s \in LFCand(X_\ell)$. Since s is a prefix of z by Lemma 4, we can compute $lcp(s \cdot val(X_r), d)$ as follows:

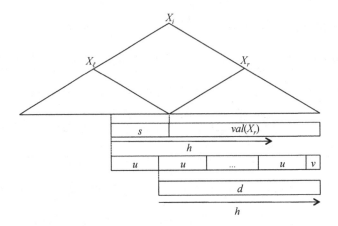

Fig. 5. Lemma 8: Case where $h = |d|$ and $|s \cdot val(X_r)| \le 2|d|$. Since $s \cdot val(X_r) = u^k v$ and $d = u^{k-1} v$, d is deleted from D_i and $s \cdot val(X_r)$ is added to D_i.

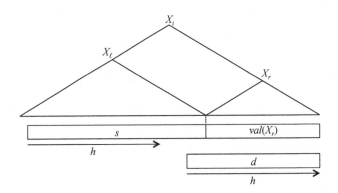

Fig. 6. Lemma 8: Case where $h = |d|$ and $|s \cdot val(X_r)| > 2|d|$. We add $s \cdot val(X_r)$ to D_i, and $s \cdot val(X_r)$ becomes the longest member of D_i.

$$lcp(s \cdot val(X_r), d) = \begin{cases} lcp(z, d) & \text{if } lcp(z, d) < |s|, \\ |s| + lcp(X_r, d[|s| + 1..|d|]) & \text{if } lcp(z, d) \ge |s|. \end{cases}$$

To compute the above lcp values using the *FM* function, for each variable X_i of S we create a new production $X_{n+i} = X_i X_i$, and hence the number of variables increases to $2n$. In addition, we construct a new SLP of size $O(n)$ that derives z in $O(n)$ time using Lemma 2. Let Z be the variable such that $val(Z) = z$. It holds that

$$lcp(z, d) = \min\{lcp(Z, X_{n+i}[|X_i| - |d| + 1..|X_{n+i}|]), |d|\} \text{ and}$$

$$lcp(X_r, d[|s|+1..|d|]) = \min\{lcp(X_r, X_{n+r}[|X_r| - |d| + |s| + 1..|X_{n+r}|]), |d| - |s|\}.$$

See also Fig. 7 and Fig. 8.

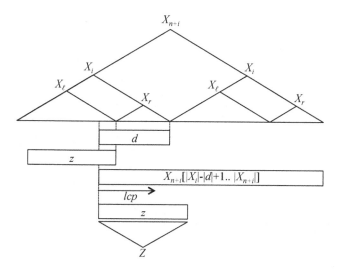

Fig. 7. Lemma 8: $lcp(z, d) = \min\{lcp(Z, X_{n+i}[|X_i| - |d| + 1..|X_{n+i}|]), |d|\}$.

By using Lemma 7, we preprocess, in $O(n^3)$ time and $O(n^2)$ space, the SLP consisting of these variables so that the query $FM(X_i, X_j, k)$ for answering $lcp(X_i[k..|X_i|], X_j)$ is supported in $O(n^2)$ time. Therefore $lcp(s \cdot val(X_r), d)$ can be computed in $O(n^2)$ time for each $s \in LFCand(X_\ell)$. Since there exist $O(\log N)$ elements in $LFCand(X_\ell)$, we can compute $LFCand(X_i)$ in $O(n^3 + n^2 \log N) = O(n^3)$ time. The total space complexity is $O(n^2)$.   □

Since there are n productions in a given SLP, using Lemma 8 we can compute $LFCand(X_n)$ for the last variable X_n in a total of $O(n^4)$ time. The main result of this paper follows.

Theorem 1. *Given an SLP $\mathcal{S}$ of size n representing a string w, we can compute $LF(w)$ in $O(n^4 + mn^3 h)$ time and $O(n^2)$ space, where m is the number of factors in $LF(w)$ and h is the height of the derivation tree of $\mathcal{S}$.*

Proof. Let $LF(w) = \ell_1^{p_1} \cdots \ell_m^{p_m}$. First, using Lemma 8 we compute $LFCand$ for all variables in $\mathcal{S}$ in $O(n^4)$ time. Next we will compute the Lyndon factors from right to left. Suppose that we have already computed $\ell_{j+1}^{p_{j+1}} \cdots \ell_m^{p_m}$, and we are computing the jth Lyndon factor $\ell_j^{p_j}$. Using Lemma 2, we construct in $O(n)$ time a new SLP of size $O(n)$ describing $w[1..|w| - \sum_{k=j+1}^{m} p_k|\ell_k|]$, which is the prefix of w obtained by removing the suffix $\ell_{j+1}^{p_{j+1}} \cdots \ell_m^{p_m}$ from w. Here we note that the new SLP actually has $O(h)$ new variables since $w[1..|w| - \sum_{k=j+1}^{m} p_k|\ell_k|]$ can be represented by a sequence of $O(h)$ variables in $\mathcal{S}$. Let Y be the last variable of the new SLP. Since $LFCand$ for all variables in $\mathcal{S}$ have already been computed, it is enough to compute $LFCand$ for $O(h)$ new variables. Hence using Lemma 8, we compute a sorted list of $LFCand(Y) = LFCand(w[1..|w| - \sum_{k=j+1}^{m} p_k|\ell_k|])$

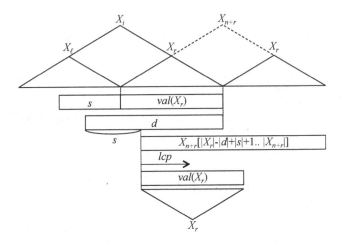

Fig. 8. Lemma 8: $lcp(X_r, d[|s| + 1..|d|]) = \min\{lcp(X_r, X_{n+r}[|X_r| - |d| + |s| + 1..|X_{n+r}|]), |d| - |s|\}$

in a total of $O(n^3h)$ time. It follows from Lemma 5 that the shortest element of $LFCand(Y)$ is $\ell_j^{p_j}$, the jth Lyndon factor of w. Note that each string in $LFCand(Y)$ is represented by its length, and so far we only know the total length $p_j|\ell_j|$ of the jth Lyndon factor. Since ℓ_j is border free, $|\ell_j|$ is the shortest period of $\ell_j^{p_j}$. We construct a new SLP of size $O(n)$ describing $\ell_j^{p_j}$, and compute $|\ell_j|$ in $O(n^3 \log N)$ time using Lemma 3. We repeat the above procedure m times, and hence $LF(w)$ can be computed in a total of $O(n^4 + m(n^3h + n^3 \log N)) = O(n^4 + mn^3h)$ time. To compute each Lyndon factor of $LF(w)$, we need $O(n^2)$ space for Lemma 3 and Lemma 8. Since $LFCand(X_i)$ for each variable X_i requires $O(\log N)$ space, the total space complexity is $O(n^2 + n \log N) = O(n^2)$. $\square$

4   Conclusions and Open Problem

Lyndon words and Lyndon factorization are important concepts of combinatorics on words, with various applications. Given a string in terms of an SLP of size n, we showed how to compute the Lyndon factorization of the string in $O(n^4 + mn^3h)$ time using $O(n^2)$ space, where m is the size of the Lyndon factorization and h is the height of the SLP. Since the decompressed string length N can be exponential w.r.t. n, m and h, our algorithm can be useful for highly compressive strings.

An interesting open problem is to compute the Lyndon factorization from a given LZ78 encoding [26]. Each LZ78 factor is a concatenation of the longest previous factor and a single character. Hence, it can be seen as a special class of SLPs, and this property would lead us to a much simpler and/or more efficient solution to the problem. Noting the number s of the LZ78 factors is $\Omega(\sqrt{N})$, a question is whether we can solve this problem in $o(s^2) + O(m)$ time.

References

1. Apostolico, A., Crochemore, M.: Fast parallel Lyndon factorization with applications. Mathematical Systems Theory 28(2), 89–108 (1995)
2. Breslauer, D., Grossi, R., Mignosi, F.: Simple real-time constant-space string matching. In: Giancarlo, R., Manzini, G. (eds.) CPM 2011. LNCS, vol. 6661, pp. 173–183. Springer, Heidelberg (2011)
3. Brlek, S., Lachaud, J.O., Provençal, X., Reutenauer, C.: Lyndon + Christoffel = digitally convex. Pattern Recognition 42(10), 2239–2246 (2009)
4. Chemillier, M.: Periodic musical sequences and Lyndon words. Soft Comput. 8(9), 611–616 (2004)
5. Chen, K.T., Fox, R.H., Lyndon, R.C.: Free differential calculus. iv. The quotient groups of the lower central series. Annals of Mathematics 68(1), 81–95 (1958)
6. Crochemore, M., Perrin, D.: Two-way string matching. J. ACM 38(3), 651–675 (1991)
7. Daykin, J.W., Iliopoulos, C.S., Smyth, W.F.: Parallel RAM algorithms for factorizing words. Theor. Comput. Sci. 127(1), 53–67 (1994)
8. Delgrange, O., Rivals, E.: STAR: an algorithm to search for tandem approximate repeats. Bioinformatics 20(16), 2812–2820 (2004)
9. Duval, J.P.: Factorizing words over an ordered alphabet. J. Algorithms 4(4), 363–381 (1983)
10. Fredricksen, H., Maiorana, J.: Necklaces of beads in k colors and k-ary de Bruijn sequences. Discrete Mathematics 23(3), 207–210 (1978)
11. Gawrychowski, P.: Optimal pattern matching in LZW compressed strings. In: Proc. SODA 2011, pp. 362–372 (2011)
12. Gawrychowski, P.: Pattern matching in Lempel-Ziv compressed strings: Fast, simple, and deterministic. In: Demetrescu, C., Halldórsson, M.M. (eds.) ESA 2011. LNCS, vol. 6942, pp. 421–432. Springer, Heidelberg (2011)
13. Gawrychowski, P.: Faster algorithm for computing the edit distance between SLP-compressed strings. In: Calderón-Benavides, L., González-Caro, C., Chávez, E., Ziviani, N. (eds.) SPIRE 2012. LNCS, vol. 7608, pp. 229–236. Springer, Heidelberg (2012)
14. Gil, J.Y., Scott, D.A.: A bijective string sorting transform. CoRR abs/1201.3077 (2012)
15. Goto, K., Bannai, H., Inenaga, S., Takeda, M.: Fast q-gram mining on SLP compressed strings. Journal of Discrete Algorithms 18, 89–99 (2013)
16. Hermelin, D., Landau, G.M., Landau, S., Weimann, O.: A unified algorithm for accelerating edit-distance computation via text-compression. In: Proc. STACS 2009, pp. 529–540 (2009)
17. Kufleitner, M.: On bijective variants of the Burrows-Wheeler transform. In: Proc. PSC 2009, pp. 65–79 (2009)
18. Lifshits, Y.: Solving classical string problems an compressed texts. In: Combinatorial and Algorithmic Foundations of Pattern and Association Discovery. Dagstuhl Seminar Proceedings, vol. 06201 (2006)
19. Lifshits, Y.: Processing compressed texts: A tractability border. In: Ma, B., Zhang, K. (eds.) CPM 2007. LNCS, vol. 4580, pp. 228–240. Springer, Heidelberg (2007)
20. Lyndon, R.C.: On Burnside's problem. Transactions of the American Mathematical Society 77, 202–215 (1954)
21. Miyazaki, M., Shinohara, A., Takeda, M.: An improved pattern matching algorithm for strings in terms of straight-line programs. In: Hein, J., Apostolico, A. (eds.) CPM 1997. LNCS, vol. 1264, pp. 1–11. Springer, Heidelberg (1997)

22. Mucha, M.: Lyndon words and short superstrings. In: Proc. SODA 2013, pp. 958–972 (2013)
23. Neuburger, S., Sokol, D.: Succinct 2D dictionary matching. Algorithmica, 1–23 (2012), 10.1007/s00453-012-9615-9
24. Provençal, X.: Minimal non-convex words. Theor. Comput. Sci. 412(27), 3002–3009 (2011)
25. Yamamoto, T., Bannai, H., Inenaga, S., Takeda, M.: Faster subsequence and don't-care pattern matching on compressed texts. In: Giancarlo, R., Manzini, G. (eds.) CPM 2011. LNCS, vol. 6661, pp. 309–322. Springer, Heidelberg (2011)
26. Ziv, J., Lempel, A.: Compression of individual sequences via variable-length coding. IEEE Transactions on Information Theory 24(5), 530–536 (1978)

Approximation of Grammar-Based Compression via Recompression[*]

Artur Jeż[1,2,**]

[1] Max Planck Institute für Informatik,
Campus E1 4, DE-66123 Saarbrücken, Germany
[2] Institute of Computer Science, University of Wrocław
ul. Joliot-Curie 15, 50-383 Wrocław, Poland
aje@cs.uni.wroc.pl

Abstract. We present a simple linear-time algorithm constructing a context-free grammar of size $\mathcal{O}(g \log(N/g))$ for the input string of size N, where g the size of the optimal grammar generating this string. The algorithm works for arbitrary size alphabets, but the running time is linear assuming that the alphabet Σ of the input string is a subset of $\{1, \ldots, N^c\}$ for some constant c. Algorithms with such an approximation guarantees and running time are known, the novelty of this paper is the particular simplicity of the algorithm as well as the analysis of the algorithm, which uses a general technique of recompression recently introduced by the author. Furthermore, contrary to the previous results, this work does not use the LZ representation of the input string in the construction, nor in the analysis.

Keywords: Grammar-based compression, Construction of the smallest grammar, SLP.

1   Introduction

Grammar Based Compression. This paper presents an alternative linear-time approximation algorithm for the construction of the smallest grammar (CFG) generating a given string T. There are three known algorithms with an approximation ratio $\mathcal{O}(\log(N/g))$, where N is the input-string length and g is the size of the optimal grammar [14,1,15]. The novelty of the proposed algorithm is its apparent simplicity (it uses only local replacement of strings) and an analysis that uses the recompression technique developed recently by the author. In particular, neither the algorithm, nor its analysis relate to the LZ-compression, which was the case for previously known algorithms.

In the grammar-based compression text is represented by a context-free grammar generating exactly one string. The idea behind this approach is that a CFG can compactly represent the structure of the text, even if this structure is not apparent. Furthermore, the natural hierarchical definition of the CFGs make such

---

[*] The full version of this paper is available at http://arxiv.org/abs/1301.5842
[**] Supported by NCN grant number 2011/01/D/ST6/07164, 2011–2014

J. Fischer and P. Sanders (Eds.): CPM 2013, LNCS 7922, pp. 165–176, 2013.

a representation suitable for algorithms, in which case the string operations can be performed on the compressed representation, without the need of the explicit decompression [3,5,9,13,4,1]. Lastly, there is a close connection between block-based compression methods and the grammar compression. To be more specific, it fairly easy to rewrite the LZW definition as a context free grammar (with just a multiplicative constant factor overhead), LZ77 can also be presented in this way, but this is much less obvious (and introduces a $\log(N/\ell)$ blow-up, where ℓ is the size of the LZ77 representation) [14,1].

The main drawback of the grammar-based compression is that producing the smallest CFG for a text is difficult: the decision problem is NP-hard [16] and the size of the grammar cannot be approximated within a constant factor [1]. Furthermore, the connection with *addition chains* makes any algorithm with an approximation guarantee $o(\log N/\log\log N)$ unlikely [1].

Approximation. The two first algorithms with approximation ratio $\mathcal{O}(\log(N/g))$ were due to Rytter [14] and independently Charikar et al. [1]. They both applied the LZ77 compression to the input string and transformed the obtained LZ77 representation to a grammar. The main idea was to require that the derivation tree of the intermediate constructed grammar was balanced, the former algorithm assumed AVL-condition, while the latter imposed that for a rule $X \to YZ$ the lengths of words generated by Y and Z are within a certain multiplicative constant factor from each other.

Sakamoto [15] proposed a different approach, based on RePair [10], a practically implemented and used algorithm for grammar-based compression. His algorithm iteratively replaced pairs of different letters and maximal blocks of letters (a^ℓ is a *maximal block* if that cannot be extended by a to either side). A special pairing of the letters was devised, so that it is 'synchronising': for any two appearances of the same string w in the instance we can represent w as $w = w_1 w_2 w_3$, where $w_1, w_3 = \mathcal{O}(1)$ and w_2 in both appearances are compressed in the same way (though w_1 and w_3 in those appearances can be compressed differently). The analysis considered the LZ77 representation of the text and proving that due to 'synchronisation' the factors of LZ77 are compressed very similarly as the text to which they refer.

However, to the author's best knowledge and understanding, the presented analysis [15] is incomplete, as the cost of nonterminals introduced for the representation of maximal blocks is not bounded in the paper; the bound that the author was able to obtain using there presented approach is $\mathcal{O}(\log(N/g)^2)$.

Proposed Approach: Recompression. In this paper another algorithm is proposed, using the general approach of recompression, developed by the author, based on iterative application of two replacement schemes performed on the text T:

pair compression of ab For two different symbols (i.e. letters or nonterminals) a, b such that substring ab appears in T replace each of ab in T by a fresh nonterminal c.

a's block compression For each maximal block a^ℓ appearing in T, where a is a letter or a nonterminal, replace all a^ℓs in T by a fresh nonterminal a_ℓ.

In one phase, all pairs (letters) appearing in the current text are listed in P (L, respectively). Then pair compression is applied to an appropriately chosen subset of P and all blocks of symbols from L, then the phase ends. If each pair and each block is compressed, the length of T drops by half; in reality the text length drops by some smaller, but constant, factor per phase. For the sake of simplicity, we treat all nonterminals introduced by the algorithm as letters.

In author's previous work it was shown that such an approach can be efficiently applied to text represented in a grammar compressed form [6,5,7]. In this paper a somehow opposite direction is followed: the recompression method is employed to the input string. This yields a simple linear-time algorithm: Performing one phase in $\mathcal{O}(|T|)$ running time is relatively easy, since the length of T drops by a constant factor in each phase, the $\mathcal{O}(N)$ running time is obtained.

However, the more interesting is the analysis, and not the algorithm itself: it is performed by applying the recompression to the optimal grammar G for the input text. In this way, the current G always generates the current string kept by the algorithm and the number of nonterminals introduced during the construction can be calculated in terms of $|G| = g$. A straightforward analysis yields that the generated grammar is of size $\mathcal{O}(g \log N)$, when the recompression-based algorithm is combined with a naive algorithm generating a grammar of size $\mathcal{O}(N)$, the resulting algorithm outputs a grammar of size $\mathcal{O}(g \log(N/g) + g)$.

We believe that the proposed algorithm is interesting, as it is very simple and its analysis for the first time does not rely on LZ77 representation of the string. Potentially this can help in both design of an algorithm with a better approximation ratio and in showing a logarithmic lower bound: Observe that the size ℓ of the LZ77 representation is not larger than g, so it might be that some algorithm produces a grammar of size $o(g \log(N/g))$, even though this is of size $\Omega(\ell \log(N/\ell))$. Secondly, as the analysis 'considers' the optimal grammar, it may be much easier to observe, where any approximation algorithm performs badly, and so try to approach a logarithmic lower bound.

Unlike Rytter's [14] and Charikar's et al. [1] algorithms, the grammar returned by the here-proposed algorithm is not balanced in any sense and it can have height $\omega(\log N)$. This is disadvantageous, as many many data structures assume some balanced form of the grammar. On the other hand, there is no evidence that an optimal grammar is balanced and the only algorithms changing an SLP into balanced once are adaptations of translations from LZ to SLP, in particular they introduce an $\mathcal{O}(\log(N/g))$ blow-up in size. Thus beating the $\mathcal{O}(\log(N/g))$ approximation ratio might involve a construction of an unbalanced grammar.

Comparison with Sakamoto's algorithm. The general approach is similar to Sakamoto's method, however, the pairing of letters seems more natural. Also, the construction of nonterminals for blocks of letters is different, the author failed to show that the bound actually holds for the variant proposed by Sakamoto. It should be noted that the analysis presented in this paper the calculation of nonterminals used due to pair compression is fairly easy, while estimating the number used for block compression is non-obvious. Also, the connection to the addition chains suggests that the compression of blocks is the difficult part of the smallest grammar problem.

Computational model. When Σ can be identified with $\{1, 2, \ldots, N^c\}$ for some constant c (i.e. RadixSort can be performed on it in linear time) the presented algorithm runs in linear time. The same applies to previous algorithms: when a LZ77 representation is created using a suffix-tree the linear-time construction for it assumes that the alphabet consists of integers and that these can be sorted in linear time [2]. While Sakamoto's method was designed for constant-size Σ, it generalises to $\Sigma = \{1, 2, \ldots, N^c\}$ with ease.

2   The Algorithm

The input sequence is $T \in \Sigma^*$ and N is its initial length; it is represented as a doubly-linked list. We treat nonterminals of the constructed grammar in the same way as the original letters and use common set Σ for both of them.

The smallest grammar generating T is denoted by G and its size, measured as the length of the productions, is g. The crucial part of the analysis is the modification of G, so that it always derives T (so in some sense according to the compression performed on T). In this way at each step the cost of new nonterminals introduced by the algorithm can be related to G and moreover we can amortise better the cost of the operations.

In the following, the terms nonterminal, rules, etc. always regard to the optimal grammar G (or its transformed version). Still, we need to estimate the number of productions in the constructed grammar, to avoid confusion, we say about *representing* a letter a and the *cost of representation*, i.e. whenever a replaces string w, the representation of a is the subgrammar generating w from a and the cost of representation is the size of this subgrammar. Note that when we replace a pair ab by c then the cost of representation is constant (i.e. 2), however, when a_ℓ replaces a block a^ℓ, the cost is much higher (roughly: $\mathcal{O}(\log \ell)$, but we try to amortise it whenever possible).

Blocks Compression. The blocks compression is very simple to implement: We read T, for a block of as of length greater than 1 we create a record (a, ℓ, p), where ℓ is a length of the block, and p is the pointer to the first letter in this block. We then sort these records lexicographically using RadixSort (ignoring the last component). There are only $\mathcal{O}(|T|)$ records and we can assume that Σ can be identified with an interval, this is all done in $\mathcal{O}(|T|)$. Now, for a fixed letter a, the consecutive tuples with the first coordinate a correspond to all blocks of a, ordered by the size. It is easy to replace them in $\mathcal{O}(|T|)$ time with new letters.

We need to represent the a_ℓ replacing a^ℓ: suppose we are to replace the blocks $a^{\ell_1}, a^{\ell_2}, \ldots, a^{\ell_k}$, where $1 < \ell_1 < \ell_2 < \cdots < \ell_k$, let $\ell_0 = 0$. Then we first represent each block of length $\ell_{i+1} - \ell_i$ using the the binary expansion: for each 2^j, where $1 < j \leq \log(\ell_k - \ell_{k-1})$ let $a_{2^j} \to a_{2^{j-1}} a_{2^{j-1}}$ and $a_1 \to a$. Then each $a_{\ell_i - \ell_{i-1}}$ can be represented as a concatenation of some of $a_1, a_2, \ldots, a_{2^j}$, based on the

Algorithm 1. TtoG: outline

```
1: while |T| > 1 do
2:     L ← list of letters in T
3:     for each a ∈ L do                              ▷ Blocks compression
4:         compress maximal blocks of a                      ▷ O(|T|)
5:     P ← list of pairs
6:     find partition of Σ into Σ_ℓ and Σ_r
7:     ▷ Try to maximize the occurrences from Σ_ℓΣ_r in T. O(|T|) time, see Lemma 2
8:     for ab ∈ P ∩ Σ_ℓΣ_r do                        ▷ These pairs do not overlap
9:         compress pair ab                                  ▷ Pair compression
10: return the constructed grammar
```

binary notation of $\ell_i - \ell_{i-1}$. Next, $a_{\ell_{i+1}}$ is represented as $a_{\ell_{i+1}} \to a_{\ell_{i+1}-\ell_i} a_{\ell_i}$. The representation cost is $\mathcal{O}(k + \max_{i+1}^{k} \log(\ell_i - \ell_{i-1}))$.

Since no two maximal blocks of the same letter can be next to each other, after the block compression there are no blocks of length greater than 1 in T.

Lemma 1. *In line 5 there are no two consecutive identical letters in T.*

Pair Compression. The pair compression is performed similarly as the block compression. However, since the pairs can overlap, compressing all pairs at the same time is not possible. Still, we can find a subset of non-overlapping pairs in T such that a constant fraction of T is covered by appearances of these pairs. This subset is defined by a *partition* of Σ into Σ_ℓ and Σ_r and choosing the pairs with the first letter in Σ_ℓ and the second in Σ_r.

Observe that if each element of Σ is randomly assigned to Σ_ℓ or Σ_r (with equal probabilities) then a fixed appearance of ab has a $1/4$ chance of being in $\Sigma_\ell\Sigma_r$. Hence the expected number of pairs in this partition is $(|T| - 1)/4$, and so there is a partition covering at least this amount of pairs in T. A simple derandomisation (using the expected-value approach) yields the following:

Lemma 2. *For T in $\mathcal{O}(|T|)$ time we can find in line 6 a partition of Σ into Σ_ℓ, Σ_r such that number of appearances of pairs $ab \in \Sigma_\ell\Sigma_r$ in T is at least $(|T| - 1)/4$ (or 1, if this less than 1). In the same running time we can provide, for each $ab \in P \cap \Sigma_\ell\Sigma_r$, a lists of pointers to appearances of ab in T.*

When for each pair $ab \in \Sigma_\ell\Sigma_r$ the list of its appearances in T is provided, the replacement of pairs is done going through and the list and replacing each of the pair, which is done in linear time. Note, that as Σ_ℓ, Σ_r are disjoint, the considered pairs cannot overlap. This ends the description of the construction.

Size and Running Time. From Lemma 2 it follows that

Lemma 3. *In each phase $|T|$ is reduced by a constant factor. In particular, TtoG runs in linear time.*

Note that there are known very fast implementations of algorithms creating an SLP based on pair compression [8].

3   Size of the Grammar: SLPs and Recompression

To bound cost of representing the letters introduced by TtoG we perform a mental experiment, in which we start with the smallest grammar G generating (the input) T and then modify it so that it generates T after each operation performed on G. Thus if we take together the rules of the current G and the representations introduced already by TtoG, we obtain a grammar generating the input text. Hence we can think of this process as gradually transforming G into the grammar returned by TtoG. The main advantage of this approach is that at each step of TtoG we can bound the cost of representation in terms of $|G|$.

We assume that G is a *Straight Line Programme*, however, we relax the notion a bit: i.e. its nonterminals are numbered $X_1, \ldots, X_m$ and each rule is of the form $X_i \to w \in \Sigma^*$ or $X_i \to uX_jv$ or $X_i \to uX_jvX_kw$, where $u, v, w \in \Sigma^*$ and $j, k < i$ (note that there is no bound on the lengths of u, v and w). Every CFG generating a unique string can be transformed into such a form, with the size increased only by a constant factor. We call the letters (strings) appearing in the productions the *explicit letters* (*strings*, respectively). The unique string derived by X_i is denoted by $\mathrm{val}(X_i)$; as already promised, the grammar G shall satisfy the condition $\mathrm{val}(X_m) = T$.

Idea of the Cost Analysis. With each explicit letter in rules of G we associate a unit *credit*. When a letter is removed from a rule, it *releases* its credit, on the other hand, when new letters appear in the rules for any reasons, their credit needs to be *paid*. Several operations on G increase the credit of G, we say that they *introduce* this additional credit. The main idea of the analysis is that when a pair compression (say ab to c) is performed, the released credit (at least 2) is enough to pay for the new letters (appearances of c) and there is a surplus, which can be associated with c, i.e. the representation cost of c is also paid by the credit. Similar approach works for blocks.

Since representation cost can be fully paid by credit, instead of trying to estimate it, we just calculate the total amount of introduced credit. This upper-bounds the cost of representation.

Pair Compression. A pair of letters ab has a *crossing appearance* in a non-terminal X_i (with a rule $X_i \to \alpha_i$) if ab is in $\mathrm{val}(X_i)$ but this appearance does not come from an explicit appearance of ab in α_i nor it is generated by any of the nonterminals in α_i. A pair is *non-crossing* if it has no crossing appearance. Unless explicitly written, we use this notion only to pairs of *different* letters.

By $PC_{ab\to c}(w)$ we denote the text obtained from w by replacing each ab by a letter c (we assume that $a \neq b$). We say that a procedure *properly implements the pair compression* of ab to c, if $\mathrm{val}(X'_m) = PC_{ab\to c}(\mathrm{val}(X_m))$. When a pair ab is noncrossing, implementing the pair compression is easy, as it is enough to replace each explicit ab with c, we refer to this procedure as $\mathsf{PairCompNCr}(ab, c)$.

In order to distinguish between the nonterminals, grammar, etc. before and after the application of compression of ab (or, in general, any procedure) we use 'primed' letters, i.e. X_i', G', T' for the nonterminals, grammar and text after this compression and 'unprimed', i.e. X_i, G T for the ones before.

Lemma 4. *If ab is a noncrossing pair, then* $\mathsf{PairCompNCr}(ab, c)$ *properly implements the compression of ab. The credit and cost of representing the new letter c is paid by the released credit. If any other pair $a'b'$ appearing in T was noncrossing in G, it is in G'.*

The proof follows by a simple case inspection.

It is left to assure that the pairs from $\Sigma_\ell \Sigma_r$ are all noncrossing. Intuitively, there are three types of 'bad' situations:

- there is a nonterminal X_i such that $\mathrm{val}(X_i)$ begins with b and aX_i appears in one of the rules;
- there is a nonterminal X_i such that $\mathrm{val}(X_i)$ ends with a and X_ib appears in one of the rules;
- there are nonterminals X_i, X_j such that $\mathrm{val}(X_i)$ ends with a and $\mathrm{val}(X_j)$ begins with b and X_iX_j appears in one of the rules.

Consider the first case, let $bw = \mathrm{val}(X_i)$. Then we modify the rule for X_i so that $\mathrm{val}(X_i) = w$ and replace each X_i in the rules by bX_i, we call this action the *left-popping b from X_i*. Similar operations can be done for other cases: when $\mathrm{val}(X_j)$ ends with a and X_jb appears in G then we right-pop this a; in the last case, when $\mathrm{val}(X_j)$ ends with b and $\mathrm{val}(X_i)$ begins with a, we left-pop this b and right-pop the a. Such operations can be performed for many letters in parallel: $\mathsf{Pop}(\Sigma_\ell, \Sigma_r)$ 'uncrosses' all pairs from the set $\Sigma_\ell \Sigma_r$, assuming that Σ_ℓ and Σ_r are disjoint subsets of Σ; in particular, it takes care of all three listed above cases.

Algorithm 2. $\mathsf{Pop}(\Sigma_\ell, \Sigma_r)$

1: **for** $i \leftarrow 1 .. m - 1$ **do**
2:     let the production for X_i be $X_i \to \alpha_i$
3:     **if** the first symbol of α_i is $b \in \Sigma_r$ **then**
4:         remove this b from α_i
5:         replace X_i in G's productions by bX_i
6:         **if** $\mathrm{val}(X_i) = \epsilon$ **then**
7:             remove X_i from G's productions
8:     Do the symmetric right-popping

Lemma 5. *After application of* $\mathsf{Pop}(\Sigma_\ell, \Sigma_r)$, *where $\Sigma_\ell \cap \Sigma_r = \emptyset$, each pair $ab \in \Sigma_\ell \Sigma_r$ is non-crossing. Furthermore, $\mathrm{val}(X_m') = \mathrm{val}(X_m)$. The credit of G increases by at most $\mathcal{O}(m)$.*

The first claim follows by a case inspection and is identical as in author's earlier work using the technique [7,5,6]. The second and third are obvious.

In order to compress pairs from $\Sigma_\ell \Sigma_r$ it is enough to first uncross them all and then compress them, we refer to this procedure as $\mathsf{PairComp}(\Sigma_\ell, \Sigma_r)$.

Lemma 6. PairComp *implements pair compression for each* $ab \in \Sigma_\ell \Sigma_r$. *The credit of G is increased by* $\mathcal{O}(m)$.

Blocks Compression. Similar notions and analysis are applied for blocks. Consider appearances of maximal a-blocks in T and their derivation by G. Then a block a^ℓ has a *crossing appearance* in X_i with a rule $X_i \to \alpha_i$, if it is contained in $\mathrm{val}(X_i)$ but this appearance is not generated by the explicit as in the rule nor in the substrings generated by the nonterminals in α_i. If as blocks have no crossing appearances, then a *has no crossing blocks*. As for noncrossing pairs, the compression of a blocks, when it has no crossing blocks, is easy: it is enough to replace every explicit appearance of maximal block a^ℓ with a_ℓ.

Algorithm 3. BlockCompNCr(a)

1: **for** each a^{ℓ_m} **do**
2:     replace each maximal a^{ℓ_m} in G by a_{ℓ_m}

Lemma 7. *Suppose that a has no crossing blocks. Then* BlockCompNCr(a) *properly compresses a's blocks.*

Furthermore, if a letter b from T had no crossing blocks in G, it does not have them in G'.

It is left to ensure that no letter has a crossing block. The solution is similar to Pop, this time though we need to remove the whole prefix and suffix from $\mathrm{val}(X_i)$ instead of a single letter: suppose that a has a crossing block because aX_i appears in the rule and $\mathrm{val}(X_i)$ begins with a. Left-popping a does not solve the problem, as it might be that $\mathrm{val}(X_i)$ still begins with a. Thus, we keep on left-popping until the first letter of $\mathrm{val}(X_i)$ is not a, i.e. we remove the a-prefix of $\mathrm{val}(X_i)$.

Lemma 8. *After* RemCrBlocks *no letter has a crossing block.*

The proof is fairly obvious: if a^ℓ was left-popped from X_i then the only letter to the left of X_i is a and $\mathrm{val}(X_i)$ does not start with a, similar argument applies to the ending letter.

Algorithm 4. RemCrBlocks: removing crossing blocks

1: **for** $i \leftarrow 1..m-1$ **do**
2:     let a, b be the first and last letter of $\mathrm{val}(X_i)$
3:     let ℓ_i, r_i be the length of the a-prefix and b-suffix of $\mathrm{val}(X_i)$
4:                           ▷ If $\mathrm{val}(X_i) \in a^*$ then $r_i = 0$ and $\ell_i = |\mathrm{val}(X_i)|$
5:     remove a^{ℓ_i} from the beginning and b^{r_i} from the end of the rule
6:     replace X_i by $a^{\ell_i} X_i b^{r_i}$ in the rules
7:     **if** $\mathrm{val}(X_i) = \epsilon$ **then** remove X_i from the rules

---

So the compression of all blocks of letters is done by first running RemCr-Blocks and then compressing each of the block. We do not compress blocks of letters that are introduced in this way. We call the resulting procedure Block-Comp. Concerning the number of credit, the arbitrary long blocks popped by RemCrBlocks are compressed (each into a single letter) and so only 4 credit per rule is introduced.

Lemma 9. BlockComp *properly compresses blocks of each letter a appearing in T before its application and introduces at most $\mathcal{O}(m)$ credit.*

Calculating the Cost of Representing Letters in Block Compression. While the credits were enough to pay the cost of representing letters introducing during the pair construction, this is not the case for block compression. The appropriate analysis is presented in this section. The overall plan is as follows: firstly, we define a scheme of representing the letters based on the grammar G and the way G is changed by BlockComp. For such a representation schema, we show that the cost of representation is $\mathcal{O}(g \log N)$. Lastly, it is proved that the actual cost of representing the letters by TtoG is smaller than the one based on G, hence it is also $\mathcal{O}(g \log N)$.

Representing blocks of letters using G. In most cases the new blocks are obtained by concatenating letters a that appear explicitly in the grammar and in such a case the credit can be used to pay for the creation of the new rule. This does not apply when the new block is obtained by concatenating two different blocks of a (popped from nonterminals) inside a rule. However, this cannot happen too often.

We create a new symbol for each a block that is either popped from a nonterminal or is in a rule at the end of the BlockComp. Such a block is a *power* if a^ℓ was popped from some X_i which was then removed or it is obtained by concatenation of two different powers of a^ℓ inside a rule (and perhaps some other explicit letters a). The second case appears only when in the rule $X_i \to u X_j v X_k w$ the popped suffix of X_j and popped prefix of X_k are blocks of the same letter, say a, and furthermore $v \in a^*$. For each block a^ℓ that is not a power we may uniquely identify another block a^k (which may be a power or ϵ) such that a^ℓ was obtained by concatenating explicit letters to a^k.

We represent the blocks as follows:

- for a block a^ℓ that is a power we express a_ℓ using the binary expansion: i.e. introduce a_{2^j} for $j \leq \log \ell$ representing a^{2^j} and then represent a_ℓ as a concatenation of appropriate a_{2^j}s, depending on the binary notation of ℓ. This costs $\mathcal{O}(1 + \log \ell)$;
- for a block a^ℓ that is obtained by concatenating $\ell - k$ explicit letters to a^k we express a_ℓ as $a_k a \ldots a$, the cost $\mathcal{O}(\ell - k)$ is covered by the credit released by the explicit letters.

Cost of G-based representation. We estimate the cost of representing the letters based on G. The idea is that each nonterminal X_i can represent a block of length at most $\log N$ and so it should be enough to spend that amount of cost on the representation.

Lemma 10. *The total cost of representing powers is $\mathcal{O}(g \log N)$.*

The general idea of the proof is that each nonterminal X_i can represent a block of length at most $\log N$ and so it should be enough to spend that amount of cost on the representation. The actual proof is technical, as we need to take into the account that block compression is performed several times on $\mathrm{val}(X_i)$; this proof appears only in the full version of this paper.

Comparing the cost of representation induced by G and the one of TtoG. We show that the cost of representing letters by actual strategy used by TtoG yields a cost at most as much as the one used by the strategy defined for G (note that the latter cost includes the credit released by explicit letters). Imagine all blocks represented by grammar-based schema as a directed weighted graph, the weights of edges correspond to the cost of representing a letter:

- when a^ℓ is a power, the node labelled with a_ℓ has an edge to ϵ with weight $1 + \log \ell$ (recall that this is the cost of representing this power);
- when a_ℓ is represented as a concatenation of $\ell - k$ letters to a_k, the node a_ℓ has an edge to a_k of weight $\ell - k$ (this is the cost of representing this block, note that it was paid by the credit on the $\ell - k$ explicit letters a).

Then the sum of the weight of such a graph is a cost of representing the blocks using the grammar schema (up to a constant factor). We transform this graph (not increasing the sum of weights) into a similar one that corresponds to the actual representation cost of the TtoG: recall that when representing $a_{\ell_1} \leq a_{\ell_2} \leq \cdots \leq a_{\ell_k}$ we represented $a_{\ell_{i+1}}$ by concatenating $a_{\ell_{i+1}-\ell_i}$ to a_{ℓ_i}, which had a cost $\mathcal{O}(1 + \log(\ell_{i+1} - \ell_i))$. So the representation can be depicted as a graph with edges from $a_{\ell_{i+1}}$ to a_{ℓ_i} with weight $1 + \log(\ell_{i+1} - \ell_i)$.

We sort the nodes according to the increasing length of the powers (we can remove duplicates, redirecting incoming edges to the other copy). For each node a_ℓ, we redirect its edge to its direct predecessor a_k and label it with a cost $1 + \log(\ell - k)$. This cannot increase the cost. All blocks represented in TtoG appear in T and so they were also represented by the G-based representation. On the other hand, some of the blocks represented by G perhaps were not represented by TtoG. For such a block a^ℓ we remove its node a_ℓ and redirect its unique incoming edge to its predecessor, the weight cannot increase in this way.

Improved Analysis. The naive algorithm, which simply represents the word as $X_1 \to w$ results in a grammar of size N. We merge the naive approach with the recompression-based algorithm: if at the beginning of a phase i TtoG already paid k for representation of the letters and the remaining text has size $|T|$ then a grammar for the input string of the total size $k + |T| + 1$ can be easily given and we choose the minimum over all possible i. We call the corresponding algorithm TtoGImp. Additionally, we show that when $|T| \approx g$ then the cost of representing letters is $\mathcal{O}(g \log(N/g))$ and so corresponding grammar considered by TtoGImp is of size $\mathcal{O}(g + g \log(N/g))$, and so, the grammar returned by TtoGImp is also of this size.

Algorithm 5. TtoGImp: improved version outline

1: $i \leftarrow 0$
2: **while** $|T| > 1$ **do**
3:     $size[i] \leftarrow |T| +$ so-far cost of representing letters ▷ Cost of grammar in phase i
4:     $i \leftarrow i + 1$ ▷ Number of the phase
5:     $L \leftarrow$ list of letters in T ▷ The compression is done as in TtoG
6:     **for** each $a \in L$ **do**
7:         compress maximal blocks of a
8:     $P \leftarrow$ list of pairs
9:     find partition of Σ into Σ_ℓ and Σ_r
10:     **for** $ab \in P \cap \Sigma_\ell \Sigma_r$ **do**
11:         compress pair ab
12: output grammar G_i, for which $size[i]$ is smallest

Lemma 11. *If at the beginning of the phase $|T| \geq g$ then so far the cost of representing letters by TtoGImp is $\mathcal{O}(g + g \log(N/g))$.*

Let t_1 and t_2 be the lengths of $|T|$ at the beginning of two consecutive phases, such that $t_1 > g \geq t_2$. By Lemma 11 the cost of representing letters before the $|T|$ was reduced to t_2 letters is $\mathcal{O}(g + g \log(N/g))$. Using similar techniques the cost of representation in this phase can be upper-bounded by the same value:

Lemma 12. *Consider a phase, such that at its beginning T has length t_1 and after it it has length t_2, where $t_1 > g \geq t_2$. Then the cost of representing letters introduced during this phase is at most $\mathcal{O}(g + g \log(N/g))$.*

From Lemmata 11 and 12 it follows that

Theorem 1. *The TtoGImp returns a grammar of size at most $\mathcal{O}(g + g \log(N/g))$, where g is the size of the optimal grammar for the input text.*

Acknowledgements. I would like to thank Paweł Gawrychowski for introducing me to the topic, for pointing out the relevant literature [12,11] and discussions; Markus Lohrey for suggesting the topic of this paper and bringing the idea of applying the recompression to the smallest grammar.

References

1. Charikar, M., Lehman, E., Liu, D., Panigrahy, R., Prabhakaran, M., Sahai, A., Shelat, A.: The smallest grammar problem. IEEE Transactions on Information Theory 51(7), 2554–2576 (2005)

2. Farach-Colton, M., Ferragina, P., Muthukrishnan, S.: On the sorting-complexity of suffix tree construction. J. ACM 47(6), 987–1011 (2000)
3. Gawrychowski, P.: Pattern matching in Lempel-Ziv compressed strings: Fast, simple, and deterministic. In: Demetrescu, C., Halldórsson, M.M. (eds.) ESA 2011. LNCS, vol. 6942, pp. 421–432. Springer, Heidelberg (2011)
4. Gąsieniec, L., Karpiński, M., Plandowski, W., Rytter, W.: Efficient algorithms for Lempel-Ziv encoding. In: Karlsson, R., Lingas, A. (eds.) SWAT 1996. LNCS, vol. 1097, pp. 392–403. Springer, Heidelberg (1996)
5. Jeż, A.: Faster fully compressed pattern matching by recompression. In: Czumaj, A., Mehlhorn, K., Pitts, A., Wattenhofer, R. (eds.) ICALP 2012, Part I. LNCS, vol. 7391, pp. 533–544. Springer, Heidelberg (2012)
6. Jeż, A.: The complexity of compressed membership problems for finite automata. Theory of Computing Systems, 1–34 (2013), http://dx.doi.org/10.1007/s00224-013-9443-6
7. Jeż, A.: Recompression: a simple and powerful technique for word equations. In: Portier, N., Wilke, T. (eds.) 30th International Symposium on Theoretical Aspects of Computer Science (STACS 2013). Leibniz International Proceedings in Informatics (LIPIcs), vol. 20, pp. 233–244. Schloss Dagstuhl–Leibniz-Zentrum fuer Informatik, Dagstuhl (2013), http://drops.dagstuhl.de/opus/volltexte/2013/3937
8. Kärkkäinen, J., Mikkola, P., Kempa, D.: Grammar precompression speeds up Burrows–Wheeler compression. In: Calderón-Benavides, L., González-Caro, C., Chávez, E., Ziviani, N. (eds.) SPIRE 2012. LNCS, vol. 7608, pp. 330–335. Springer, Heidelberg (2012)
9. Karpiński, M., Rytter, W., Shinohara, A.: Pattern-matching for strings with short descriptions. In: Galil, Z., Ukkonen, E. (eds.) CPM 1995. LNCS, vol. 937, pp. 205–214. Springer, Heidelberg (1995)
10. Larsson, N.J., Moffat, A.: Offline dictionary-based compression. In: Data Compression Conference, pp. 296–305. IEEE Computer Society (1999)
11. Lohrey, M., Mathissen, C.: Compressed membership in automata with compressed labels. In: Kulikov, A., Vereshchagin, N. (eds.) CSR 2011. LNCS, vol. 6651, pp. 275–288. Springer, Heidelberg (2011)
12. Mehlhorn, K., Sundar, R., Uhrig, C.: Maintaining dynamic sequences under equality tests in polylogarithmic time. Algorithmica 17(2), 183–198 (1997)
13. Plandowski, W.: Testing equivalence of morphisms on context-free languages. In: van Leeuwen, J. (ed.) ESA 1994. LNCS, vol. 855, pp. 460–470. Springer, Heidelberg (1994)
14. Rytter, W.: Application of Lempel-Ziv factorization to the approximation of grammar-based compression. Theor. Comput. Sci. 302(1-3), 211–222 (2003)
15. Sakamoto, H.: A fully linear-time approximation algorithm for grammar-based compression. J. Discrete Algorithms 3(2-4), 416–430 (2005)
16. Storer, J.A., Szymanski, T.G.: The macro model for data compression. In: Lipton, R.J., Burkhard, W.A., Savitch, W.J., Friedman, E.P., Aho, A.V. (eds.) STOC, pp. 30–39. ACM (1978)

Fast Algorithm for Partial Covers in Words

Tomasz Kociumaka[1], Solon P. Pissis[4,5,*], Jakub Radoszewski[1],
Wojciech Rytter[1,2,**], and Tomasz Waleń[3,1]

[1] Faculty of Mathematics, Informatics and Mechanics,
University of Warsaw, Warsaw, Poland
{kociumaka,jrad,rytter,walen}@mimuw.edu.pl
[2] Faculty of Mathematics and Computer Science,
Copernicus University, Toruń, Poland
[3] Laboratory of Bioinformatics and Protein Engineering,
International Institute of Molecular and Cell Biology in Warsaw, Poland
[4] Laboratory of Molecular Systematics and Evolutionary Genetics,
Florida Museum of Natural History, University of Florida, USA
[5] Scientific Computing Group (Exelixis Lab & HPC Infrastructure),
Heidelberg Institute for Theoretical Studies (HITS gGmbH), Germany
solon.pissis@h-its.org

Abstract. A factor u of a word w is a *cover* of w if every position in w lies within some occurrence of u in w. A word w covered by u thus generalizes the idea of a *repetition*, that is, a word composed of exact concatenations of u. In this article we introduce a new notion of *partial cover*, which can be viewed as a relaxed variant of cover, that is, a factor covering at least a given number of positions in w. Our main result is an $O(n \log n)$-time algorithm for computing the shortest partial covers of a word of length n.

1 Introduction

The notion of periodicity in words and its many variants have been well-studied in numerous fields like combinatorics on words, pattern matching, data compression, automata theory, formal language theory, and molecular biology. However the classic notion of periodicity is too restrictive to provide a description of a word such as abaababaaba, which is covered by copies of aba, yet not exactly periodic. To fill this gap, the idea of *quasiperiodicity* was introduced [1]. In a periodic word, the occurrences of the single periods do not overlap. In contrast, the occurrences of a quasiperiod in a quasiperiodic word may overlap. Quasiperiodicity thus enables the detection of repetitive structures that would be ignored by the classic characterization of periods.

The most well-known formalization of quasiperiodicity is the cover of word. A factor u of a word w is said to be a *cover* of w if $u \neq w$, and every position in w lies within some occurrence of u in w. Equivalently, we say that u *covers* w.

---

* Supported by the NSF–funded iPlant Collaborative (NSF grant #DBI-0735191).
** Supported by grant no. N206 566740 of the National Science Centre.

Note that a cover of w must also be a *border* — both prefix and suffix — of w. Thus, in the above example, aba is the shortest cover of abaababaaba.

A linear-time algorithm for computing the shortest cover of a word was proposed by Apostolico et al. [2], and a linear-time algorithm for computing all the covers of a word was proposed by Moore & Smyth [3]. Breslauer [4] gave an online linear-time algorithm computing the *minimal cover array* of a word — a data structure specifying the shortest cover of every prefix of the word. Li & Smyth [5] provided a linear-time algorithm for computing the *maximal cover array* of a word, and showed that, analogous to the border array [6], it actually determines the structure of *all* the covers of every prefix of the word.

Still it remains unlikely that an arbitrary word, even over the binary alphabet, has a cover; for example, abaaababaabaaaababaa is a word that not only has no cover, but whose every prefix also has no cover. In this article we provide a natural form of quasiperiodicity. We introduce the notion of *partial covers*, that is, factors covering at least a given number of positions in w. Recently, Flouri et al. [7] suggested a related notion of *enhanced covers* which are additionally required to be borders of the word.

Partial covers can be viewed as a relaxed variant of covers alternative to approximate covers [8]. The approximate covers require each position to lie within an approximate occurrence of the cover. This allows for small irregularities within each fragment of a word. On the other hand partial covers require exact occurrences but drop the condition that all positions need to be covered. This allows some fragments to be completely irregular as long as the total length of such fragments is small. The significant advantage of partial covers is that they enjoy a more combinatorial structure, and consequently the algorithms solving the most natural problems are much more efficient than those concerning approximate covers, where the time complexity rarely drops below quadratic and some problems are even NP-hard.

Let $Covered(v, w)$ denote the number of positions in w covered by occurrences of the word v in w; we call this value the *cover index* of v within w. For example, $Covered(\text{aba}, \text{aababab}) = 5$. We primarily concentrate on the following problem, but the tools we develop can be used to answer various questions concerning partial covers.

PARTIALCOVERS problem
  Input: a word w and a positive integer $\alpha \leq |w|$.
  Output: all shortest factors v such that $Covered(v, w) \geq \alpha$.

Example 1. Let $w = \text{bcccacccaccaccb}$ and $\alpha = 11$. Then the only shortest partial covers are ccac and cacc.

Our contribution. The following summarizes our main result.

Theorem 1. *The* PARTIALCOVERS *problem can be solved in* $O(n \log n)$ *time and* $O(n)$ *space, where* $n = |w|$.

We extensively use suffix trees, for an exposition see [6,9]. A suffix tree is a compact trie of suffixes, the nodes of the trie which become nodes of the suffix tree are called *explicit* nodes, while the other nodes are called *implicit*. Each edge of the suffix tree can be viewed as an *upward* maximal path of implicit nodes starting with an explicit node. Moreover, each node belongs to a unique path of that kind. Then, each node of the trie can be represented in the suffix tree by the edge it belongs to and an index within the corresponding path. Such a representation of the unique node in the trie corresponding to a factor is called the *locus* of that factor. Our algorithm finds the loci of the shortest partial covers.

Informal Structure of the Algorithm. The algorithm first augments the suffix tree of w, and a linear number of implicit extra nodes become explicit. Then, for each node of the augmented tree, two integer values are computed. They allow for determining the size of the covered area for each implicit node by a simple formula, since limited to a single edge of the augmented suffix tree, these values form an arithmetic progression.

2   Augmented and Annotated Suffix Trees

Let w be a word of length n over a totally ordered alphabet Σ. Then the suffix tree T of w can be constructed in $O(n \log |\Sigma|)$ time [10,11]. For an explicit or implicit node v of T, we denote by $\hat{v}$ the word obtained by spelling the characters on a path from the root to v. We also denote $|v| = |\hat{v}|$. The leaves of T play an auxiliary role and do not correspond to factors, instead they are labeled with the starting positions of the suffixes.

We define the *Cover Suffix Tree* of w, denoted by $CST(w)$, as an *augmented* — new nodes are added — suffix tree in which the nodes are *annotated* with information relevant to covers. $CST(w)$ is similar to the data structure named $MAST$ (see [12,13]).

For a set X of integers and $x \in X$, we define

$$next_X(x) = \min\{y \in X, y > x\},$$

and we assume $next_X(x) = \infty$ if $x = \max X$. By $Occ(v)$ we denote the set of starting positions of occurrences of $\hat{v}$ in w. For any $i \in Occ(v)$, we define:

$$\delta(i, v) = next_{Occ(v)}(i) - i.$$

Note that $\delta(i, v) = \infty$ if i is the last occurrence of $\hat{v}$. Additionally, we define:

$$cv(v) = Covered(\hat{v}, w), \quad \Delta(v) = \big| \{i \in Occ(v) : \delta(i, v) \geq |v|\} \big|;$$

see, for example, Fig. 1.

In $CST(w)$, we introduce additional explicit nodes called *extra nodes*, which correspond to halves of square factors in w, i.e. we make v explicit if $\hat{v}\hat{v}$ is a factor of w. Moreover we annotate all explicit nodes (including extra nodes) with the values cv, Δ; see, for example, Fig. 2. The number of extra nodes is linear [14], so $CST(w)$ takes $O(n)$ space.

bcccacccaccaccb

1 2 3 4 5 6 7 8 9 10 11 12 13 14 15

Fig. 1. Let $w = $ bcccacccaccaccb and let v be the node corresponding to cacc. We have $Occ(v) = \{4, 8, 11\}$, $cv(v) = 11$, $\Delta(v) = 2$.

Lemma 1. *Let $v_1, v_2, \ldots, v_k$ be the consecutive implicit nodes on the edge from an explicit node v of $CST(w)$ to its explicit parent. Then*

$$(cv(v_1),\ cv(v_2),\ cv(v_3),\ \ldots,\ cv(v_k)) =$$
$$(cv(v) - \Delta(v),\ cv(v) - 2\Delta(v),\ cv(v) - 3\Delta(v),\ \ldots, c(v) - k \cdot \Delta(v)).$$

Proof. Consider any v_i, $1 \le i \le k$. Note that $Occ(v_i) = Occ(v)$, since otherwise v_i would be an explicit node of $CST(w)$. Also note that if any two occurrences of $\hat{v}$ in w overlap, then the corresponding occurrences of $\hat{v}_i$ overlap. Otherwise the path from v to v_i (excluding v) would contain an extra node. Hence, when we go up from v (before reaching its parent) the size of the covered area decreases at each step by $\Delta(v)$.   □

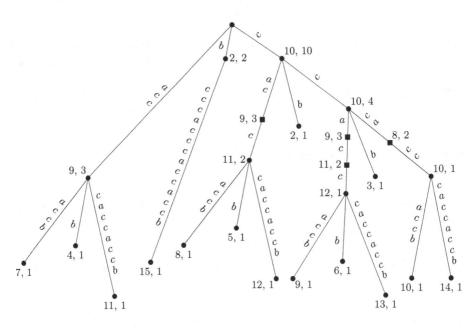

Fig. 2. $CST(w)$ for $w = $ bcccacccaccaccb. It contains four extra nodes that are denoted by squares in the figure. Each node is annotated with $cv(v), \Delta(v)$. Leaves are omitted for clarity.

Example 2. Consider the word w from Fig. 2. The word cccacc corresponds to an explicit node of $CST(w)$; we denote it by v. We have $cv(v) = 10$ and $\Delta(v) = 1$ since the two occurrences of the factor cccacc in w overlap. The word cccac corresponds to an implicit node v' and $cv(v') = 10 - 1 = 9$. Now the word ccca corresponds to an extra node v'' of $CST(w)$. Its occurrences are adjacent in w and $cv(v'') = 8$, $\Delta(v'') = 2$. The word ccc corresponds to an implicit node v''' and $cv(v''') = 8 - 2 = 6$.

As a consequence of Lemma 1 we obtain the following result.

Lemma 2. *Assume we are given $CST(w)$. Then we can compute:*

(1) for any α, the loci of the shortest partial covers in linear time;
(2) given the locus of a factor u in the suffix tree $CST(w)$, the cover index $Covered(u, w)$ in $O(1)$ time.

Proof. Part (2) is a direct consequence of Lemma 1. As for part (1), for each edge of $CST(w)$, leading from v to its parent v', we need to find minimum $|v| \geq j > |v'|$ for which $cv(v) - \Delta(v) \cdot (|v| - j) \geq \alpha$. Such a linear inequality can be solved in constant time.   □

Due to this fact the efficiency of the PARTIALCOVERS problem (Theorem 1) relies on the complexity of $CST(w)$ construction.

3   Extension of Disjoint-Set Data Structure

In this section we extend the classic disjoint-set data structure to compute the *change lists* of the sets being merged, as defined below. First let us extend the *next* notation. For a partition $\mathcal{P} = \{P_1, \ldots, P_k\}$ of $U = \{1, \ldots, n\}$, we define

$$next_{\mathcal{P}}(x) = next_{P_i}(x) \text{ where } x \in P_i.$$

Now for two partitions $\mathcal{P}, \mathcal{P}'$ let us define the *change list* (see also Fig. 3) by

$$ChangeList(\mathcal{P}, \mathcal{P}') = \{(x, next_{\mathcal{P}'}(x)) : next_{\mathcal{P}}(x) \neq next_{\mathcal{P}'}(x)\}.$$

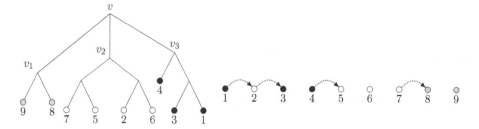

Fig. 3. Let $\mathcal{P}$ be the partition of $\{1, \ldots, 9\}$ whose classes consist of leaves in the subtrees rooted at children of v, $\mathcal{P} = \{\{1, 3, 4\}, \{2, 5, 6, 7\}, \{8, 9\}\}$, and let $\mathcal{P}' = \{\{1, \ldots, 9\}\}$. Then $ChangeList(\mathcal{P}, \mathcal{P}') = \{(1, 2), (2, 3), (4, 5), (7, 8)\}$ (depicted by dotted arrows).

We say that $(\mathcal{P}, id)$ is a partition of U *labeled* by L if $\mathcal{P}$ is a partition of U and $id : \mathcal{P} \to L$ is a one-to-one (injective) mapping. A label $\ell \in L$ is called *valid* if $id(P) = \ell$ for some $P \in \mathcal{P}$ and *free* otherwise.

Lemma 3. *Let $n \le k$ be positive integers such that k is of magnitude $\Theta(n)$. There exists a data structure of size $O(n)$, which maintains a partition $(\mathcal{P}, id)$ of $\{1, \dots, n\}$ labeled by $L = \{1, \dots, k\}$. Initially $\mathcal{P}$ is a partition into singletons with $id(\{x\}) = x$. The data structure supports the following operations:*

- *$Find(x)$ for $x \in \{1, \dots, n\}$ gives the label of $P \in \mathcal{P}$ containing x.*
- *$Union(I, \ell)$ for a set I of valid labels and a free label ℓ replaces all $P \in \mathcal{P}$ with labels in I by their set-theoretic union with the label ℓ. The change list of the corresponding modification of $\mathcal{P}$ is returned.*

Any valid sequence of Union operations is performed in $O(n \log n)$ time and $O(n)$ space in total. A single Find operation takes $O(1)$ time.

Proof. Note that these are actually standard disjoint-set data structure operations except for the fact that we require *Union* to return the change list.

We use an approach similar to Brodal and Pedersen [15] (who use the results of [16]) originally devised for computation of maximal quasiperiodicities.

Theorem 3 of [15] states that a subset X of a linearly ordered universe can be stored in a height-balanced tree of linear size supporting the following operations:

$X.MultiInsert(Y)$: insert all elements of Y to X,
$X.MultiPred(Y)$: return all (y, x) for $y \in Y$ and $x = \max\{z \in X, z < y\}$,
$X.MultiSucc(Y)$: return all (y, x) for $y \in Y$ and $x = \min\{z \in X, z > y\}$,

in $O\left(|Y| \max\left(1, \log \frac{|X|}{|Y|}\right)\right)$ time.

In the data structure we store each $P \in \mathcal{P}$ as a height-balanced tree. Additionally, we store several auxiliary arrays, whose semantics follows. For each $x \in \{1, \dots, n\}$ we maintain a value $next[x] = next_{\mathcal{P}}(x)$ and a pointer $tree[x]$ to the tree representing P such that $x \in P$. For each $P \in \mathcal{P}$ (technically for each tree representing $P \in \mathcal{P}$) we store $id[P]$ and for each $\ell \in L$ we store $id^{-1}[\ell]$, a pointer to the corresponding tree (null for free labels).

Answering *Find* is trivial as it suffices to follow the *tree* pointer and return the id value. The *Union* operation is perfomed according to the pseudocode given below (for brevity we write P_i instead of $id^{-1}[i]$).

Claim. The *Union* operation correctly computes the change list and updates the data structure.

Proof. If (a, b) is in the change list, then a and b come from different sets P_i, in particular at least one of them does not come from P_{i_0}. Depending on which one it is, the pair (a, b) is found by *MultiPred* or *MultiSucc* operation. On the other hand, while computing C, the table *next* is not updated yet (i.e. corresponds to the state before *Union* operation) while S is already updated. Consequently the pairs inserted to C indeed belong to the change list. Once C is proved to be the change list, it is clear that *next* is updated correctly. For the other components of the data structure, correctness of updates is evident.   □

Function $Union(I, \ell)$

   $i_0 := \mathrm{argmax}\{|P_i| : i \in I\}; \ S := P_{i_0};$

   foreach $i \in I \setminus \{i_0\}$ **do**

      foreach $x \in P_i$ **do** $tree[x] := S;$

      $S.MultiInsert(P_i);$

   $C := \emptyset;$

   foreach $i \in I \setminus \{i_0\}$ **do**

      foreach $(b, a) \in S.MultiPred(P_i)$ **do**

         if $next[a] \neq b$ **then** $C := C \cup \{(a, b)\};$

      foreach $(a, b) \in S.MultiSucc(P_i)$ **do**

         if $next[a] \neq b$ **then** $C := C \cup \{(a, b)\};$

      $id^{-1}[i] := \mathrm{null};$

   $id[S] := \ell; \ id^{-1}[\ell] := S;$

   foreach $(x, y) \in C$ **do** $next[x] := y;$

   return $C;$

Claim. Any sequence of *Union* operations takes $O(n \log n)$ time in total.

Proof. Let us introduce a potential function $\Phi(\mathcal{P}) = \sum_{P \in \mathcal{P}} |P| \log |P|$. We shall prove that the running time of a single *Union* operation is proportional to the increase in potential. Clearly

$$0 \leq \Phi(\mathcal{P}) = \sum_{P \in \mathcal{P}} |P| \log |P| \leq \sum_{P \in \mathcal{P}} |P| \log n = n \log n,$$

so this suffices to obtain a desired $O(n \log n)$ bound.

Let us consider a *Union* operation that merges partition classes of sizes $p_1 \geq p_2 \geq \ldots \geq p_k$ to a single class of size $p = \sum_{i=1}^{k} p_i$. The most time-consuming steps of the algorithm are the operations on height-balanced trees, which, for single i, run in $O\left(\max\left(p_i, p_i \log \frac{p}{p_i}\right)\right)$ time. These operations are not performed for the largest set and for the remaining ones we have $p_i < \frac{1}{2}p$ (i.e. $\log \frac{p}{p_i} \geq 1$). This lets us bound the time complexity of the *Union* operations as follows:

$$\sum_{i=2}^{k} \max\left(p_i, p_i \log \frac{p}{p_i}\right) = \sum_{i=2}^{k} p_i \log \frac{p}{p_i} \leq \sum_{i=1}^{k} p_i \log \frac{p}{p_i} =$$

$$\sum_{i=1}^{k} p_i (\log p - \log p_i) = p \log p - \sum_{i=1}^{k} p_i \log p_i,$$

which is equal to the potential growth. □

This concludes the proof of Lemma 3. □

4 $O(n \log n)$-time Construction of $CST(w)$

The suffix tree of w augmented with extra nodes is called the *skeleton* of $CST(w)$, which we denote by $sCST(w)$. The following lemma follows from the fact that all square factors can be computed in linear time [17,18], and the nodes corresponding to them (a linear number) can be inserted into the suffix tree easily in $O(n \log n)$ time.

Lemma 4. $sCST(w)$ *can be constructed in $O(n \log n)$ time.*

We introduce auxiliary notions related to covered area of nodes:

$$cv_h(v) = \sum_{\substack{i \in Occ(v) \\ \delta(i,v) < h}} \delta(i,v), \quad \Delta_h(v) = |\{i \in Occ(v) \; : \; h \leq \delta(i,v)\}|.$$

Observation 1. $cv(v) = cv_{|v|}(v) + \Delta_{|v|}(v) \cdot |v|, \; \Delta(v) = \Delta_{|v|}(v).$

In the course of the algorithm some nodes will have their values c, Δ already computed; we call them *processed nodes*. Whenever v will be processed, so will its descendants.

The algorithm processes inner nodes v of $sCST(w)$ in the order of non-increasing height $|v|$. We maintain the partition $\mathcal{P}$ of $\{1, \ldots, n\}$ given by sets of leaves of subtrees rooted at *peak nodes*. Initially the peak nodes are the leaves of $sCST(w)$. Each time we process v all its children are peak nodes. Consequently, after processing v they are no longer peak nodes and v becomes a new peak node; see, for example, Fig. 4. The sets in the partition are labeled with identifiers of the corresponding peak nodes. Recall that leaves are labeled with the starting positions of the corresponding suffixes. We allow any labeling of the remaining nodes as long as each node of $sCST(w)$ has a distinct label of magnitude $O(n)$. We maintain the following technical invariant.

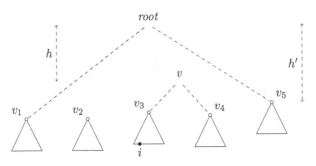

Fig. 4. One stage of the algorithm, where the peak nodes are $v_1, \ldots, v_5$ while the currently processed node is v. If $i \in List[d]$ and $v_3 = Find(i)$, then $d = \delta(i, v_3) = Dist[i]$. The current partition is $\mathcal{P} = \{Leaves(v_1), Leaves(v_2), Leaves(v_3), Leaves(v_4), Leaves(v_5)\}$. After v is processed, the partition changes to $\mathcal{P} = \{Leaves(v_1), Leaves(v_2), Leaves(v), Leaves(v_5)\}$. The *Union* operation merges $Leaves(v_4), Leaves(v_3)$ and returns the corresponding change list.

Invariant(h)

(A) For each peak node z we store:

$$cv'[z] = cv_h(z), \ \Delta'[z] = \Delta_h(z).$$

(B) For each $i \in \{1, \ldots, n\}$ we store $Dist[i] = \delta(i, Find(i))$.
(C) For each $d < h$ we store $List[d] = \{i : Dist[i] = d\}$.

Algorithm COMPUTECST (w)

   $T := sCST(w);$
   $\mathcal{P} :=$ partition of $\{1, \ldots, n\}$ into singletons;
   foreach $v :$ a leaf of T **do** $cv'[v] := 0;\ \Delta'[v] := 1;$
   $h := n + 1;$
   foreach $v :$ an inner node of T, in non-increasing order of $|v|$ **do**
      $Lift(h, |v|);\ h := |v|;$
      {Now part (A) of Invariant(h) is satisfied}
      $cv'[v] := \sum_{u \in children(v)} cv'[u];$
      $\Delta'[v] := \sum_{u \in children(v)} \Delta'[u];$
      $ChangeList(v) := Union(children(v), v);$
      foreach $(p, q) \in ChangeList(v)$ **do** $LocalCorrect(p, q, v);$
      $cv[v] := cv'[v] + \Delta'[v] \cdot |v|;\ \Delta[v] := \Delta'[v];$
   return T together with values of $cv, \Delta;$

In the algorithm, h is the smallest height (the smallest value of $|z|$) among the current set of peak nodes z; the height is not defined for leaves, so we start with $h = n + 1$.

Description of the $Lift(h_{old}, h_{new})$ **Operation.** The procedure $Lift$ is of auxiliary nature but plays an important preparatory role in processing the current node. According to part (A) of our invariant, for all peak nodes z we know the values: $cv'[z] = cv_{h_{old}}(z),\ \Delta'[z] = \Delta_{h_{old}}(z)$. Now we have to change h_{old} to h_{new} and guarantee validity of the invariant: $cv'[z] = cv_{h_{new}}(z),\ \Delta'[z] = \Delta_{h_{new}}(z)$. This is exactly what the following operation does.

Function $Lift(h_{old}, h_{new})$

   for $h := h_{old} - 1$ **downto** h_{new} **do**
      foreach i in $List[h]$ **do**
         $v := Find(i);$
         $\Delta'[v] := \Delta'[v] + 1;\ cv'[v] := cv'[v] - h;$

Description of the *LocalCorrect*(p, q, v) **Operation.** Here we assume that $\hat{v}$ occurs at positions $p < q$ and that these are consecutive occurrences. Moreover, we assume that these occurrences are followed by distinct characters, i.e. $(p, q) \in$ *ChangeList*(v). The *LocalCorrect* procedure updates $Dist[p]$ to make part (B) of the invariant hold for p again. The data structure *List* is updated accordingly so that (C) remains satisfied.

Function *LocalCorrect*(p, q, v)

    $d := q - p; \; d' := Dist[p];$

    if $d' < |v|$ **then** $cv'[v] := cv'[v] - d'$ **else** $\Delta'[v] := \Delta'[v] - 1;$

    if $d < |v|$ **then** $cv'[v] := cv'[v] + d$ **else** $\Delta'[v] := \Delta'[v] + 1;$

    $Dist[p] := d;$

    $remove(i, List[d']); \; insert(i, List[d]);$

Complexity of the Algorithm. In the course of the algorithm we compute *ChangeList*(v) for each $v \in T$. Due to Lemma 3 we have:

$$\sum_{v \in T} |ChangeList(v)| \; = \; O(n \log n).$$

Consequently we perform $O(n \log n)$ operations *LocalCorrect*. In each of them at most one element is added to a list $List[d]$ for some d. Hence the total number of insertions to these lists is also $O(n \log n)$.

    The cost of each operation *Lift* is proportional to the total size of lists $List[h]$ processed in this operation. As for each h the list $List[h]$ is processed once and the total number of insertions into lists is $O(n \log n)$, the total cost of all operations *Lift* is also $O(n \log n)$. This proves the following fact which, together with Lemma 3, implies our main result (Theorem 1).

Lemma 5. *Algorithm* COMPUTECST *computes* $CST(w)$ *in* $O(n \log n)$ *time and* $O(n)$ *space, where* $n = |w|$.

5 Final Remarks

We have presented an algorithm which constructs a data structure, called the *Cover Suffix Tree*, in $O(n \log n)$ time and $O(n)$ space. In the algorithm, to simplify its presentation, we used all halves of square factors as extra nodes. However, it suffices to consider primitive square halves only and all such nodes can be shown to be necessary for Lemma 1 to hold. As such, they can be introduced on the fly (in the *Lift* operation) without using the algorithms of [17,18].

    The Cover Suffix Tree has been developed in order to solve the PARTIALCOVERS problem, but it gives a well-structured description of the cover indices of all factors. Consequently, various queries related to partial covers can be answered efficiently. For example, with the Cover Suffix Tree one can solve in linear time a problem symmetric to PARTIALCOVERS: given constraints on factors of w (e.g. on their length), find a factor that maximizes the number of positions covered.

An interesting open problem is to reduce the construction time to $O(n)$. This could be difficult, though, since this would yield alternative linear-time algorithms finding primitively rooted squares and computing seeds (for a definition see [19]); and the only known linear-time algorithms for these problems are rather complex.

References

1. Apostolico, A., Ehrenfeucht, A.: Efficient detection of quasiperiodicities in strings. Theor. Comput. Sci. 119(2), 247–265 (1993)
2. Apostolico, A., Farach, M., Iliopoulos, C.S.: Optimal superprimitivity testing for strings. Inf. Process. Lett. 39(1), 17–20 (1991)
3. Moore, D., Smyth, W.F.: An optimal algorithm to compute all the covers of a string. Inf. Process. Lett. 50(5), 239–246 (1994)
4. Breslauer, D.: An on-line string superprimitivity test. Inf. Process. Lett. 44(6), 345–347 (1992)
5. Li, Y., Smyth, W.F.: Computing the cover array in linear time. Algorithmica 32(1), 95–106 (2002)
6. Crochemore, M., Hancart, C., Lecroq, T.: Algorithms on Strings. Cambridge University Press (2007)
7. Flouri, T., Iliopoulos, C.S., Kociumaka, T., Pissis, S.P., Puglisi, S.J., Smyth, W.F., Tyczyński, W.: New and efficient approaches to the quasiperiodic characterisation of a string. In: Holub, J., Žďárek, J. (eds.) PSC, pp. 75–88. Czech Technical University in Prague, Czech Republic (2012)
8. Sim, J.S., Park, K., Kim, S., Lee, J.: Finding approximate covers of strings. Journal of Korea Information Science Society 29(1), 16–21 (2002)
9. Crochemore, M., Rytter, W.: Jewels of Stringology. World Scientific (2003)
10. Farach, M.: Optimal suffix tree construction with large alphabets. In: FOCS, pp. 137–143 (1997)
11. Ukkonen, E.: On-line construction of suffix trees. Algorithmica 14(3), 249–260 (1995)
12. Apostolico, A., Preparata, F.P.: Data structures and algorithms for the string statistics problem. Algorithmica 15(5), 481–494 (1996)
13. Brodal, G.S., Lyngsø, R.B., Östlin, A., Pedersen, C.N.S.: Solving the string statistics problem in time $\mathcal{O}(n \log n)$. In: Widmayer, P., Triguero, F., Morales, R., Hennessy, M., Eidenbenz, S., Conejo, R. (eds.) ICALP 2002. LNCS, vol. 2380, pp. 728–739. Springer, Heidelberg (2002)
14. Fraenkel, A.S., Simpson, J.: How many squares can a string contain? J. Comb. Theory, Ser. A 82(1), 112–120 (1998)
15. Brodal, G.S., Pedersen, C.N.S.: Finding maximal quasiperiodicities in strings. In: Giancarlo, R., Sankoff, D. (eds.) CPM 2000. LNCS, vol. 1848, pp. 397–411. Springer, Heidelberg (2000)
16. Brown, M.R., Tarjan, R.E.: A fast merging algorithm. J. ACM 26(2), 211–226 (1979)

17. Gusfield, D., Stoye, J.: Linear time algorithms for finding and representing all the tandem repeats in a string. J. Comput. Syst. Sci. 69(4), 525–546 (2004)
18. Crochemore, M., Iliopoulos, C.S., Kubica, M., Radoszewski, J., Rytter, W., Waleń, T.: Extracting powers and periods in a string from its runs structure. In: Chavez, E., Lonardi, S. (eds.) SPIRE 2010. LNCS, vol. 6393, pp. 258–269. Springer, Heidelberg (2010)
19. Kociumaka, T., Kubica, M., Radoszewski, J., Rytter, W., Waleń, T.: A linear time algorithm for seeds computation. In: Rabani, Y. (ed.) SODA, pp. 1095–1112. SIAM (2012)

Linear Time Lempel-Ziv Factorization: Simple, Fast, Small*

Juha Kärkkäinen, Dominik Kempa, and Simon J. Puglisi

Department of Computer Science,
University of Helsinki
Helsinki, Finland
{firstname.lastname}@cs.helsinki.fi

Abstract. Computing the LZ factorization (or LZ77 parsing) of a string is a computational bottleneck in many diverse applications, including data compression, text indexing, and pattern discovery. We describe new linear time LZ factorization algorithms, some of which require only $2n \log n + O(\log n)$ bits of working space to factorize a string of length n. These are the most space efficient linear time algorithms to date, using $n \log n$ bits less space than any previous linear time algorithm. The algorithms are also simple to implement, very fast in practice, and amenable to streaming implementation.

1 Introduction

In the 35 years since its discovery the LZ77 factorization of a string — named after its authors Abraham Lempel and Jacob Ziv, and the year 1977 in which it was published — has been applied all over computer science. The first uses of LZ77 were in data compression, and to this day it lies at the heart of efficient and widely used file compressors, like `gzip` and `7zip`. LZ77 is also important as a *measure* of compressibility. For example, its size is a lower bound on the size of the smallest context-free grammar that represents a string [2]. Our particular motivation is the construction of compressed full-text indexes [15], several recent and powerful instances of which are based on LZ77 [8,7,14]. In all these applications (and in most of the many others we have not listed) computation of the factorization is a time- and space-bottleneck in practice.

Related work. There exists a variety of worstcase linear time algorithms to compute the LZ factorization [1,9,16]. All of them require at least $3n \log n$ bits of working space[1] in the worstcase. The most space efficient linear time algorithm is due to Chen et al. [3]. By overwriting the suffix array it achieves a working space of $(2n + s) \log n$ bits, where s is the maximal size of the stack used in the algorithm. However, in the worstcase $s = \Theta(n)$. Another space efficient solution

---

* This research is partially supported by Academy of Finland grants 118653 (ALGO-DAN) and 250345 (CoECGR).
[1] Working space excludes input string, output factorization, and $O(\log n)$ terms.

requiring $(2n+\sqrt{n})\log n$ bits of space in the worstcase is from [6] but it computes only the lengths of LZ77 factorization phrases. It can be extended to compute the full parsing at the cost of extra $n \log n$ bits.

All of these algorithms rely on the suffix array, which can be constructed in $O(n)$ time and using $(1 + \epsilon)n \log n$ bits of space (in addition to the input string but including the output of size $n \log n$ bits) [11]. This raises the question of whether the space complexity of linear time LZ77 factorization can be reduced from $3n \log n$ bits. In this paper, we answer the question in the affirmative by describing a linear time algorithm using $2n \log n$ bits.

In terms of practical performance, the fastest linear time LZ factorization algorithms are the very recent ones by Goto and Bannai [9], all using at least $3n \log n$ bits of working space. Other candidates for the fastest algorithms are described by Kempa and Puglisi [13]. Due to nearly simultaneous publication, no comparison between them exists so far. Experiments in this paper put the algorithms of Kempa and Puglisi slightly ahead. Their algorithms are also very space efficient; one of them uses $2n \log n + n$ bits of working space and others even less. However, their worstcase time complexity is $\Theta(n \log \sigma)$ for an alphabet of size σ. More details about these algorithms are given in Section 2.

Our contribution. We describe two linear time algorithms for LZ factorization. The first algorithm uses $3n \log n$ bits of working space and can be seen as a reorganization of an algorithm by Goto and Bannai [9]. However, this reorganization makes it smaller and faster. In our experiments, this is the fastest of all algorithms when the input is not highly repetitive.

The second algorithm reduces the working space to $2n \log n$ bits, which is at least $n \log n$ bits less than any previous linear time algorithm uses in the worstcase. The space reduction does not come at a great cost in performance. The algorithm is the fastest on some inputs and never far behind the fastest. It relies on novel combinatorial observations that might be of independent interest.

Both algorithms share several nice features. They are simple and easy to implement; they are alphabet-independent, using only character comparisons to access the input; and they make just one sequential pass over the suffix array, enabling streaming from disk. Our experiments show that streaming not only reduces the working space by a further $n \log n$ bits, but also speeds up the computation when the time for reading inputs from disk is taken into account.

2   Preliminaries

Strings. Throughout we consider a string $X = X[1..n] = X[1]X[2]\ldots X[n]$ of $|X| = n$ symbols drawn from an ordered alphabet of size σ.

For $i = 1, \ldots, n$ we write $X[i..n]$ to denote the *suffix* of X of length $n - i + 1$, that is $X[i..n] = X[i]X[i + 1]\ldots X[n]$. We will often refer to suffix $X[i..n]$ simply as "suffix i". Similarly, we write $X[1..i]$ to denote the *prefix* of X of length i. We write $X[i..j]$ to represent the *substring* $X[i]X[i + 1]\ldots X[j]$ of X that starts at position i and ends at position j. Let $lcp(i, j)$ denote the length of the longest-common-prefix of suffix i and suffix j. For example, in the string $X = zzzzzipzip$,

$\mathsf{lcp}(2,5) = 1 = |z|$, and $\mathsf{lcp}(5,8) = 3 = |zip|$. For technical reasons we define $\mathsf{lcp}(i,0) = \mathsf{lcp}(0,i) = 0$ for all i.

Suffix Arrays. The suffix array SA is an array $\mathsf{SA}[1..n]$ containing a permutation of the integers $1..n$ such that $\mathsf{X}[\mathsf{SA}[1]..n] < \mathsf{X}[\mathsf{SA}[2]..n] < \cdots < \mathsf{X}[\mathsf{SA}[n]..n]$. In other words, $\mathsf{SA}[j] = i$ iff $\mathsf{X}[i..n]$ is the jth suffix of X in ascending lexicographical order. The inverse suffix array ISA is the inverse permutation of SA, that is $\mathsf{ISA}[i] = j$ iff $\mathsf{SA}[j] = i$. Conceptually, $\mathsf{ISA}[i]$ tells us the position of suffix i in SA.

The array $\Phi[0..n]$ (see [12]) is defined by $\Phi[i] = \mathsf{SA}[\mathsf{ISA}[i]-1]$, that is, the suffix $\Phi[i]$ is the immediate lexicographical predecessor of the suffix i. For completeness and for technical reasons we define $\Phi[\mathsf{SA}[1]] = 0$ and $\Phi[0] = \mathsf{SA}[n]$ so that Φ forms a permutation with one cycle.

LZ77. The LZ77 factorization uses the notion of a *longest previous factor* (LPF). The LPF at position i in X is a pair (p_i, ℓ_i) such that, $p_i < i$, $\mathsf{X}[p_i..p_i + \ell_i - 1] = \mathsf{X}[i..i + \ell_i - 1]$ and $\ell_i > 0$ is maximized. In other words, $\mathsf{X}[i..i + \ell_i - 1]$ is the longest prefix of $\mathsf{X}[i..n]$ which also occurs at some position $p_i < i$ in X. If $\mathsf{X}[i]$ is the leftmost occurrence of a symbol in X then such a pair does not exist. In this case we define $p_i = \mathsf{X}[i]$ and $\ell_i = 0$. Note that there may be more than one potential p_i, and we do not care which one is used.

The LZ77 factorization (or LZ77 parsing) of a string X is then just a greedy, left-to-right parsing of X into longest previous factors. More precisely, if the jth LZ factor (or *phrase*) in the parsing is to start at position i, then we output (p_i, ℓ_i) (to represent the jth phrase), and then the $(j + 1)$th phrase starts at position $i + \ell_i$, unless $\ell_i = 0$, in which case the next phrase starts at position $i + 1$. We call a factor (p_i, ℓ_i) *normal* if it satisfies $l_i > 0$ and *special* otherwise. The number of phrases in the factorization is denoted by z.

For the example string $\mathsf{X} = zzzzzipzip$, the LZ77 factorization produces:

$$(z,0), (1,4), (i,0), (p,0), (5,3).$$

The second and fifth factors are normal, and the other three are special.

NSV/PSV. The LPF pairs can be computed using *next and previous smaller values* (NSV/PSV) defined as

$$\mathsf{NSV}_{\mathsf{lex}}[i] = \min\{j \in [i+1..n] \mid \mathsf{SA}[j] < \mathsf{SA}[i]\}$$
$$\mathsf{PSV}_{\mathsf{lex}}[i] = \max\{j \in [1..i-1] \mid \mathsf{SA}[j] < \mathsf{SA}[i]\}.$$

If the set on the right hand side is empty, we set the value to 0. Further define

$$\mathsf{NSV}_{\mathsf{text}}[i] = \mathsf{SA}[\mathsf{NSV}_{\mathsf{lex}}[\mathsf{ISA}[i]]] \tag{1}$$
$$\mathsf{PSV}_{\mathsf{text}}[i] = \mathsf{SA}[\mathsf{PSV}_{\mathsf{lex}}[\mathsf{ISA}[i]]]. \tag{2}$$

If $\mathsf{NSV}_{\mathsf{lex}}[\mathsf{ISA}[i]] = 0$ ($\mathsf{PSV}_{\mathsf{lex}}[\mathsf{ISA}[i]] = 0$) we set $\mathsf{NSV}_{\mathsf{text}}[i] = 0$ ($\mathsf{PSV}_{\mathsf{text}}[i] = 0$).

If (p_i, ℓ_i) is a normal factor, then either $p_i = \mathsf{NSV}_{\mathsf{text}}[i]$ or $p_i = \mathsf{PSV}_{\mathsf{text}}[i]$ is always a valid choice for p_i [4]. To choose between the two (and to compute the ℓ_i component), we have to compute $\mathsf{lcp}(i, \mathsf{NSV}_{\mathsf{text}}[i])$ and $\mathsf{lcp}(i, \mathsf{PSV}_{\mathsf{text}}[i])$ and choose the larger of the two. This is given as a procedure **LZ-Factor** in Fig. 1.

Lazy LZ Factorization. The fastest LZ factorization algorithms in practice are from recent papers by Kempa and Puglisi [13] and Goto and Bannai [9]. A common feature between them is a lazy evaluation of LCP values: $\mathsf{lcp}(i, \mathsf{NSV}_{\text{text}}[i])$ and $\mathsf{lcp}(i, \mathsf{PSV}_{\text{text}}[i])$ are computed only when i is a starting position of a phrase. The values are computed by a plain character-by-character comparison of the suffixes, but it is easy to see that the total time complexity is $O(n)$. This is in contrast to most previous algorithms that compute the LCP values for every suffix using more complicated techniques. The new algorithms in this paper use lazy evaluation too.

Goto and Bannai [9] describe algorithms that compute and store the full set of NSV/PSV values. One of their algorithms, BGT, computes the $\mathsf{NSV}_{\text{text}}$ and $\mathsf{PSV}_{\text{text}}$ arrays with the help of the Φ array. The LZ factorization is then easily computed by repeatedly calling `LZ-Factor`. Two other algorithms, BGS and BGL, compute the $\mathsf{NSV}_{\text{lex}}$ and $\mathsf{PSV}_{\text{lex}}$ arrays and use them together with SA and ISA to simulate $\mathsf{NSV}_{\text{text}}$ and $\mathsf{PSV}_{\text{text}}$ as in Eqs. (1) and (2). All three algorithms run in linear time and they use $3n \log n$ (BGT), $4n \log n$ (BGL) and $(4n+s) \log n$ (BGS) bits of working space, where s is the size of the stack used by BGS. In the worst case $s = \Theta(n)$. The algorithms for computing the NSV/PSV values are not new but come from [16] (BGT) and from [4] (BGL and BGS). However, the use of lazy LCP evaluation makes the algorithms of Goto and Bannai faster in practice than earlier algorithms.

Kempa and Puglisi [13] extend the lazy evaluation to the NSV/PSV values too. Using ISA and a small data structure that allows arbitrary NSV/PSV queries over SA to be answered quickly, they compute $\mathsf{NSV}_{\text{text}}[i]$ and $\mathsf{PSV}_{\text{text}}[i]$ only when i is a starting position of a phrase. The approach requires $(2+1/b)n \log n$ bits of working space and $O(n + zb + z \log(n/b))$ time, where b is a parameter controlling a space-time tradeoff in the NSV/PSV data structure. If we set $b = \log n$, and given $z = O(n/ \log_\sigma n)$, then in the worstcase the algorithm requires $O(n \log \sigma)$ time, and $2n \log n + n$ bits of space. Despite the superlinear time complexity, this algorithm (ISA9) is both faster and more space efficient than earlier linear time algorithms. Kempa and Puglisi also show how to reduce the space to $(1 + \epsilon)n \log n + n + O(\sigma \log n)$ bits by storing a succinct representation of ISA (algorithms ISA6r and ISA6s). Because of the lazy evaluation, these algorithms are especially fast when the resulting LZ factorization is small.

Optimized Parsing. Fig. 1 shows two versions of the basic parsing procedure. The standard version is essentially how the computation is done in all prior implementations using lazy LZ factorization. The optimized version is the first, small contribution of this paper. It is based on the observation that $\mathsf{lcp}(nsv, psv) = \min(\mathsf{lcp}(i, nsv), \mathsf{lcp}(i, psv))$ and performs $\mathsf{lcp}(nsv, psv)$ fewer symbol comparisons than the standard version.

3 $3n \log n$-Bit Algorithm

Our first algorithm is closely related to the algorithms of Goto and Bannai [9], particularly BGT and BGS. It first computes the $\mathsf{NSV}_{\text{text}}$ and $\mathsf{PSV}_{\text{text}}$ arrays and

Procedure LZ-Factor(i, nsv, psv)
1: $\ell_{nsv} \leftarrow \mathsf{lcp}(i, nsv)$
2: $\ell_{psv} \leftarrow \mathsf{lcp}(i, psv)$
3: **if** $\ell_{nsv} > \ell_{psv}$ **then**
4:     $(p, \ell) \leftarrow (nsv, \ell_{nsv})$
5: **else**
6:     $(p, \ell) \leftarrow (psv, \ell_{psv})$
7: **if** $\ell = 0$ **then** $p \leftarrow \mathsf{X}[i]$
8: **output** factor (p, ℓ)
9: **return** $i + \max(\ell, 1)$

Procedure LZ-Factor(i, nsv, psv)
1: $\ell \leftarrow \mathsf{lcp}(nsv, psv)$
2: **if** $\mathsf{X}[i + \ell] = \mathsf{X}[nsv + \ell]$ **then**
3:     $\ell \leftarrow \ell + 1$
4:     $(p, \ell) \leftarrow (nsv, \ell + \mathsf{lcp}(i + \ell, nsv + \ell))$
5: **else**
6:     $(p, \ell) \leftarrow (psv, \ell + \mathsf{lcp}(i + \ell, psv + \ell))$
7: **if** $\ell = 0$ **then** $p \leftarrow \mathsf{X}[i]$
8: **output** factor (p, ℓ)
9: **return** $i + \max(\ell, 1)$

Fig. 1. The standard (left) and optimized (right) versions of the basic procedure for computing a phrase starting at a position i given $nsv = \mathsf{NSV}_{\text{text}}[i]$ and $psv = \mathsf{PSV}_{\text{text}}[i]$. The return value is the starting position of the next phrase.

uses them for lazy LZ factorization similarly to the BGT algorithm (lines 11–13 in Fig. 2). However, the NSV/PSV values are computed using the technique of the BGS algorithm, which comes originally from [4].

The NSV/PSV computation scans the suffix array while maintaining a stack of suffixes, which are always in double ascending order: both in ascending lexicographical order and in ascending order of text position. The following are equivalent characterizations of the stack content after processing suffix $\mathsf{SA}[i]$:

– $\mathsf{SA}[i]$, $\mathsf{PSV}_{\text{text}}[\mathsf{SA}[i]]$, $\mathsf{PSV}_{\text{text}}[\mathsf{PSV}_{\text{text}}[\mathsf{SA}[i]]]$, ..., 0
– 0 and all $\mathsf{SA}[k]$, $k \in [1..i]$, such that $\mathsf{SA}[k] = \min \mathsf{SA}[k..i]$
– 0 and all $\mathsf{SA}[k]$, $k \in [1..i]$, such that $\mathsf{NSV}_{\text{text}}[\mathsf{SA}[k]] \notin \mathsf{SA}[k + 1..i]$.

Our version of this NSV/PSV computation is shown on lines 1–10 in Fig. 2. It differs from the BGS algorithm of Goto and Bannai in the following ways:

1. We write the NSV/PSV values to the text ordered arrays $\mathsf{NSV}_{\text{text}}$ and $\mathsf{PSV}_{\text{text}}$ instead of the lexicographically ordered arrays $\mathsf{NSV}_{\text{lex}}$ and $\mathsf{PSV}_{\text{lex}}$. Because of this, the second phase of the algorithm does not need the SA and ISA arrays.
2. BGS uses a dynamically growing separate stack while we overwrite the suffix array with the stack. This is possible because the stack is never larger than the already scanned part of SA, which we do not need any more (see above). The worst case size of the stack is $\Theta(n)$ (but it is almost always much smaller in practice).
3. Similar to the algorithms of Goto and Bannai, we store the arrays $\mathsf{PSV}_{\text{text}}$ and $\mathsf{NSV}_{\text{text}}$ interleaved so that the values $\mathsf{PSV}_{\text{text}}[i]$ and $\mathsf{NSV}_{\text{text}}[i]$ are next to each other. We compute the PSV value when popping from the stack instead of when pushing to the stack as BGS does. This way $\mathsf{PSV}_{\text{text}}[i]$ and $\mathsf{NSV}_{\text{text}}[i]$ are computed and written at the same time which can reduce the number of cache misses.

Because of these differences, our algorithm uses between $n \log n$ and $2n \log n$ bits less space and is significantly faster than BGS, which is the fastest of the algorithms in [9].

Algorithm KKP3
1: SA[0] ← 0   // bottom of stack
2: SA[$n+1$] ← 0   // empties the stack at end
3: top ← 0   // top of stack
4: **for** i ← 1 **to** $n+1$ **do**
5:       **while** SA[top] > SA[i] **do**
6:            NSV$_{\text{text}}$[SA[top]] ← SA[i]
7:            PSV$_{\text{text}}$[SA[top]] ← SA[$top-1$]
8:            top ← $top-1$ // pop from stack
9:       top ← $top+1$
10:      SA[top] ← SA[i]   // push to stack
11: i ← 1
12: **while** $i \le n$ **do**
13:      i ← LZ-Factor(i, NSV$_{\text{text}}$[i], PSV$_{\text{text}}$[i])

Fig. 2. LZ factorization using $3n \log n$ bits of working space (the arrays SA, NSV$_{\text{text}}$ and PSV$_{\text{text}}$)

4 $2n \log n$-Bit Algorithm

Our second algorithm reduces space by computing and storing only the NSV values at first. It then computes the PSV values from the NSV values on the fly. As a side effect, the algorithm also computes the Φ array! This is a surprising reversal of direction compared to some algorithms that compute NSV and PSV values from Φ [16,9].

For $t \in [0..n]$, let $\mathcal{X}_t = \{X[i..n] \mid i \le t\}$ be the set of suffixes starting at or before position t. Let Φ_t be Φ restricted to $\mathcal{X}_t$, that is, for $i \in [1..t]$, suffix $\Phi_t[i]$ is the immediate lexicographical predecessor of suffix i among the suffixes in $\mathcal{X}_t$. In particular, $\Phi_n = \Phi$. As with the full Φ, we make Φ_t a complete unicyclic permutation by setting $\Phi_t[i_{\min}] = 0$ and $\Phi_t[0] = i_{\max}$, where $i_{\min}$ and $i_{\max}$ are the lexicographically smallest and largest suffixes in $\mathcal{X}_t$. We also set $\Phi_0[0] = 0$. A useful way to view Φ_t is as a circular linked list storing $\mathcal{X}_t$ in the descending lexicographical order with $\Phi_t[0]$ as the head of the list.

Now consider computing Φ_t given Φ_{t-1}. We need to insert a new suffix t into the list, which can be done using standard insertion into a singly-linked list provided we know the position. It is easy to see that t should be inserted between NSV$_{\text{text}}$[t] and PSV$_{\text{text}}$[t]. Thus

$$\Phi_t[i] = \begin{cases} t & \text{if } i = \text{NSV}_{\text{text}}[t] \\ \text{PSV}_{\text{text}}[t] & \text{if } i = t \\ \Phi_{t-1}[i] & \text{otherwise} \end{cases}$$

and furthermore

$$\text{PSV}_{\text{text}}[t] = \Phi_{t-1}[\text{NSV}_{\text{text}}[t]] \ .$$

The pseudocode for the algorithm is given in Fig 3. The NSV values are computed essentially the same way as in the first algorithm (lines 1–9) and stored in the array Φ. In the second phase, the algorithm maintains the invariant that after t rounds of the loop on lines 12–18, $\Phi[0..t] = \Phi_t$ and $\Phi[t+1..n] = \mathsf{NSV}_{\text{text}}[t+1..n]$.

Algorithm KKP2

```
1:  SA[0] ← 0      // bottom of stack
2:  SA[n + 1] ← 0    // empties the stack at end
3:  top ← 0        // top of stack
4:  for i ← 1 to n + 1 do
5:      while SA[top] > SA[i] do
6:          Φ[SA[top]] ← SA[i]   // Φ[SA[top]] = NSVtext[SA[top]]
7:          top ← top − 1       // pop from stack
8:      top ← top + 1
9:      SA[top] ← SA[i]        // push to stack
10: Φ[0] ← 0
11: next ← 1
12: for t ← 1 to n do
13:     nsv ← Φ[t]
14:     psv ← Φ[nsv]
15:     if t = next then
16:         next ← LZ-Factor(t, nsv, psv)
17:     Φ[t] ← psv
18:     Φ[nsv] ← t
```

Fig. 3. LZ factorization using $2n \log n$ bits of working space (the arrays SA and Φ)

An interesting observation about the algorithm is that the second phase computes Φ from $\mathsf{NSV}_{\text{text}}$ without any additional information. Since the suffix array can be computed from Φ, the $\mathsf{NSV}_{\text{text}}$ array alone contains sufficient information to reconstruct the suffix array.

5   Getting Rid of the Stack

The above algorithms overwrite the suffix array with the stack, which can be undesirable. First, we might need the suffix array later for another purpose. Second, since the algorithms make just one sequential pass over the suffix array, we could stream the suffix array from disk to further reduce the memory usage. In this section, we describe variants of our algorithms that do not overwrite SA (and still make just one pass over it).

The idea, already used in the BGL algorithm of Goto and Bannai [9], is to replace the stack with $\mathsf{PSV}_{\text{text}}$ pointers. As observed in Section 3, if j is the suffix on the top of the stack, then the next suffixes in the stack are $\mathsf{PSV}_{\text{text}}[j]$, $\mathsf{PSV}_{\text{text}}[\mathsf{PSV}_{\text{text}}[j]]$, etcetera. Thus given $\mathsf{PSV}_{\text{text}}$ we do not need an explicit stack at all. Both of our algorithms can be modified to exploit this:

- In KKP3, we need to compute the $\mathsf{PSV}_{\text{text}}$ values when pushing on the stack rather than when popping. The body of the main loop (lines 5–10 in Fig. 2) now becomes:

$$\begin{aligned}
&\textbf{while } top > \mathsf{SA}[i] \textbf{ do} \\
&\quad \mathsf{NSV}_{\text{text}}[top] \leftarrow \mathsf{SA}[i] \\
&\quad top \leftarrow \mathsf{PSV}_{\text{text}}[top] \\
&\mathsf{PSV}_{\text{text}}[\mathsf{SA}[i]] \leftarrow top \\
&top \leftarrow \mathsf{SA}[i]
\end{aligned}$$

- KKP2 needs to be modified to compute $\mathsf{PSV}_{\text{text}}$ values first instead of $\mathsf{NSV}_{\text{text}}$ values. The $\mathsf{PSV}_{\text{text}}$-first version is symmetric to the $\mathsf{NSV}_{\text{text}}$-first algorithm. In particular, Φ_t is replaced by the inverse permutation Φ_t^{-1}. The algorithm is shown in Fig. 4.

Algorithm KKP2n

```
1:  top ← 0        // top of stack
2:  for i ← 1 to n do
3:      while top > SA[i] do
4:          top ← Φ⁻¹[top]     // pop from stack
5:          Φ⁻¹[SA[i]] ← top   // Φ⁻¹[SA[i]] = PSVtext[SA[i]]
6:          top ← SA[i]        // push to stack
7:  Φ⁻¹[0] ← 0
8:  next ← 1
9:  for t ← 1 to n do
10:     psv ← Φ⁻¹[t]
11:     nsv ← Φ⁻¹[psv]
12:     if t = next then
13:         next ← LZ-Factor(t, nsv, psv)
14:     Φ⁻¹[t] ← nsv
15:     Φ⁻¹[psv] ← t
```

Fig. 4. LZ factorization using $2n \log n$ bits of working space (the arrays SA and Φ^{-1}) without an explicit stack. The SA remains intact after the computation.

The versions without an explicit stack are slightly slower because of the non-locality of stack operations. A faster way to avoid overwriting SA would be to use a separate stack. However, the stack can grow as big as n (for example when $X = a^{n-1}b$) which increases the worst case space requirement by $n \log n$ bits. We can get the best of both alternatives by adding a fixed size stack buffer to the stackless version. The buffer holds the top part of the stack to speed up stack operations. When the buffer gets full, the bottom half of its contents is discarded, and when the buffer gets empty, it is filled half way using the PSV pointers. This version is called KKP2b.

All the algorithm variants have linear time complexity.

Table 1. Files used in the experiments. They are from the standard (S) Pizza&Chili corpus (http://pizzachili.dcc.uchile.cl/texts.html) and from the repetitive (R) Pizza&Chili corpus (http://pizzachili.dcc.uchile.cl/repcorpus.html). We truncated all files to 150MiB. The repetitive corpus files are either multiple versions of similar data (R) or artificially generated (A). The value of n/z (the average length of a phrase in the LZ factorization) is included as a measure of repetitiveness.

Name	Abbr.	σ	n/z	Source	Description
proteins	pro	25	9.57	S	Swissprot database
english	eng	220	13.77	S	Gutenberg Project
dna	dna	16	14.65	S	Human genome
sources	src	228	17.67	S	Linux and GCC sources
coreutils	cor	236	110	R/R	GNU Coreutils sources
cere	cer	5	112	R/R	Baking yeast genomes
kernel	ker	160	214	R/R	Linux Kernel sources
einstein.en	ein	124	3634	R/R	Wikipedia articles
tm29	tm	2	2912K	R/A	Thue-Morse sequence
rs.13	rs	2	3024K	R/A	Run-Rich String sequence

6   Experimental Results

We implemented the algorithms described in this paper and compared their performance in practice to algorithms from [13] and [9]. The main experiment measured the time to compute the LZ factorization of the text. All algorithms take the text and the suffix array as an input hence we omit the time to compute SA. The data sets used in experiments are described in detail in Table 1. All algorithms use the optimized version of LZ-Factor (Fig. 1), which slightly reduces the time (e.g. for KKP3 by 2% on non-repetitive files). The implementations are available at http://www.cs.helsinki.fi/group/pads/.

Experiments Setup. We performed experiments on a 2.4GHz Intel Core i5 CPU equipped with 3072KiB L2 cache and 4GiB of main memory. The machine had no other significant CPU tasks running and only a single thread of execution was used. The OS was Linux (Ubuntu 10.04, 64bit) running kernel 2.6.32. All programs were compiled using g++ version 4.4.3 with -O3 -static -DNDEBUG options. For each combination of algorithm and test file we report the median runtime from five executions.

Discussion. The LZ factorization times are shown in the top part of Table 2. In nearly all cases algorithms introduced in this paper outperform the algorithms from [9] (which are, to our knowledge, the fastest up-to-date linear time LZ factorization algorithms) while using the same or less space. In particular the KKP2 algorithms are always faster and simultaneously use at least $n \log n$ bits less space. A notably big difference is observed for non-repetitive data, where KKP3 significantly dominates all prior solutions.

Table 2. Time and space consumption for computing LZ factorization/LPF array. The timing values were obtained with the standard C `clock` function and are scaled to seconds per gigabyte. The times do not include any reading from or writing to disk. The second column summarizes the practical working space (excluding the output in case of LZ factorization) of each algorithm assuming byte alphabet and 32-bit integers. Note that LPF-online computes only the ℓ_i component of LPF array. If this is sufficient, KKP2-LPF can be modified (without affecting the speed) to use only $9n$ bytes.

	Alg.	Mem	pro	eng	dna	src	cor	cer	ker	ein	tm	rs
	KKP3	$13n$	74.5	75.7	81.7	50.5	43.6	63.2	45.7	56.9	38.2	77.8
	KKP2	$9n$	83.9	80.6	92.7	54.7	40.2	53.3	41.6	43.6	35.1	49.0
	KKP2b	$9n$	84.1	80.6	92.7	54.8	40.2	53.2	41.5	43.5	35.1	49.4
LZ factorization	KKP2n	$9n$	88.1	84.6	97.3	56.1	40.6	57.7	42.2	47.6	38.7	52.0
	ISA6r	$6n$	-	-	-	-	43.3	51.8	39.2	31.1	34.2	34.8
	ISA6s	$6n$	198.0	171.0	175.2	115.0	49.4	56.3	45.7	37.1	39.6	40.8
	ISA9	$9n$	92.7	83.9	86.1	59.3	41.9	53.0	42.8	45.2	36.4	51.8
	iBGS	$17n$	99.8	93.2	97.5	69.3	51.5	65.5	52.9	60.0	44.1	59.5
	iBGL	$17n$	123.2	108.6	113.4	77.8	52.2	66.1	53.0	58.6	44.2	59.5
	iBGT	$13n$	171.4	153.9	188.0	99.8	55.4	84.1	56.2	52.8	44.4	56.5
	KKP3-LPF	$13n$	115.5	112.9	133.5	71.1	56.0	88.0	58.0	63.5	49.2	82.8
LPF	KKP2-LPF	$13n$	140.3	132.4	167.2	83.6	54.6	82.6	55.6	51.3	41.1	58.0
	iOG	$13n$	210.1	188.0	243.7	121.3	66.8	104	66.4	60.6	50.3	62.7
	LPF-online	$13n$	160.4	162.3	187.2	114.2	103	137	109	127	100	148

The new algorithms (e.g. KKP2b) also dominate in most cases the general purpose practical algorithms from [13] (ISA9 and ISA6s), while offering stronger worst case time guarantees, but are a frame slower (and use about 50% more space in practice) than ISA6r for highly repetitive data.

The comparison of KKP2n to KKP2 reveals the expected slowdown (up to 16%) due to the non-local stack simulation. However, this effect is almost completely eliminated by buffering the top part of the stack (KKP2b). With a 256KiB buffer we obtained runtimes almost identical to KKP2 ($< 1\%$ difference in all cases). We observed a similar effect when applying this optimization to the KKP3 algorithm but, for brevity, we only present the results for KKP2.

Full LPF array. All our algorithms can be modified to compute the full LPF array, i.e. the set of longest previous factors (p_i, ℓ_i) for $i \in [1..n]$ in linear time. After obtaining $\mathsf{NSV}_{\text{text}}$ and $\mathsf{PSV}_{\text{text}}$ values, instead of repeatedly calling `LZ-Factor` to compute the LZ factorization, we compute all previous factors using the algorithm of Crochemore and Ilie [4, Fig. 2]. We compared this approach to the fastest algorithms for computing LPF array by Ohlebusch and Gog [16] (with the interleaving optimization from [9]) and LPF-online from [5] (see [13] and [16] for comparison). For LCP array computation we use the fastest version of Φ algorithm consuming $13n$ bytes of space [12].

As shown in Table 2, modified KKP2 algorithm consistently outperforms old methods. The LPF variant of KKP3 is even faster, when input is not repetitive.

Table 3. Times for computing LZ factorization, taking into account the disk reading time. The values are wallclock times scaled to seconds per gigabyte. KKP1s is a version of KKP2b that streams the suffix array from disk, and so requires only $n \log n$ bits of working space.

Alg.	pro	eng	dna	src	cor	cer	ker	ein	tm	rs
KKP1s	106.5	100.2	109.0	86.6	71.0	74.9	68.4	67.6	66.1	66.1
KKP2b	150.6	143.7	155.7	117.8	103.6	115.9	102.9	107.3	96.8	111.6

Streaming. As explained in Section 5 our new algorithms can be implemented so that SA is only accessed sequentially in a read-only manner, allowing it to be streamed from the disk. Furthermore, all algorithms (including full LPF variants) can stream the output, which is produced in order, directly to disk. The streaming versions of KKP2b and KKP2b-LPF, called KKP1s and KKP1s-LPF, use only $n \log n$ bits of working space in addition to the text and small stack and disk buffers. We have implemented KKP1s and compared its performance to KKP2b under the assumption that SA is stored on the disk and the disk reading time is included in the total runtime. Reading from the disk was performed with the standard C `fread` function, either as a single read (KKP2b) or using a 32KiB buffer (KKP1s).

Surprisingly, in such setting, KKP1s is significantly faster than KKP2b, as shown in Table 3. Further investigation revealed that the advantage of the streaming algorithm is apparently due to the implementation of I/O in the Linux operating system. The Linux kernel performs implicit asynchronous read ahead operations when a file is accessed sequentially, allowing an overlap of I/O and CPU computation (see [17]).

7 Future Work

We have reduced the working memory of linear time LZ factorization to $2n \log n$ bits, but one wonders if only $(1 + \epsilon)n \log n$ bits (for an arbitrary constant ϵ) is enough, as it is for suffix array construction [11]. In [13] working space of $(1 + \epsilon)n \log n + n$ bits is achieved, but at the price of $O(n \log \sigma)$ runtime. We are also exploring even more space efficient (but slower) approaches [10].

Our streaming algorithms are a first step towards exploiting external memory in LZ factorization. We are currently exploring semi-external variants of these algorithms that keep little else than the input string in memory. This is achieved by permuting the NSV/PSV values from lex order to text order using external memory. Fully external memory as well as parallel and distributed approaches would also be of high interest, especially given the recent pattern matching indexes which use LZ77.

Finally, another problem is to find a scalable way to accurately estimate the size of the LZ factorization in lieu of actually computing it. Such a tool would be useful for entropy estimation, and to guide the selection of appropriate compressors and compressed indexes when managing massive data sets.

Acknowledgments. We thank Keisuke Goto and Hideo Bannai for an early copy of their paper [9].

References

1. Al-Hafeedh, A., Crochemore, M., Ilie, L., Kopylova, E., Smyth, W., Tischler, G., Yusufu, M.: A comparison of index-based Lempel-Ziv LZ77 factorization algorithms. ACM Comput. Surv. 45(1), 5:1–5:17 (2012)
2. Charikar, M., Lehman, E., Liu, D., Panigrhy, R., Prabhakaran, M., Sahai, A., Shelat, A.: The smallest grammar problem. IEEE Transactions on Information Theory 51(7), 2554–2576 (2005)
3. Chen, G., Puglisi, S.J., Smyth, W.F.: Fast and practical algorithms for computing all the runs in a string. In: Ma, B., Zhang, K. (eds.) CPM 2007. LNCS, vol. 4580, pp. 307–315. Springer, Heidelberg (2007)
4. Crochemore, M., Ilie, L.: Computing longest previous factor in linear time and applications. Information Processing Letters 106(2), 75–80 (2008)
5. Crochemore, M., Ilie, L., Iliopoulos, C.S., Kubica, M., Rytter, W., Waleń, T.: LPF computation revisited. In: Fiala, J., Kratochvíl, J., Miller, M. (eds.) IWOCA 2009. LNCS, vol. 5874, pp. 158–169. Springer, Heidelberg (2009)
6. Crochemore, M., Ilie, L., Smyth, W.F.: A simple algorithm for computing the Lempel-Ziv factorization. In: DCC 2008, pp. 482–488. IEEE Computer Society (2008)
7. Gagie, T., Gawrychowski, P., Kärkkäinen, J., Nekrich, Y., Puglisi, S.J.: A faster grammar-based self-index. In: Dediu, A.-H., Martín-Vide, C. (eds.) LATA 2012. LNCS, vol. 7183, pp. 240–251. Springer, Heidelberg (2012)
8. Gagie, T., Gawrychowski, P., Puglisi, S.J.: Faster approximate pattern matching in compressed repetitive texts. In: Asano, T., Nakano, S.-i., Okamoto, Y., Watanabe, O. (eds.) ISAAC 2011. LNCS, vol. 7074, pp. 653–662. Springer, Heidelberg (2011)
9. Goto, K., Bannai, H.: Simpler and faster Lempel Ziv factorization. In: DCC 2013, pp. 133–142. IEEE Computer Society (2013)
10. Kärkkäinen, J., Kempa, D., Puglisi, S.J.: Lightweight Lempel-Ziv parsing. In: Bonifaci, V. (ed.) SEA 2013. LNCS, vol. 7933, pp. 139–150. Springer, Heidelberg (2013)
11. Kärkkäinen, J., Sanders, P., Burkhardt, S.: Linear work suffix array construction. Journal of the ACM 53(6), 918–936 (2006)
12. Kärkkäinen, J., Manzini, G., Puglisi, S.J.: Permuted longest-common-prefix array. In: Kucherov, G., Ukkonen, E. (eds.) CPM 2009 Lille. LNCS, vol. 5577, pp. 181–192. Springer, Heidelberg (2009)
13. Kempa, D., Puglisi, S.J.: Lempel-Ziv factorization: simple, fast, practical. In: Zeh, N., Sanders, P. (eds.) ALENEX 2013, pp. 103–112. SIAM (2013)
14. Kreft, S., Navarro, G.: Self-indexing based on LZ77. In: Giancarlo, R., Manzini, G. (eds.) CPM 2011. LNCS, vol. 6661, pp. 41–54. Springer, Heidelberg (2011)
15. Navarro, G., Mäkinen, V.: Compressed full-text indexes. ACM Computing Surveys 39(1), article 2 (2007)
16. Ohlebusch, E., Gog, S.: Lempel-Ziv factorization revisited. In: Giancarlo, R., Manzini, G. (eds.) CPM 2011. LNCS, vol. 6661, pp. 15–26. Springer, Heidelberg (2011)
17. Wu, F.: Sequential file prefetching in Linux. In: Wiseman, Y., Jiang, S. (eds.) Advanced Operating Systems and Kernel Applications: Techniques and Technologies, ch. 11, pp. 217–236. IGI Global (2009)

External Memory Generalized Suffix and LCP Arrays Construction

Felipe A. Louza[1], Guilherme P. Telles[2], and Cristina Dutra De Aguiar Ciferri[1]

[1] Institute of Mathematics and Computer Science, University of São Paulo, São Carlos, SP, Brazil
{louza,cdac}@icmc.usp.br
[2] Institute of Computing, University of Campinas, Campinas, SP, Brazil
gpt@ic.unicamp.br

Abstract. A suffix array is a data structure that, together with the LCP array, allows solving many string processing problems in a very efficient fashion. In this article we introduce eGSA, the first external memory algorithm to construct both generalized suffix and LCP arrays for sets of strings. Our algorithm relies on a combination of buffers, induced sorting and a heap. Performance tests with real DNA sequence sets of size up to 8.5 GB showed that eGSA can indeed be applied to sets of large sequences with efficient running time on a low-cost machine. Compared to the algorithm that most closely resembles eGSA purpose, eSAIS, eGSA reduced the time spent to construct the arrays by a factor of 2.5−4.8.

Keywords: generalized suffix array, generalized LCP array, external memory algorithms, text indexes, DNA indexing.

1 Introduction

Suffix arrays [1] play an important role in several string processing tasks, from pattern matching to data compression and information retrieval [2]. The suffix array combined with the longest common prefix (LCP) array provides a powerful data structure to solve many string processing problems in optimal time and space [3].

Many algorithms have been proposed for internal memory suffix arrays construction, including linear ones [4, 5]. This is also the case for LCP arrays construction [6, 7]. These algorithms are limited by internal memory size, but there is a significant number of applications that deal with a huge amount of strings that may impair the use of existing algorithms for internal memory suffix arrays construction. An example is sequence comparisons and pattern searching in molecular sequences, a field where databases have been growing at exponential rate for years. For instance, GenBank[1] has more than 150 million sequences with more than 140 billion characters.

---

[1] ftp://ftp.ncbi.nih.gov/genbank/gbrel.txt

J. Fischer and P. Sanders (Eds.): CPM 2013, LNCS 7922, pp. 201–210, 2013.
© Springer-Verlag Berlin Heidelberg 2013

Several algorithms have been developed for external memory suffix array construction [8–10]. More recently, Bingmann *et al.* [11] proposed eSAIS, the first external memory algorithm that constructs both suffix and LCP arrays for a single string.

Suffix arrays were also generalized to index sets of strings [12]. In the literature, works that use generalized suffix arrays do not use external memory, *e.g.* [13, 14]. On the other hand, works that investigate suffix arrays on external memory are not specifically aimed to index sets of strings. In this article we fill this gap. We introduce eGSA, an external memory algorithm to construct both generalized suffix and LCP arrays for a set of strings. We also show that eGSA can indeed be applied to sets of large sequences efficiently.

The rest of the article is organized as follows. Section 2 introduces concepts and notation, Section 3 describes the proposed algorithm, Section 4 shows performance tests that validate the algorithm, and Section 5 concludes the article and highlights future work.

2 Background

Let Σ be an ordered alphabet of symbols and let \$ be a symbol not in Σ that precedes every symbol in Σ. We denote the reflexive and transitive closure of Σ by Σ^* and the concatenation of strings or symbols by the dot operator ($\cdot$). We define $\Sigma^\$ = \{T \cdot \$ | T \in \Sigma^*\}$.

Let $T = T[1]T[2]\ldots T[n]$ be any string of length n. A substring of T is denoted $T[i, j] = T[i]\ldots T[j]$, $1 \leq i \leq j \leq n$. A prefix of T is a substring $T[1, k]$ and a suffix is a substring $T[k, n]$. We denote a suffix starting with a symbol α as an α-suffix.

A suffix array for a string T is an array of integers that provides the lexicographic order for all suffixes of T. We use the symbol $<$ for the lexicographic order relation between strings. Formally, a suffix array for a string $T \in \Sigma^\$$ of size n, called *SA*, is an array of integers $SA = [i_1, i_2, \ldots, i_n]$ such that $T[i_1, n] < T[i_2, n] < \ldots < T[i_n, n]$. The function $pos(T[k, n])$ maps the position of $T[k, n]$ in *SA*. We define an α-bucket as a block of a partition of *SA* that contains only α-suffixes.

Let $lcp(S, T)$ be the length of the longest common prefix of S and T, where $S, T \in \Sigma^\$$. The *LCP* array for T is an array of integers such that $LCP[i] = lcp(T[SA[i], n], T[SA[i-1], n])$ and $LCP[0] = 0$.

The *Burrows-Wheeler transform* of a string allows its efficient compression [15]. It may be stored in an array such that $BWT[i] = T[SA[i] - 1]$ if $SA[i] \neq 1$ or $BWT[i] = \$$ otherwise. *BWT* has a close relationship to *SA* and can be trivially obtained from it [10].

Suffix arrays have been generalized to index sets of strings. Given a set of k strings $\mathcal{T} = \{T_1, \ldots, T_k\}$ from $\Sigma^\$$ with lengths $n_1, \ldots, n_k$, the generalized suffix array of $\mathcal{T}$, denoted *GSA*, is an array of pairs of integers (i, j) that specifies the lexicographic order of all suffixes $T_i[j, n_i]$ of strings in $\mathcal{T}$. An order relation is defined for the tail suffixes $T_i[n_i - 1, n_i] = \$$ as $T_i[n_i - 1, n_i] < T_j[n_j - 1, n_j]$ if

$i < j$. *LCP* and *BWT* can also be generalized for sets of strings. The generalized *LCP* will be denoted *GLCP*.

3   Proposed Algorithm: eGSA

Our algorithm is called eGSA (*External Generalized Suffix and LCP Arrays Construction Algorithm*) and is based on the two-phase, multiway merge-sort presented by Garcia-Molina *et al.* [16]. The input for eGSA is a set of k strings $\mathcal{T} = \{T_1, \ldots, T_k\}$ with lengths $n_1, \ldots, n_k$ stored in the external memory and the output, which is written to external memory, is composed both by *GSA* and *GLCP* arrays for $\mathcal{T}$.

In a glance, eGSA works as follows. In the first phase it sorts the suffixes of each T_i in internal memory, obtaining SA_i, LCP_i and other auxiliary arrays, that are written to external memory. In the second phase, eGSA uses internal memory buffers to merge the previously computed arrays, obtaining *GSA* and *GLCP*. We detail each phase next.

3.1   Phase 1: Sorting

The first phase of eGSA builds SA_i and LCP_i for every $T_i \in \mathcal{T}$, using any internal memory algorithm for suffix sorting [4, 5] and *lcp* computing [6, 7]. Note that, if there is not enough internal memory available for this phase, we can use any external memory algorithm to construct them. Furthermore, two other arrays are computed, BWT_i and PRE_i, which are used to improve the second phase. The computation of all these arrays is performed in internal memory. At the end of this first phase, the arrays are written to external memory in a sequential fashion.

The prefix array for T_i, PRE_i, is defined by Barsky *et al.* [17] such that $PRE_i[j]$ is the prefix of $T_i[SA_i[j], n_i]$ of length p, for a configurable constant p. We notice that the probability that $PRE_i[j]$ is equal to $PRE_i[j+1]$ is large, since the suffixes are sorted in SA_i. Then, to avoid redundancy, we adopt a different strategy, similar to the lef-justified approach in [18], and construct PRE_i through non-overlapping substrings as $PRE_i[j] = T_i[SA_i[j] + h_j, SA_i[j] + h_j + p]$, where $h_j = min(LCP_i[j], h_{j-1} + p)$ and $h_0 = 0$.

Figure 1 shows the output structures of this phase of eGSA for $T_1 = GATAGA\$$ and $p = 3$. The last column is merely illustrative and shows the suffixes $T_1[SA_1[j], n_1]$. For simplicity when $SA_1[j] + h_j + p > n_1 = 7$ we consider $T[SA_1[j] + h_j + p] = \$$.

3.2   Phase 2: Merging

The second phase of eGSA merges the arrays computed in the first phase to obtain *GSA* and *GLCP* for $\mathcal{T}$, as follows.

Let $R_i = \langle SA_i, LCP_i, BWT_i, PRE_i \rangle$. Each R_i is partitioned into r_i blocks $R_i^1, \ldots, R_i^{r_i}$, having b consecutive elements from each array except perhaps for

j	$SA_1[j]$	$LCP_1[j]$	$BWT_1[j]$	$PRE_1[j]$	$T_1[SA[j], n_1]$
1	6	-	A	$$$	**$**
2	5	0	G	A$$	**A$**
3	3	1	T	GA$	**AGA$**
4	1	1	G	TAG	**ATAGA$**
5	4	0	A	GA$	**GA$**
6	0	2	$	TAG	**GATAGA$**
7	2	0	A	TAG	**TAGA$**

Fig. 1. Output structures of phase 1, for $T_1 = GATAGA\$$ and $p = 3$

$R_i^{r_i}$. For each R_i the algorithm uses two internal memory buffers: a string buffer S_i, which stores substrings up to s symbols of T_i, and a partition buffer B_i, which stores a block R_i^j. $B_i[j]$ is composed of $\langle SA_i[k], LCP_i[k], BWT_i[k], PRE_i[k]\rangle$, for $j = k \bmod b$. We also use two other buffers. The output buffer stores d elements from the *GSA* and *GLCP* arrays. The induced buffer has size c and stores information that is necessary in the inducing strategy, to be discussed below. The values of s, b, d and c determine the amount of internal memory used in this phase.

Each block R_i^1 is initially loaded into its buffer B_i. Then the heading elements of each buffer B_i are inserted into a binary heap. The smallest suffix in the heap is moved to the output buffer and replaced by the next element from the same buffer B_i. This operation is repeated until all partition blocks are empty. When the output buffer is full, it is written to external memory.

The most sensitive operation in this phase is the comparison of elements from each buffer. Using a naïve approach may require too many random disk accesses. This is due the fact that for every SA_i involved in the comparison, the corresponding suffixes must be accessed in external memory, loaded into string buffers and then compared.

To reduce disk accesses, we propose an enhanced comparison method composed of three strategies: (i) prefix assembling; (ii) *lcp* comparisons; and (iii) inducing suffixes. These strategies are described below.

Prefix Assembly. Let j be the index of the smallest element in the buffer B_i. PRE_i is used to load the initial prefix of $T_i[SA_i[j], n_i]$ into S_i with no disk accesses, just concatenating previous $PRE_i[k]$, for $k = 1, 2, \ldots, j$. As j changes, buffer S_i is updated such that $S_i[1, h_j + p + 1] = S_i[1, h_j] \cdot PRE_i[j] \cdot \#$, where $h_j = min(LCP_i[j], h_{j-1} + p)$, $h_0 = 0$, and $\#$ is an end-of-buffer marker not in Σ. Thus, if a string comparison does not involve more than $h_j + p$ symbols, a disk access is not necessary. Otherwise $\#$ is reached and T_i is accessed in the external memory.

The last column of Figure 1 illustrates in bold the prefixes recovered by prefix assembly. For instance, if $j = 5$ then $h_5 = 0$ and S_1 stores *GA$*. Next, when $j = 6$ then $h_6 = min(LCP_i[6], h_5 + p) = min(2, 0 + 3) = 2$, and $S_1[3, 3 + 3 - 1] = S_1[3, 5]$ receives $PRE_i[5] = TAG$. In this case, $S_1 = S_1[1, 2] \cdot S_1[3, 5] \cdot \# = GA \cdot TAG \cdot \# = GATAG\#$.

LCP Comparisons. Let X, Y and Z be nodes in the binary heap storing $B_a[i]$, $B_b[j]$ and $B_c[k]$, respectively. Suppose that node X is the parent of Y and Z. As $X < Y$ and $X < Z$, then $T_a[SA_a[i], n_a] < T_b[SA_b[j], n_b]$ and $T_a[SA_a[i], n_a] < T_c[SA_c[k], n_c]$. The *lcp* values can be used to speed up suffix comparisons in the heap [19]. The following lemma formalizes this relation. The proof is simple and will be omitted.

Lemma 1. *Let S_1, S_2 and S_3 be strings, such that $S_1 < S_2$ and $S_1 < S_3$. If $lcp(S_1, S_2) > lcp(S_1, S_3)$ then $S_2 < S_3$. If $lcp(S_1, S_2) < lcp(S_1, S_3)$ then $S_2 > S_3$. Otherwise, if $lcp(S_1, S_2) = lcp(S_1, S_3) = l$ then $lcp(S_2, S_3) \geq l$.*

The order of Y and Z can be determined using Lemma 1, and if $lcp(X, Y) = lcp(X, Z)$ then $lcp(Y, Z) \geq lcp(X, Y) = l$, and Y and Z can be compared directly starting from position l. As X is removed from the heap, $B_a[i]$ is moved to the output buffer and X is replaced by another node W storing $B_a[i+1]$. The order of W with respect to its children can also be determined by Lemma 1 along the heap, as the right position for W is searched. The *lcp* values in the heap are updated as nodes are swapped. Hence, using *lcp* values many direct comparisons of strings that are in the external memory are avoided.

Inducing Suffixes. *Induced sorting* is the determination of the order of unsorted suffixes from already sorted suffixes that is used by many internal memory algorithms [4]. We apply an induced sorting approach based on the following lemma, whose proof is straightforward.

Lemma 2. *Let Suff be the set of all suffixes of $\mathcal{T}$, $T_i \in \mathcal{T}$, $1 \leq j \leq n_i$ and $\alpha \in \Sigma$. If $T_i[j, n_i]$ is the smallest element of Suff (w.r.t. the lexicographic order) then $T_i[j - 1, n_i] = \alpha \cdot T_i[j, n_i]$ is the smallest α-suffix of Suff $\setminus \{T_i[j, n_i]\}$.*

Lemma 2 can be used to sort the suffixes of T_i as follows. *Suff* starts with every suffix of T_i and as the smallest suffix $T_i[j, n_i] = \alpha \cdot T_i[j + 1, n_i]$ is found, $T_i[j, n_i]$ is removed from *Suff* and inserted into the smallest available position of the α-bucket. Then $T_i[j - 1, n_i] = \beta \cdot T_i[j, n_i]$ is induced to the smallest available position in the β-bucket, $\alpha, \beta \in \Sigma$.

Note that if $\alpha > \beta$ the suffix $T_i[j - 1, n_i]$ was already sorted. Moreover, the induced suffixes $T_i[j - 1, n_i] = \beta \cdot T_i[j, n_i]$ cannot be removed from *Suff* because they must induce suffixes $T_i[j - 2, n_i]$ as well. To this end, when the smallest β-suffix $T_i[j - 1, n_i]$ is the smallest suffix in *Suff*, the β-bucket is read starting from the second element. As the suffixes $T_i[j - 2, n_i]$ are analyzed to be induced, the suffixes $T_i[j - 1, n_i]$ are removed from *Suff*. Also, if $\alpha = \beta$, then reading induced suffixes from the β-bucket can cause the induction of already induced suffixes. So no induction is done when $\alpha \geq \beta$.

However, this approach is not efficient to sort a single string T_i, since it is always necessary to find the smallest suffix $T_i[j, n_i]$. But in a merge algorithm, the smallest suffix is one of those remaining in buffer B_i, and can be determined efficiently using the heap. Let *Suff* be the set of all suffixes of $\mathcal{T}$ and suppose that $B_i[k]$ is at the root of the heap. Then $T_i[j, n_i]$ is the smallest suffix in *Suff*, and using the approach described previously we can induce $T_i[j-1, n_i]$ if $\alpha < \beta$. For this, we use BWT_i to determine if $T_i[j] < T_i[j-1]$ and whether or not $T_i[j-1, n_i]$ can be induced.

The smallest suffixes are moved to the output buffer, and the induced suffixes are written to the induced buffer, which is written to external memory as it gets full. When the smallest β-suffix $T_i[j-1, n_i]$ is the smallest in $Suff$, the β-bucket is read from external memory, and induces other suffixes as necessary. Note that there is no need to compare the induced suffixes in this step, it is sufficient only to follow the order imposed by the β-bucket in the heap.

The LCP values of the induced suffixes must also be induced, since they are not calculated when the induced suffixes are not compared in the heap. Let $T_a[i, n_a]$ be a suffix that induces an α-suffix and let $T_b[j, n_b]$ be the suffix that induces the following α-suffix. Then $LCP(T_a[i-1, n_a], T_b[j-1, n_b]) = LCP(T_a[i, n_a], T_b[j, n_b]) + 1$. But since the suffixes $T_a[i, n_a]$ and $T_b[j, n_b]$ may not be consecutive in GSA, the value of $LCP(T_a[i, n_a], T_b[j, n_b])$ may not be obtained directly. For that, let the range minimum query on $GLCP$ be $rmq(i, j) = \min_{i < k \leq j}\{GLCP[k]\}$. Since $T_a[i, n_a]$ and $T_b[j, n_b]$ are already sorted, $LCP(T_a[j, n_a], T_b[j, n_b]) = rmq(pos(T_a[j, n_a]) + 1, pos(T_b[j, n_b]))$. The rmq values may be computed as $GLCP$ is moved to the output buffer storing the min function for each $\alpha \in \Sigma^*$.

Therefore, when a suffix $T_i[j, n_i]$ is induced in the second phase, its corresponding LCP is also induced from the rmq values. As induced suffixes may also induce, the corresponding LCP must be stored in the induced buffer together with the induced suffixes in their α-bucket. As the induced suffixes are recovered from the external memory, the LCP must also be recovered to update rmq.

4   Performance Evaluation

The performance of eGSA was analyzed through tests with real DNA sequences from the genomes of (1) Human, (2) Medaka, (3) Zebrafish, (4) Cow, (5) Mouse and (6) Chicken, which were obtained from the Ensembl genome database[2]. We generated 5 datasets, described in Table 1. We preprocessed these datasets to remove the character N (unknown). Each character in a dataset uses one byte. The mean and maximum LCP values provide an approximation of suffix sorting difficulty [9].

Our algorithm was implemented in ANSI/C. The construction of the suffix and LCP arrays in the first phase of eGSA was performed by the *inducing+saislite* algorithm [6], which uses approximately $9 \times |T_i|$ bytes. We used $p = 23$ for the size of PRE_i. The buffers S_i, B_i, output and induced were set to use 200 KB, 10 MB, 64 MB and 16 MB of internal memory, respectively. We remark that eGSA uses 1 byte for each character in S_i. The output produced by eGSA was validated using a trivial checking algorithm. The source code is freely available from http://code.google.com/p/egsa/.

---

[2] http://www.ensembl.org/

Table 1. Datasets used in our experiments. Column 2 indicates the genomes that compose each dataset. Column 3 shows the number of strings (i.e. chromosomes). Columns 4 and 5 show the computed mean and maximum *lcp* values. Column 6 reports the dataset size.

Dataset	Genomes	Number of strings	mean LCP	max. LCP	Input size (GB)
1	2	24	19	2,573	0.54
2	6	30	17	5,476	0.92
3	3, 6	56	58	71,314	2.18
4	2, 3, 4	80	44	71,314	4.26
5	1, 4, 5, 6	105	59	168,246	8.50

We compared our algorithm with the eSAIS algorithm [11], which is the fastest algorithm to date that computes both suffix and LCP arrays in external memory. However, eSAIS is aimed at indexing only one string T_i. To use this algorithm to index a set of strings, we concatenated all strings in $\mathcal{T}$, replacing $ of each T_i by a new terminal symbol $\$_i$, such that $\$_i < \$_j$ if $i < j$ and $\$_i < \alpha$ for each $\alpha \in \Sigma$. This approach limits the number k of strings that can be indexed. For DNA sequences, using 1 byte for each character, k is limited by $256 - |\{A, C, G, T\}| = 252$. We are aware of the existence of the algorithms by Bauer *et al.* [20, 21] that aim at indexing sets of fixed size, small strings in external memory. However, we did not consider comparing them with eGSA because they solve a different problem.

The eGSA was compiled by GNU gcc compiler, version 4.6.3, with optimizing option -O3. The experiments were conducted in the Linux Ubuntu 12.04/64 bits operating system, running on an Intel Core i7 2.67 GHz processor 8MB L2 cache, 12 GB of internal memory and a 1 TB SATA hard disk with 5900 RPM and 64MB cache. The amount of internal memory usage across the experiments was restricted to 4 GB.

Table 2 shows the experimental results of eGSA and eSAIS execution. Although the comparison is not totally fair because eSAIS was not designed for multiple strings, eGSA have consistently outperformed eSAIS by a factor of 2.5−4.8 in time (columns μs/input byte). Then we may safely conclude that eGSA is an efficient algorithm for generalized suffix and *LCP* arrays construction on external memory. Moreover, phase 2 of eGSA used only 1.1 GB of internal memory for dataset 5.

In the same fashion, we can analyze eGSA through efficiency, that is the proportion of time for which the CPU busy, not waiting for I/O. As shown in Table 2, the efficiency decreases for dataset 5, while still more efficient than eSAIS. This is not caused only by the dataset size but also by the maximum *lcp* value for the dataset. We can see that the CPU time ratio for both algorithms on datasets 4 and 5 is close, what indicates that I/O is probably related to efficiency loss. We believe that adjusting buffer sizes may improve efficiency, and that, in particular, the string buffer size s is more closely related to this issue. As with many other external memory algorithms, buffer size adjust is often necessary.

Table 2. Results for the comparison of eGSA and eSAIS. Columns 2 and 3 report the running time in microseconds per input byte. Columns 4 and 5 report the total running time (wallclock) in seconds. Columns 6 and 7 report the total cputime time not accounting for the time of I/O. Columns 8 and 9 report the efficiency of each algorithm, that is the proportion of cputime by wallclock. Finally, column 10 reports the ratio of eSAIS cputime by eGSA cputime.

Dataset	μs/input byte		wallclock (sec)		cputime (sec)		efficiency		cputime ratio
	eSAIS	eGSA	eSAIS	eGSA	eSAIS	eGSA	eSAIS	eGSA	eSAIS/eGSA
1	5.86	1.72	3,413	1,005	1,236	687	0.36	0.68	1.80
2	5.97	1.24	5,883	1,228	2,110	715	0.36	0.58	2.95
3	6.23	2.27	14,596	5,314	4,385	3,349	0.30	0.63	1.31
4	6.41	2.31	29,383	10,590	8,542	7,566	0.29	0.71	1.13
5	7.24	2.79	66,106	25,502	16,652	13,003	0.25	0.51	1.28

Furthermore, we also registered that the proportion of induced suffixes is 37.4% on the average, what shows that inducing suffixes is a major improvement strategy in eGSA.

The theoretical cost of phase 1 of eGSA is dominated by the algorithms used to construct SA_i and LCP_i. In phase 2, the number of node swaps in the heap is bounded by $N \log k$, where N is the sum of the k string lengths. Each node swap requires comparing a number of characters that is equal to the maximum value of lcp for $\mathcal{T}$ ($maxlcp$). The cost of this phase is dominated by the $(N \log k)\, maxlcp$ comparisons, beyond I/O operations.

5   Conclusions and Future Work

In this article we proposed eGSA, which is the first external memory algorithm to construct both generalized suffix and LCP arrays for a set of strings. The proposed algorithm was validated through performance tests using real DNA sequences from different species, which were combined in datasets with different number of strings and data volume.

The results showed that eGSA is efficient. Compared to the eSAIS algorithm, the algorithm that most closely resembles eGSA purpose, eGSA reduced the time spent to construct the arrays by a factor of 2.5−4.8.

Another advantage of eGSA is that it may be employed to build generalized suffix and LCP arrays from suffix and LCP arrays that have already been computed individually for strings in a dataset. Moreover, eGSA may be used to construct the core data structures used by LOF-SA search algorithms [18] and to build generalized suffix trees in external memory [17]. Furthermore, it may be applied to construct the Longest Previous Factor array, which is used in text compression and for detecting motifs and repeats [22].

We are currently extending the eGSA algorithm to also construct a generalized *Burrows-Wheeler transform* of a set of strings. Another future work is redesigning the algorithm for multiple disks, one for write operations and the others for read operations. The data structures and algorithms used in our approach suggest that the running time of eGSA is subquadratic, but a remaining task is to formalize the asymptotic analysis both for memory and I/O operations and to compare them with experimental results for this and other datasets.

Acknowledgments. This work has been supported by the Brazilian agencies FAPESP, CNPq and CAPES.

References

1. Manber, U., Myers, E.W.: Suffix arrays: A new method for on-line string searches. SIAM J. Comput. 22(5), 935–948 (1993)
2. Gusfield, D.: Algorithms on strings, trees, and sequences: computer science and computational biology. Cambridge University Press, New York (1997)
3. Kärkkäinen, J., Manzini, G., Puglisi, S.J.: Permuted longest-common-prefix array. In: Kucherov, G., Ukkonen, E. (eds.) CPM 2009 Lille. LNCS, vol. 5577, pp. 181–192. Springer, Heidelberg (2009)
4. Puglisi, S.J., Smyth, W.F., Turpin, A.H.: A taxonomy of suffix array construction algorithms. ACM Computing Surveys 39(2), 1–31 (2007)
5. Nong, G., Zhang, S., Chan, W.H.: Linear suffix array construction by almost pure induced-sorting. In: Proc. Data Compression Conference, pp. 193–202 (2009)
6. Fischer, J.: Inducing the LCP-array. In: Dehne, F., Iacono, J., Sack, J.-R. (eds.) WADS 2011. LNCS, vol. 6844, pp. 374–385. Springer, Heidelberg (2011)
7. Gog, S., Ohlebusch, E.: Fast and lightweight lcp-array construction algorithms. In: Proc. Meeting on Algorithm Engineering & Experiments, pp. 25–34 (2011)
8. Crauser, A., Ferragina, P.: A theoretical and experimental study on the construction of suffix arrays in external memory. Algorithmica 32(1), 1–35 (2002)
9. Dementiev, R., Kärkkäinen, J., Mehnert, J., Sanders, P.: Better external memory suffix array construction. ACM J. of Experimental Algorithmics 12 (2008)
10. Ferragina, P., Gagie, T., Manzini, G.: Lightweight data indexing and compression in external memory. Algorithmica 63(3), 707–730 (2012)
11. Bingmann, T., Fischer, J., Osipov, V.: Inducing suffix and lcp arrays in external memory. In: Proc. Meeting on Algorithm Engineering & Experiments, pp. 88–103 (2013)
12. Shi, F.: Suffix arrays for multiple strings: A method for on-line multiple string searches. In: Jaffar, J., Yap, R.H.C. (eds.) ASIAN 1996. LNCS, vol. 1179, pp. 11–22. Springer, Heidelberg (1996)
13. Pinho, A., Ferreira, P., Garcia, S., Rodrigues, J.: On finding minimal absent words. BMC bioinformatics 10, 137 (2009)
14. Arnold, M., Ohlebusch, E.: Linear time algorithms for generalizations of the longest common substring problem. Algorithmica 60(4), 806–818 (2011)
15. Burrows, M., Wheeler, D.: A block-sorting lossless data compression algorithm. Systems Research (1994)
16. Garcia-Molina, H., Widom, J., Ullman, J.D.: Database System Implementation. Prentice-Hall, Inc., Upper Saddle River (1999)
17. Barsky, M., Stege, U., Thomo, A., Upton, C.: A new method for indexing genomes using on-disk suffix trees. Proc. ACM International Conference on Information and Knowledge Management 236(1-2), 649 (2008)
18. Sinha, R., Puglisi, S.J., Moffat, A., Turpin, A.: Improving suffix array locality for fast pattern matching on disk. Proc. ACM SIGMOD, 661–672 (2008)

19. Ng, W., Kakehi, K.: Merging string sequences by longest common prefixes. Information Processing Society of Japan Digital Courier 4, 69–78 (2008)
20. Bauer, M.J., Cox, A.J., Rosone, G.: Lightweight algorithms for constructing and inverting the bwt of string collections. Theoretical Computer Science (2012) (in press)
21. Bauer, M.J., Cox, A.J., Rosone, G., Sciortino, M.: Lightweight LCP Construction for Next-Generation Sequencing Datasets. In: Raphael, B., Tang, J. (eds.) WABI 2012. LNCS, vol. 7534, pp. 326–337. Springer, Heidelberg (2012)
22. Crochemore, M., Ilie, L., Iliopoulos, C.S., Kubica, M., Rytter, W., Wale, T.: Computing the longest previous factor. European J. of Combinatorics 34(1), 15–26 (2013)

Efficient All Path Score Computations on Grid Graphs

Ury Matarazzo*, Dekel Tsur*, and Michal Ziv-Ukelson**

Department of Computer Science Ben-Gurion University of the Negev, Israel

Abstract. We study the *Integer-weighted Grid All Paths Scores* (IGAPS) problem, which is given a grid graph, to compute the maximum weights of paths between every pair of a vertex on the first row of the graph and a vertex on the last row of the graph. We also consider a variant of this problem, periodic IGAPS, where the input grid graph is periodic and infinite. For these problems, we consider both the general (dense) and the sparse cases.

For the sparse IGAPS problem with 0-1 weights, we give an $O(r \log^3(n^2/r))$ time algorithm, where r is the number of (diagonal) edges of weight 1. Our result improves upon the previous $O(n\sqrt{r})$ result by Krusche and Tiskin for this problem.

For the periodic IGAPS problem we give an $O(Cn^2)$ time algorithm, where C is the maximum weight of an edge. This improves upon the previous $O(C^2n^2)$ algorithm of Tiskin. We also show a reduction from periodic IGAPS to IGAPS. This reduction yields $o(n^2)$ algorithms for this problem.

1 Introduction

String comparison is a fundamental problem in computer science that has applications in computational biology, computer vision, and other areas. String comparison is often performed using *string alignment*: The characters of two input strings are aligned to each other, and a *scoring function* gives a score to the alignment according to pairs of the aligned characters and unaligned characters. The goal of the string alignment problem is to seek an alignment that maximizes (or minimizes) the score. In this paper we consider maximal scores to be optimal, but minimization problems can be solved symmetrically. The problem can be solved in $O(n^2)$ time [27], where n is the sum of lengths of A and B. Common scoring functions are the *edit distance* score, and the *LCS* (longest common subsequence) score.

A *grid graph* is a directed graph $G = (V, E)$ whose vertex set is $V = \{(i,j) : 0 \le i \le m, 0 \le j \le n\}$, and whose edge set consists of three types:

---

* The research of D.T. and U.M was partially supported by ISF grant 981/11 and by the Frankel Center for Computer Science at Ben Gurion University of the Negev.

** The research of M.Z.U. was partially supported by ISF grant 478/10 and by the Frankel Center for Computer Science at Ben Gurion University of the Negev.

J. Fischer and P. Sanders (Eds.): CPM 2013, LNCS 7922, pp. 211–222, 2013.

1. Diagonal edges $((i, j), (i + 1, j + 1))$ for all $0 \leq i < m, 0 \leq j < n$.
2. Horizontal edges $((i, j), (i, j + 1))$ for all $0 \leq i \leq m, 0 \leq j < n$.
3. Vertical edges $((i, j), (i + 1, j))$ for all $0 \leq i < m, 0 \leq j \leq n$.

In the *Grid All Paths Scores* (GAPS) problem, the input is a grid graph and the goal is to compute the maximum weights of paths between every pair of a vertex on the first row of the graph and a vertex on the last row of the graph. For simplicity of presentation, we will assume in some parts of this paper that $m = n$.

The *Integer GAPS* (IGAPS) problem is a special case of GAPS in which the weights of the edges are integers in the range 0 to C, and additionally, the weights of all the horizontal (resp., vertical) edges between two columns (resp., rows) of vertices are equal. The *Binary GAPS* (BGAPS) problem is a special case of IGAPS in which the horizontal and vertical edges have weight 0, and diagonal edges have weight 0 or 1.

The alignment problem on strings A and B can be represented by using an $(|A| + 1) \times (|B| + 1)$ grid graph, known as the *alignment grid graph* (cf. [21]). Vertical (respectively, horizontal) edges correspond to alignment of a character in A (respectively, B) with a gap, and diagonal edges correspond to alignment of two characters in A and B. Each edge of the graph has a weight. A path from the j-th vertex on row i to the j'-th vertex on row i' corresponds to an alignment of $A[i..i']$ and $B[j..j']$.

The GAPS problem was introduced by Apostolico et al. [3] in order to obtain fast parallel algorithms for LCS computation. It has since been studied in several additional papers [1, 2, 7, 11–15, 21–23]. Schmidt [21] showed that the GAPS problem can be solved in $O(n^2 \log n)$ time. In the same paper, Schmidt showed that IGAPS can be solved in $O(Cn^2)$ time. An $O(n^2)$ algorithm based on a similar approach for the BGAPS problem was also given by Alves et al. [1] and Tiskin [23]. Tiskin [22, p. 60] gave an $O(n^2(\log \log n / \log n)^2)$ time algorithm for a special case of BGAPS, in which the grid graph corresponds to an LCS problem on two strings. Tiskin also showed that IGAPS can be reduced to BGAPS. However, this reduction increases the size of the grid graph by a factor of C^2. Thus, the time for solving IGAPS with this reduction is either $O(C^2 n^2)$ (for general grid graphs) or $O(C^2 n^2 (\log \log n / \log n)^2)$ (for grid graphs that correspond to alignment problems on two strings).

A special case of the BGAPS problem is when the number of diagonal edges with weight 1 is significantly smaller than n^2. We call this problem *sparse BGAPS*. Krusche and Tiskin [12] showed that sparse BGAPS can be solved in $O(n\sqrt{r})$ time, where r is the number of edges of weight 1. For the special case of a permutation grid graph (namely, each column and each row have exactly one edge of weight 1), Tiskin [22] gave an $O(n \log^2 n)$ time algorithm. Another special case of BGAPS is when the grid graph corresponds to the LCS computation of two strings with little similarity. Landau et al. [15] gave an algorithm for this variant with time complexity $O(nL)$, where L is the LCS of the two strings.

Efficient computations and storage of GAPS provide very powerful tools that can be also used for solving many problems on strings: optimal alignment computation [5], approximate tandem repeats [17, 21], approximate non-overlapping repeats [4,9,21], common substring alignment [16,18], sparse spliced alignment [10, 20], alignment of compressed strings [6], fully-incremental string comparison [8,19,22], and other problems.

Additional types of computations are useful in some of the applications. A *periodic grid graph* is an infinite graph obtained by concatenating horizontally an infinite number of a (finite) grid graph. The *periodic IGAPS* problem is a variant of the IGAPS problem, in which the input is a periodic grid graph. Note that while there are an infinite number of vertex pairs whose maximum path score need to be computed, due to the periodicity of the graph, the output can be represented in finite space. The periodic IGAPS problem was studied by Tiskin [25] who gave an $O(C^2n^2)$ time algorithm for the problem.

1.1 Our Contribution and Road Map

In this work we address several variants of the IGAPS problem. Our contribution includes generalizations and improvements to previous results as follows (summarized in Table 1).

We start by working out some of the previously vague details from Schmidt's algorithm [21] for a special case of the IGAPS problem (the assumption in [21] is that all horizontal and vertical edges have weight w_1, and each diagonal edge has weight w_1 or w_2, for some fixed w_1 and w_2). We generalize Schmidt's algorithm to yield an $O(Cn^2)$ algorithm for the general IGAPS problem (Sections 2 and 3).

In Section 4 we consider the sparse BGAPS problem. We give an $O(r \log^3(n^2/r))$ time algorithm, which improves the previous result of Krusche and Tiskin for this problem.

Next, we turn to address the periodic IGAPS problem in Section 5. Our first result on this front is obtained by extending the $O(Cn^2)$ algorithm for IGAPS to handle the periodic variant of the problem (Section 5.1). This improves Tiskin's $O(C^2n^2)$ result for periodic IGAPS. We then show, in Section 5.2, that periodic IGAPS can be reduced to BGAPS. Therefore, we obtain an $O(C^2n^2(\log\log n/\log n)^2)$ time algorithm for periodic IGAPS (when the grid graph corresponds to an alignment problem), and an $O(r \log^3(n^2/r))$ time algorithm for periodic sparse BGAPS.

Due to space limitation, proofs were omitted.

2   Preliminaries

A *sequence* is an ordered list of integers. For a sequence S, let $S[k]$ denote the k-th element of S, and let $S[k : k']$ denote the sequence $(S[k], \ldots, S[k'])$ (if $k > k'$ then $S[k : k']$ is an empty sequence). Let merge(S_1, S_2) denote the sequence obtained from merging two sorted sequences S_1 and S_2 into one sorted sequence.

Let G be an $m \times n$ grid graph with weights on the edges. Let (i, j) denote the vertex on row i and column j of the graph. The grid graph $G[i_1..i_2, j_1..j_2]$ is the

Table 1. Results for GAPS and periodic IGAPS. The results of this paper are marked by asterisks. The results for periodic IGAPS are based on reducing the periodic problems to the non-periodic problems and using the corresponding non-periodic algorithms. The results in the third row are for the IGAPS problem when the grid graph corresponds to an alignment problem.

Type	Non-periodic	Periodic
IGAPS	$O(Cn^2)$ [21]	$O(C^2n^2)$ [25]
		$O(Cn^2)$ *
	$O(C^2n^2(\log\log n/\log n)^2)$ [22]	$O(C^2n^2(\log\log n/\log n)^2)$ *
Sparse BGAPS	$O(n\sqrt{r})$ [12]	
	$O(r\log^3(n^2/r))$ *	$O(r\log^3(n^2/r))$ *
Permutation	$O(n\log^2 n)$ [22]	$O(n\log^2 n)$ *

subgraph obtained by taking the subgraph of G induced by the vertices $\{(i,j) : i_1 \le i \le i_2, j_1 \le j \le j_2\}$ and then renumbering the vertices by subtracting i_1 from each row number and j_1 from each column number. Let G_1 and G_2 be two grid graphs with the same number of rows. The *horizontal concatenation* of G_1 and G_2 is the grid graph obtained by merging the vertices in the last column of G_1 with the vertices of the first column of G_2 (each vertex is merged with a vertex with the same row number). The *removal of a column j* in G means taking the two subgraphs $G_1 = G[0..m, 0..j]$ and $G_2 = G[0..m, j+1..n]$, and concatenating G_1 and G_2 horizontally. Vertical concatenation and removal of a row are defined analogously.

For a grid graph G, we will denote the weights of the diagonal, horizontal, and vertical edges leaving a vertex (i,j) by $W_{i,j}, W_{i,j}^H, W_{i,j}^V$, respectively. Recall that we assume that $W_{i,j}^H = W_{i',j}^H$ for all i, i', j, and $W_{i,j}^V = W_{i,j'}^V$ for all i, j, j'. We now claim that we can assume without loss of generality that all horizontal and vertical edges have weight 0. We show this by giving a reduction from the general case to the restricted case.[1] The first step of the reduction is to replace the weight of each diagonal edge leaving (i,j) by $W_{i,j}' = \max(W_{i,j}, W_{i,j}^H + W_{i,j+1}^V)$. Clearly, every path in G has the same weight under the new and original weights. The next step is to replace the weight of each diagonal edge leaving (i,j) by $W_{i,j}'' = W_{i,j}' - (W_{i,j}^H + W_{i,j+1}^V)$ and replace the weights of all horizontal and vertical edges by 0. It is easy to verify that the weight of a path from (i,j) to (i',j') in the original graph is equal to the weight of the path in the new graph plus $\sum_{k=j}^{j'-1} W_{i,k}^H + \sum_{k=i}^{i'-1} W_{k,j}^V$. Thus, we shall assume throughout the paper that $W_{i,j}^H = 0$ and $W_{i,j}^V = 0$ for all i and j.

We define $\mathrm{OPT}_i(k,j)$ to be the maximum weight of a path from $(0,k)$ to (i,j). If $k > j$ then $\mathrm{OPT}_i(k,j)$ is not defined. Define

$$\mathrm{DIFFC}_{i,j}(k) = \mathrm{OPT}_i(k,j+1) - \mathrm{OPT}_i(k,j)$$

---

[1] This reduction was suggested by an anonymous referee.

and

$$\text{DIFFR}_{i,j}(k) = \text{OPT}_{i+1}(k,j) - \text{OPT}_i(k,j).$$

Note that $\text{DIFFC}_{i,j}$ and $\text{DIFFR}_{i,j}$ are defined for $0 \le k \le j$. The $\text{DIFFC}_{m,j}$ functions give an implicit representation of the all-scores matrix of G. Thus, our goal is to show how to compute all values of these functions. The algorithm for computing the values of the $\text{DIFFC}_{m,j}$ functions also computes all values of all $\text{DIFFC}_{i,j}$ and $\text{DIFFR}_{i,j}$ functions.

We now give some properties of the $\text{DIFFC}_{i,j}$ and $\text{DIFFR}_{i,j}$ functions.

Lemma 1 (Schmidt [21]). *For every i and j, $\text{DIFFC}_{i,j}(k-1) \le \text{DIFFC}_{i,j}(k)$ and $\text{DIFFR}_{i,j}(k-1) \ge \text{DIFFR}_{i,j}(k)$ for all k.*

In the next lemma we give upper and lower bounds for $\text{DIFFC}_{i,j}$ and $\text{DIFFR}_{i,j}$.

Lemma 2. $0 \le \text{DIFFC}_{i,j}(k) \le C$ *and* $0 \le \text{DIFFR}_{i,j}(k) \le C$ *for all* $0 \le k \le j$.

In the following lemmas, we will show that $\text{DIFFC}_{i+1,j}$ and $\text{DIFFR}_{i,j+1}$ can be computed efficiently from $\text{DIFFC}_{i,j}$ and $\text{DIFFR}_{i,j}$. For every k, the values $\text{DIFFC}_{i+1,j}(k)$ and $\text{DIFFR}_{i,j+1}(k)$ depend on $\text{OPT}_{i+1}(k,j+1)$. The optimal path from $(0,k)$ to $(i+1,j+1)$ passes through either $(i+1,j)$, (i,j), or $(i,j+1)$. Thus,

$$\text{OPT}_{i+1}(k,j+1) = \max\{\text{OPT}_{i+1}(k,j), \text{OPT}_i(k,j) + W_{i,j}, \text{OPT}_i(k,j+1)\}.$$

From the equality above, we obtain the following equality for $\text{OPT}_{i+1}(k,j+1) - \text{OPT}_i(k,j)$.

Lemma 3. $\text{OPT}_{i+1}(k,j+1) - \text{OPT}_i(k,j) = \text{MAX}_{i,j}(k)$, *where*

$$\text{MAX}_{i,j}(k) = \max\{\text{DIFFR}_{i,j}(k), W_{i,j}, \text{DIFFC}_{i,j}(k)\}.$$

Lemma 4. *For* $0 \le k \le j$,

$$\text{DIFFC}_{i+1,j}(k) = \text{MAX}_{i,j}(k) - \text{DIFFR}_{i,j}(k) \tag{1}$$

$$\text{DIFFR}_{i,j+1}(k) = \text{MAX}_{i,j}(k) - \text{DIFFC}_{i,j}(k). \tag{2}$$

Recall that the functions $\text{DIFFC}_{i,j}$ and $\text{DIFFR}_{i,j}$ were defined only for $k \le j$. We now extend the definition of the $\text{DIFFC}_{i,j}$ and $\text{DIFFR}_{i,j}$ functions so that these functions will be defined for every integer k, $0 \le k \le n$. We want to extend each function in a way that preserves the monotonicity property and also preserves the correctness of Lemma 4. This is done by defining $\text{DIFFC}_{i,j}(k) = C$ and $\text{DIFFR}_{i,j}(k) = 0$ for $j < k \le n$. By Lemma 2, the monotonicity of the $\text{DIFFC}_{i,j}$ and $\text{DIFFR}_{i,j}$ functions is kept.

Lemma 5. *Equations (1) and (2) hold for all* $0 \le k \le n$.

The $\text{DIFFC}_{i,j}$ and $\text{DIFFR}_{i,j}$ functions are monotone functions with integer values in the range 0 to C. We define a compact representation for these functions. Intuitively, $\text{SC}_{i,j}$ and $\text{SR}_{i,j}$ are sequences that contain the "step" indices of the corresponding $\text{DIFFC}_{i,j}$ and $\text{DIFFR}_{i,j}$ functions, i.e., the indices in which the values of these sequences change. The elements of each such sequence are sorted in non-decreasing order. Formally,

- The sequence $SC_{i,j}$ contains $DIFFC_{i,j}(0)$ elements with value $-\infty$, $DIFFC_{i,j}(k) - DIFFC_{i,j}(k-1)$ elements with value k for every $1 \leq k \leq n$, and $C - DIFFC_{i,j}(n)$ elements with value ∞.
- The sequence $SR_{i,j}$ contains $C - DIFFR_{i,j}(0)$ elements with value $-\infty$, $DIFFR_{i,j}(k-1) - DIFFR_{i,j}(k)$ elements with value k for every $1 \leq k \leq n$, and $DIFFR_{i,j}(n)$ elements with value ∞.

Each element of $SC_{i,j}$ and $SR_{i,j}$ is called a *step*. From the extended definition of the $DIFFC_{i,j}$ and $DIFFR_{i,j}$ functions it follows that each sequence $SC_{i,j}$ and $SR_{i,j}$ has exactly C elements.

Recall that by Lemma 5, $DIFFC_{i+1,j}(k) = MAX_{i,j}(k) - DIFFR_{i,j}(k)$. Therefore, the steps of $DIFFC_{i+1,j}(k)$ depend on the steps of $MAX_{i,j}$ and $DIFFR_{i,j}$. Based on this observation, our main result of this section shows how to compute $SC_{i+1,j}$ and $SR_{i,j+1}$ from $SC_{i,j}$ and $SR_{i,j}$. For the following theorem denote $SC_{i,j}[C+1] = \infty$ and $SR_{i,j}[0] = \infty$.

Theorem 1. *Let $i_1 = 1 + C - W_{i,j}$ and $i_2 = 1 + W_{i,j}$. If $W_{i,j} = C$ or $SR_{i,j}[i_1 - 1] < SC_{i,j}[i_2]$ then*

$$SC_{i+1,j} = \text{merge}(SC_{i,j}[i_2 : C], SR_{i,j}[i_1 : C]) \tag{3}$$

$$SR_{i,j+1} = \text{merge}(SC_{i,j}[1 : i_2 - 1], SR_{i,j}[1 : i_1 - 1]), \tag{4}$$

and otherwise

$$SC_{i+1,j} = S[C+1 : 2C] \tag{5}$$

$$SR_{i,j+1} = S[1 : C], \tag{6}$$

where $S = \text{merge}(SR_{i,j}, SC_{i,j})$.

3 Algorithm for IGAPS

The algorithm for IGAPS follows directly from Theorem 1.

(1) **For** $j = 0, \ldots, n - 1$ **do** $SC_{0,j} \leftarrow (j+1, \ldots, j+1)$.
(2) **For** $i = 0, \ldots, n - 1$ **do**
(3) $SR_{i,0} \leftarrow (-\infty, \ldots, -\infty)$.
(4)   **For** $j = 1, \ldots, n - 1$ **do**
(5)   Compute $SC_{i+1,j}$ and $SR_{i,j+1}$ using Theorem 1.

By Theorem 1, the computation in line 5 takes $O(C)$ time, so the overall time complexity of the algorithm is $O(Cn^2)$.

4 Algorithm for Sparse BGAPS

In what follows, an edge of weight 1 is called *active*. A row (resp., column) of G is called *inactive* if there is no active edge that starts at this row (resp., column).

In this section we give an algorithm for sparse BGAPS. Our algorithm is based on the algorithm of Krusche and Tiskin [12]. Both algorithms use a divide and conquer approach, namely they divide the input grid graph into subgraphs and solve the problem recursively on each subgraph. There are two differences between these algorithms. First, the algorithm of Krusche and Tiskin stops the partitioning when a subgraph of the grid graph has no active edges, whereas our algorithm stops the partitioning when the number of active edges is at most the size of the subgraph. Furthermore, the conquer steps of the two algorithms are different. The rest of this section is organized as follows. We first describe a result of Tiskin [26] which is used for the conquer step of our algorithm. Then, we give an algorithm for handling the case when the number of active edges is at most the size of the subgraph. Finally, we describe the algorithm for sparse BGAPS and analyze its time complexity.

Due to the assumption of 0-1 weights, each sequence $SC_{i,j}$ or $SR_{i,j}$ contains a single element. We shall therefore refer to $SC_{i,j}$ or $SR_{i,j}$ as an integer rather than a sequence. Theorem 1 then reduces to the following simplified form.

Theorem 2 (Tiskin [24]). *If $W_{i,j} = 1$ then $SC_{i+1,j} = SR_{i,j}$ and $SR_{i,j+1} = SC_{i,j}$. Otherwise, $SC_{i+1,j} = \max(SC_{i,j}, SR_{i,j})$ and $SR_{i,j+1} = \min(SC_{i,j}, SR_{i,j})$.*

According to the definitions in Section 2, the initialization of $SC_{0,j}$ is $SC_{0,j} = j + 1$. For $SR_{i,0}$ we use the initialization $SR_{i,0} = -i$. Note that this initialization is different from the one used in Section 2. For an $m \times n$ grid graph G we define

$$\text{OUT}(G) = (SC_{m,0}, SC_{m,1}, \ldots, SC_{m,n-1}, SR_{m-1,n}, SR_{m-2,n}, \ldots, SR_{0,n}).$$

With the initialization given above and by Theorem 2, $\text{OUT}(G)$ is a permutation of $(-m + 1, -m + 2, \ldots, n)$.

The algorithm of Section 3, restricted for the case of 0-1 weights, has an interpretation as a transposition network [12] (or using a different terminology, as seaweeds [22]). We now describe this interpretation using different terminology. Start with $m + n$ balls located on the edges in the first column and first row of G. The balls are numbered by $-m + 1, -m + 2, \ldots, n$ according to their anti-clockwise order. The balls are then moved along the edges of the graph. When the horizontal and vertical edges leaving a vertex (i, j), denoted e_1 and e_2, contain each a ball, these two balls are moved to the horizontal and vertical edges entering $(i + 1, j + 1)$, denoted e_3 and e_4 (the numbers of these balls represent the values $SC_{i,j}$ and $SR_{i,j}$, respectively). If $W_{i,j} = 1$ then the ball in e_1 is moved to e_4, and the ball in e_2 is moved to e_3. In other words, the ball in e_1 moves to the right, and the ball in e_2 is moved down. We call this movement *non-crossing*. If $W_{i,j} = 0$ then the movement of the balls depends on their numbers. If the ball in e_1 has smaller number than the ball in e_2 (indicating that the corresponding paths have already crossed once) then the movement of the ball is non-crossing. Otherwise, the movement is *crossing*: the ball in e_1 is moved to e_3 and the ball in e_2 is moved to e_4. We say in this case that the paths of the balls cross. Note that in both cases, exactly one ball is moved to e_3 and one ball is moved to e_4. Thus, for each edge on the last column and last row of the graph there is a distinct ball that reaches the edge.

When two balls reach the edges leaving some vertex and the movement for this vertex is crossing, the ball entering the horizontal edge must have a number greater than the number of the ball entering the vertical edge. If afterward these two balls reach the edges leaving another vertex, then the ball entering the horizontal edge has a number *smaller* than the number of the ball entering the vertical edge. Therefore, by definition, the movement in this step is non-crossing. In other words, the paths of two balls can cross at most once. Therefore, an alternative way to define the movement of the balls in e_1 and e_2 for the case $W_{i,j} = 0$ is: If the paths of the balls crossed before, then the movement is non-crossing, and otherwise the movement is crossing. The computation of $\text{OUT}(G)$ is equivalent to computing the destination edges of all balls.

We will use the following results.

Theorem 3 (Tiskin [26]). *Let G be a grid graph obtained by horizontal or vertical concatenation of two $m \times n$ grid graphs G_1 and G_2. Given $\text{OUT}(G_1)$ and $\text{OUT}(G_2)$, $\text{OUT}(G)$ can be computed in $O(n + m \log m)$ time.*

Lemma 6 (Tiskin [22]). *Let G be an $n \times n$ grid graph. Let G' be the grid graph obtained from G by removal of inactive rows and columns. Then $\text{OUT}(G)$ can be computed from $\text{OUT}(G')$ in $O(n)$ time.*

We now give an algorithm that computes $\text{OUT}(G)$ for an $n \times n$ grid graph G with at most n active edges. This algorithm will later be used as a subroutine in our solution for the sparse BGAPS problem. The algorithm is an extension of an algorithm of Tiskin [22] for permutation grid graphs. Let G be a grid graph with at most n active edges. If G has at most $n/2$ active edges, then there are at least $n/2$ inactive rows and at least $n/2$ inactive columns. Choose $n/2$ inactive rows and $n/2$ inactive columns and remove them from G to obtain a grid graph G'. Recursively compute $\text{OUT}(G')$ and then use Lemma 6 to obtain $\text{OUT}(G)$. Otherwise, let i be the maximum index such that $G_1 = G[0..i, 0..n]$ has at most $n/2$ active edges. Note that i is well defined since $G[0..0, 0..n]$ has no active edges, and $G[0..n, 0..n]$ has more than $n/2$ active edges. Let $G_2 = G[i..i + 1, 0..n]$ and $G_3 = G[i + 1..n, 0..n]$. The usage of G_2 ensures that G_3 has at most $n/2$ active edges (note that G_1 also has at most $n/2$ edges by definition). Thus, we can remove $n/2$ inactive rows and $n/2$ inactive columns in each graph and obtain graphs G_1' and G_3'. Using recursion, $\text{OUT}(G_1')$ and $\text{OUT}(G_3')$ are computed, and then $\text{OUT}(G_1)$ and $\text{OUT}(G_3)$ are obtained. Moreover, since $\text{OUT}(G_2)$ is of size $1 \times n$, $\text{OUT}(G_2)$ can be computed in $O(n)$ time. Finally, compute $\text{OUT}(G)$ from $\text{OUT}(G_1)$, $\text{OUT}(G_2)$, and $\text{OUT}(G_3)$ using Theorem 3. The time complexity function $T_2(n)$ of the algorithm satisfies the recurrence $T(n) = 2T(n/2) + O(n \log n)$. Thus, $T(n) = O(n \log^2 n)$.

Our algorithm for sparse BGAPS is as follows. Let G be an input graph of size $n \times n$. We assume that we are given as input a list of the active edges of G. For simplicity, we assume that n is a power of two.

(1) If the number of active edges is at most n, compute $\text{OUT}(G)$ and stop.
(2) Partition G into four subgraphs $G_1 = G[0..n/2, 0..n/2]$, $G_2 = G[0..n/2, n/2..n]$, $G_3 = G[n/2..n, 0..n/2]$, and $G_4 = G[n/2..n, n/2..n]$.
(3) Recursively compute $\text{OUT}(G_i)$ for each of the subgraphs.
(4) Compute $\text{OUT}(G)$ by application of Theorem 3 three times.

Time complexity analysis. Consider the total time the algorithm spends on level j of the recursion. The size of each subgraph in this level is $n' \times n'$, where $n' = n/2^j$. Clearly, the number of graphs handled in level j is at most 4^j. Moreover, the number of subgraphs in level $j-1$ on which line 3 of the algorithm operates is at most $r/\frac{n}{2^{j-1}}$ as each such subgraph contains at least $\frac{n}{2^{j-1}}$ active edges, and these subgraphs are disjoint. It follows that the number of subgraphs in level j is at most $4\frac{r}{n/2^{j-1}}$. The time complexity of handling one subgraph in level j is either $O(n' \cdot \log^2 n')$ if step 1 is performed, and $O(n' \log n')$ if steps 2–4 are performed. Therefore, the total time of the algorithm is

$$O\left(\sum_{j=0}^{\log n} \min\left(4^j, \frac{r2^j}{n}\right) \cdot \frac{n}{2^j} \log^2 \frac{n}{2^j}\right) = O\left(r \log^3 \frac{n^2}{r}\right).$$

5   Algorithms for Periodic IGAPS

We begin with some notations. Let G be an $n \times n$ grid graph. The *periodic grid graph* G^∞ is the graph obtained by taking an infinite number of copies of G and concatenating them horizontally. The columns of G^∞ are numbered by all (positive and negative) integers.

For the periodic IGAPS problem, we use the same notation as for the non-periodic problem, with minor differences. The $\text{DIFFC}_{i,j}(k)$ and $\text{DIFFR}_{i,j}(k)$ functions are defined for all integers $k \leq j$ and these functions are extended for all integers $k > j$ as before. The step sequences are defined as follows. $\text{SC}_{i,j}$ is a sequence that contains $\min_k \text{DIFFC}_{i,j}(k)$ elements with value $-\infty$, $\text{DIFFC}_{i,j}(k) - \text{DIFFC}_{i,j}(k-1)$ elements with value k for every k, and $C - \text{DIFFC}_{i,j}(j)$ elements with value ∞. The elements of $\text{SC}_{i,j}$ are sorted in non-decreasing order. The sequence $\text{SR}_{i,j}$ is defined similarly. Theorem 1 also holds for the periodic problem.

Since Lemma 6 also holds for the periodic problem, we will assume without loss of generality that G does not contain inactive rows.

5.1   Direct Algorithm

In this section we describe a quadratic time algorithm for periodic IGAPS. The following lemma shows that the $\text{SC}_{i,j}$ and $\text{SR}_{i,j}$ sequences have a periodic property. Thus, it suffices to compute $\text{SC}_{i,j}$ and $\text{SR}_{i,j}$ only for $0 \leq j \leq n-1$.

Lemma 7. *For all i, j, l, $\text{SC}_{i,j-n}[l] = \text{SC}_{i,j}[l] - n$ and $\text{SR}_{i,j-n}[l] = \text{SR}_{i,j}[l] - n$, where $-\infty - n = -\infty$ and $\infty - n = \infty$.*

Similarly to the non-periodic problem, the algorithm processes a subgraph of the input graph (corresponding to $0 \leq j \leq n-1$) in top-to-bottom traversal order. The initialization of the $SC_{0,j}$ sequences is the same as in the non-periodic problem (namely, $SC_{0,j}$ contains C elements with value $j+1$). In the non-periodic problem, the initialization of the $SR_{i,0}$ sequences is trivial (see Section 3). In the periodic problem, it may not be easy to determine $SR_{i,0}$. However, the following lemma shows that in each row i there is at least one j for which the sequence $SR_{i,j}$ can be easily determined.

Lemma 8. *Suppose that $W_{i,j} = \max_{j'} W_{i,j'}$. Then, for all $k \leq j+1$ there is a maximum weight path P from $(0,k)$ to $(i+1, j+1)$ such that the last edge in P is diagonal or vertical.*

Lemma 9. *Suppose that $W_{i,j} = \max_{j'} W_{i,j'}$. Then, $SR_{i,j+1}$ contains $C - W_{i,j}$ elements with value $-\infty$, and then the elements of $SC_{i,j}[1 : W_{i,j}]$.*

Based on Lemma 9, the algorithm processes row i of the graph as follows. First, it finds an index j^* such that $W_{i,j^*} = \max_{j'} W_{i,j'}$, and computes SR_{i,j^*+1} according to Lemma 9. Then, it computes $SC_{i+1,j}$ and $SR_{i,j+1}$ for $j = j^*+1, \ldots, n-1$ using Theorem 1. Next, the algorithm computes $SR_{i,0}$ from $SR_{i,n}$ according to Lemma 7. Finally, it computes $SC_{i+1,j}$ and $SR_{i,j+1}$ for $j = 0, \ldots, j^*$ using Theorem 1.

5.2   Reduction to BGAPS

In this section, we show that periodic IGAPS can be reduced to BGAPS. Since Tiskin showed a reduction from integer weights to 0-1 weights [22], it suffices to show a reduction from periodic BGAPS to BGAPS. We will use $SC_{i,j}^G$ and $SR_{i,j}^G$ to denote the sequences $SC_{i,j}$ and $SR_{i,j}$ with respect to a grid graph G.

Let G be an $n \times n$ grid graph. We extend the definition of $\text{OUT}(\cdot)$ to periodic grid graphs as follows:

$$\text{OUT}(G^\infty) = (SC_{n,0}^{G^\infty}, SC_{n,1}^{G^\infty}, \ldots, SC_{n,n-1}^{G^\infty}).$$

To solve periodic BGAPS, it suffices to compute $\text{OUT}(G^\infty)$.

The grid graph G^k is the graph obtained by horizontal concatenation of k copies of G. Let j_i denote the minimum index j such that $W_{i,j} = 1$. The reduction from periodic IGAPS to GAPS is based on the following lemma.

Lemma 10. *Let $k \geq 1$. Let $0 \leq i \leq n-1$ and $0 \leq j \leq n-1$ be indices such that either*

1. *$i \leq k-1$, or*
2. *$i = k$ and $j > j_i$*

then $SC_{i,j+\alpha}^{G^k} = SC_{i,j}^{G^\infty} + \alpha$, where $\alpha = (k-1)n$. Moreover, for $0 \leq i \leq n-1$ and $0 \leq j \leq n$, if either

1. $i \leq k - 2$, or
2. $i = k - 1$ and $j > j_i$

then $\mathrm{SR}_{i,j+\alpha}^{G^k} = \mathrm{SR}_{i,j}^{G^\infty} + \alpha$.

Corollary 1. If $k \geq n + 1$ then $\mathrm{OUT}(G^\infty) = (\mathrm{SC}_{n,\alpha}^{G^k} - \alpha, \mathrm{SC}_{n,1+\alpha}^{G^k} - \alpha, \ldots, \mathrm{SC}_{n,n-1+\alpha}^{G^k} - \alpha)$.

Our goal is to show how to compute $(\mathrm{SC}_{n,\alpha}^{G^k}, \mathrm{SC}_{n,1+\alpha}^{G^k}, \ldots, \mathrm{SC}_{n,n-1+\alpha}^{G^k})$ for some $k \geq n + 1$. Define

$$\mathrm{OUT}'(G^k) = (\mathrm{SC}_{n,\alpha}^{G^k}, \mathrm{SC}_{n,1+\alpha}^{G^k}, \ldots, \mathrm{SC}_{n,n-1+\alpha}^{G^k}, \mathrm{SR}_{n-1,kn}^{G^k}, \mathrm{SR}_{n-2,kn}^{G^k}, \ldots, \mathrm{SR}_{0,kn}^{G^k}).$$

The following lemma follows from Theorem 3.

Lemma 11. Given $\mathrm{OUT}'(G^k)$, $\mathrm{OUT}'(G^{2k})$ can be computed in $O(n \log n)$ time.

Based on Corollary 1 and Lemma 11, we obtain the following algorithm for solving periodic BGAPS.

(1) Compute $\mathrm{OUT}'(G)$.
(2) Let n' be the smallest power of 2 which is greater than or equal to n.
(3) **For** $k = 1, 2, 4, \ldots, n'/2$ **do**
(4)   Compute $\mathrm{OUT}'(G^{2k})$ from $\mathrm{OUT}'(G^k)$.
(5) Output $(\mathrm{OUT}'(G^{n'})[1] - (n' - 1)n, \ldots, \mathrm{OUT}'(G^{n'})[n] - (n' - 1)n)$.

We have shown the following theorem.

Theorem 4. The periodic BGAPS problem on a grid graph G can be solved in $T(G) + O(n \log^2 n)$ time, where $T(G)$ is the time complexity of solving BGAPS on G.

References

1. Alves, C.E.R., Cáceres, E.N., Song, S.W.: An all-substrings common subsequence algorithm. Discrete Applied Mathematics 156(7), 1025–1035 (2008)
2. Apostolico, A., Atallah, M.J., Hambrusch, S.E.: New clique and independent set algorithms for circle graphs. Discrete Applied Mathematics 36(1), 1–24 (1992)
3. Apostolico, A., Atallah, M.J., Larmore, L.L., McFaddin, S.: Efficient parallel algorithms for string editing and related problems. SIAM J. on Computing 19(5), 968–988 (1990)
4. Benson, G.: A space efficient algorithm for finding the best nonoverlapping alignment score. Theoretical Computer Science 45(1&2), 357–369 (1995)
5. Crochemore, M., Landau, G.M., Ziv-Ukelson, M.: A sub-quadratic sequence alignment algorithm for unrestricted cost matrices. SIAM J. on Computing 32(5), 1654–1673 (2003)
6. Hermelin, D., Landau, G.M., Landau, S., Weimann, O.: A unified algorithm for accelerating edit-distance computation via text-compression. In: Proc. 26th Symposium on Theoretical Aspects of Computer Science (STACS), pp. 529–540 (2009)

7. Hyyrö, H.: An efficient linear space algorithm for consecutive suffix alignment under edit distance (*short preliminary paper*). In: Amir, A., Turpin, A., Moffat, A. (eds.) SPIRE 2008. LNCS, vol. 5280, pp. 155–163. Springer, Heidelberg (2008)

8. Ishida, Y., Inenaga, S., Shinohara, A., Takeda, M.: Fully incremental LCS computation. In: Liśkiewicz, M., Reischuk, R. (eds.) FCT 2005. LNCS, vol. 3623, pp. 563–574. Springer, Heidelberg (2005)

9. Kannan, S., Myers, E.W.: An algorithm for locating nonoverlapping regions of maximum alignment score. SIAM J. on Computing 25(3), 648–662 (1996)

10. Kent, C., Landau, G.M., Ziv-Ukelson, M.: On the complexity of sparse exon assembly. Journal of Computational Biology 13(5), 1013–1027 (2006)

11. Kim, S.-R., Park, K.: A dynamic edit distance table. J. of Discrete Algorithms 2(2), 303–312 (2004)

12. Krusche, P., Tiskin, A.: String comparison by transposition networks. In: Proc. of London Algorithmics Workshop, pp. 184–204 (2008)

13. Krusche, P., Tiskin, A.: New algorithms for efficient parallel string comparison. In: Proc. 22nd Symposium on Parallel Algorithms and Architectures (SPAA), pp. 209–216 (2010)

14. Landau, G.M., Myers, E.W., Schmidt, J.P.: Incremental string comparison. SIAM J. on Computing 27(2), 557–582 (1998)

15. Landau, G.M., Myers, E.W., Ziv-Ukelson, M.: Two algorithms for LCS consecutive suffix alignment. J: of Computer and System Sciences 73(7), 1095–1117 (2007)

16. Landau, G.M., Schieber, B., Ziv-Ukelson, M.: Sparse LCS common substring alignment. Information Processing Letters 88(6), 259–270 (2003)

17. Landau, G.M., Schmidt, J.P., Sokol, D.: An algorithm for approximate tandem repeats. J. of Computational Biology 8(1), 1–18 (2001)

18. Landau, G.M., Ziv-Ukelson, M.: On the common substring alignment problem. J. of Algorithms 41(2), 338–359 (2001)

19. Russo, L.M.S.: Multiplication Algorithms for Monge Matrices. In: Chavez, E., Lonardi, S. (eds.) SPIRE 2010. LNCS, vol. 6393, pp. 94–105. Springer, Heidelberg (2010)

20. Sakai, Y.: An almost quadratic time algorithm for sparse spliced alignment. Theory of Computing Systems, pp. 1–22 (2009)

21. Schmidt, J.P.: All highest scoring paths in weighted grid graphs and their application to finding all approximate repeats in strings. SIAM J. of Computing 27(4), 972–992 (1998)

22. Tiskin, A.: Semi-local string comparison: algorithmic techniques and applications. arXiv:0707.3619v16

23. Tiskin, A.: Semi-local longest common subsequences in subquadratic time. J. Discrete Algorithms 6(4), 570–581 (2008)

24. Tiskin, A.: Semi-local string comparison: Algorithmic techniques and applications. Mathematics in Computer Science 1(4), 571–603 (2008)

25. Tiskin, A.: Periodic string comparison. In: Kucherov, G., Ukkonen, E. (eds.) CPM 2009 Lille. LNCS, vol. 5577, pp. 193–206. Springer, Heidelberg (2009)

26. Tiskin, A.: Fast distance multiplication of unit-monge matrices. In: Proc. 21st Symposium on Discrete Algorithms (SODA), pp. 1287–1296 (2010)

27. Wagner, R.A., Fischer, M.J.: The string-to-string correction problem. J. of the ACM 21(1), 168–173 (1974)

Time-Space Trade-Offs
for the Longest Common Substring Problem

Tatiana Starikovskaya[1] and Hjalte Wedel Vildhøj[2]

[1] Moscow State University, Department of Mechanics and Mathematics
tat.starikovskaya@gmail.com
[2] Technical University of Denmark, DTU Compute
hwv@hwv.dk

Abstract. The Longest Common Substring problem is to compute the longest substring which occurs in at least $d \geq 2$ of m strings of total length n. In this paper we ask the question whether this problem allows a deterministic time-space trade-off using $O(n^{1+\varepsilon})$ time and $O(n^{1-\varepsilon})$ space for $0 \leq \varepsilon \leq 1$. We give a positive answer in the case of two strings ($d = m = 2$) and $0 < \varepsilon \leq 1/3$. In the general case where $2 \leq d \leq m$, we show that the problem can be solved in $O(n^{1-\varepsilon})$ space and $O(n^{1+\varepsilon} \log^2 n(d \log^2 n + d^2))$ time for any $0 \leq \varepsilon < 1/3$.

1 Introduction

The *Longest Common Substring (LCS) Problem* is among the fundamental and classic problems in combinatorial pattern matching [6]. Given two strings T_1 and T_2 of total length n, this is the problem of finding the longest substring that occurs in both strings. In 1970 Knuth conjectured that it was not possible to solve the problem in linear time [10], but today it is well-known that the LCS can be found in $O(n)$ time by constructing and traversing a suffix tree for T_1 and T_2 [6]. However, obtaining linear time comes at the cost of using $\Theta(n)$ space, which in real-world applications might be infeasible.

In this paper we explore solutions to the LCS problem that achieve sublinear, i.e., $o(n)$, space usage[1] at the expense of using superlinear time. For example, our results imply that the LCS of two strings can be found deterministically in $O(n^{4/3})$ time while using only $O(n^{2/3})$ space. We will also study the time-space trade-offs for the more general version of the LCS problem, where we are given m strings $T_1, T_2, \ldots, T_m$ of total length n, and the goal is to find the longest common substring that occurs in at least d of these strings, $2 \leq d \leq m$.

1.1 Known Solutions

For $m = d = 2$ the LCS is the longest common prefix between any pair of suffixes from T_1 and T_2. Naively comparing all pairs leads to an $O(n^2|\text{LCS}|)$ time and $O(1)$ space solution, where $|\text{LCS}|$ denotes the length of the LCS.

---

[1] We assume the input is in read-only memory and not counted in the space usage.

J. Fischer and P. Sanders (Eds.): CPM 2013, LNCS 7922, pp. 223–234, 2013.

As already mentioned we can also find the LCS in $O(n)$ time and space by finding the deepest node in the suffix tree that has a suffix from both T_1 and T_2 in its subtree. Alternatively, we can build a data structure that for any pair of suffixes can be queried for the value of their longest common prefix. Building such a data structure is known as the *Longest Common Extension (LCE) Problem* and it has several known solutions [2,7]. If a data structure for a string of length n with query time $q(n)$ and space usage $s(n)$ can be built in time $p(n)$, then this implies a solution for the LCS problem using $O(q(n)n^2 + p(n))$ time and $O(s(n))$ space. For example using the deterministic data structure of Bille et al. [2], the LCS problem can be solved in $O(n^{2(1+\varepsilon)})$ time and $O(n^{1-\varepsilon})$ space for any $0 \le \varepsilon \le 1/2$.

In the general case where $2 \le d \le m$, the LCS can still be found in $O(n)$ time and space using the suffix tree approach. Using Rabin-Karp fingerprints [9] we can also obtain an efficient randomised algorithm using sublinear space. The algorithm is based on the following useful trick: Suppose that we have an efficient algorithm for deciding if there is a substring of length i that occurs in at least d of the m strings. Moreover, assume that the algorithm outputs such a string of length i if it exists. Then we can find the LCS by repeating the algorithm $O(\log |\text{LCS}|)$ times in an exponential search for the maximum value of i. To determine if there is a substring of length i that occurs in at least d strings, we start by checking if any of the $n^{1-\varepsilon}$ first substrings of length i occurs at least d times. We can check this efficiently by storing their fingerprints in a hash table and sliding a window of length i over the strings T_j, $j = 1, \ldots, m$. For each substring we look up its fingerprint in the hash table and increment an associated counter if it is the first time we see this fingerprint in T_j. If at any time a counter exceeds d, we stop and output the window. In this way we can check all i length substrings in $O(n^\varepsilon)$ rounds each taking time $O(n)$. Thus, this gives a Monte Carlo algorithm for the general LCS problem using $O(n^{1+\varepsilon} \log |\text{LCS}|)$

Table 1. The first half summarises solutions for $d = m = 2$, and the second half summarises solutions for the general case. The complexity bounds are worst-case unless otherwise stated; w.h.p. means with probability at least $1 - 1/n^c$ for any constant c.

	Space	Time	Trade-Off Interval	Description
$d = m = 2$	$O(1)$	$O(n^2|\text{LCS}|)$		Naive solution.
	$O(n^{1-\varepsilon})$	$O(n^{2(1+\varepsilon)})$	$0 \le \varepsilon \le \frac{1}{2}$	Deterministic LCE d.s. [2]
	$O(n^{1-\varepsilon})$	$O(n^{2+\varepsilon} \log |\text{LCS}|)$ w.h.p.	$0 \le \varepsilon \le 1$	Randomised LCE d.s. [2]
	$O(n^{1-\varepsilon})$	$O(n^{1+\varepsilon})$	$0 < \varepsilon \le \frac{1}{3}$	Our solution, $d=m=2$.
$2 \le d \le m$	$O(n^{1-\varepsilon})$	$O(n^{1+\varepsilon} \log |\text{LCS}|)$	$0 \le \varepsilon \le 1$	Randomised fingerprints. Correct w.h.p.
	$O(n^{1-\varepsilon})$	$O(n^{1+\varepsilon} \log^2 n(d \log^2 n + d^2))$	$0 \le \varepsilon < \frac{1}{3}$	Our solution, $2 \le d \le m$.
	$O(n)$	$O(n)$		Suffix tree.

time and $O(n^{1-\varepsilon})$ space for all $0 \leq \varepsilon \leq 1$. From the properties of fingerprinting we know that the algorithm succeeds with high probability. The algorithm can also be turned into a Las Vegas algorithm by verifying that the fingerprinting function is collision free in $O(n^2)$ time. Table 1 summarises the solutions.

1.2   Our Results

We show the following main result:

Theorem 1. *Given m strings $T_1, T_2, \ldots, T_m$ of total length n, an integer $2 \leq d \leq m$ and a trade-off parameter ε, the longest common substring that occurs in at least d of the m strings can be found in*

(i) $O(n^{1-\varepsilon})$ space and $O(n^{1+\varepsilon})$ time for $d = m = 2$ and $0 < \varepsilon \leq \frac{1}{3}$, or in

(ii) $O(n^{1-\varepsilon})$ space and $O(n^{1+\varepsilon} \log^2 n(d \log^2 n + d^2))$ time for $2 \leq d \leq m$, $0 \leq \varepsilon < \frac{1}{3}$.

The main innovation in these results is that they are both deterministic. Moreover, our first solution improves over the randomised fingerprinting trade-off by removing the $\log |LCS|$ factor. The basis of both solutions is a sparse suffix array determining the lexicographic order on $O(n^{1-\varepsilon})$ suffixes sampled from the strings $T_1, T_2, \ldots, T_m$ using difference covers.

2   Preliminaries

Throughout the paper all logarithms are base 2, and positions in strings are numbered from 1. Notation $T[i..j]$ stands for a substring $T[i]T[i+1] \cdots T[j]$ of T, and $T[i..]$ denotes the suffix of T starting at position i. The longest common prefix of strings T_1 and T_2 is denoted by $\text{lcp}(T_1, T_2)$.

2.1   Suffix Trees

We assume a basic knowledge of *suffix trees*. In order to traverse and construct suffix trees in linear time and space, we will assume that the size of the alphabet is constant. Thus, the suffix tree for a set of strings $\mathcal{S}$, denoted $ST(\mathcal{S})$, together with suffix links, can be built in $O(n)$ time and space, where n is the total length of strings in $\mathcal{S}$ [6]. We remind that a suffix link of a node labelled by a string ℓ points to the node labelled by $\ell[2..]$ and that suffix links exist for all inner nodes of a suffix tree. We need the following lemma:

Lemma 1. *Let $ST(\mathcal{S})$ be the suffix tree for a set of strings $\mathcal{S}$, and $\mathcal{A}$ be a set of all nodes (explicit or implicit) of $ST(\mathcal{S})$ labelled by substrings of another string T. I.e., the labels of the nodes in $\mathcal{A}$ are exactly all common substrings of T and strings from $\mathcal{S}$. Then $ST(\mathcal{S})$ can be traversed in $O(|T|)$ time so that*

(i) All nodes visited during the traversal will belong to $\mathcal{A}$, and
(ii) Every node in $\mathcal{A}$ will have at least one visited descendant.

Proof. We first explain how the tree is traversed. We traverse $ST(\mathcal{S})$ with T starting at the root. If a mismatch occurs or the end of T is reached at a node v (either explicit or implicit) labelled by a string ℓ we first jump to a node v' labelled by $\ell[2..]$. We do that in three steps: 1) walk up to the higher end u of the edge v belongs to; 2) follow the suffix link from u to a node u'; 3) descend from u' to v' comparing only the first characters of the labels of the edges with the corresponding characters of $\ell[2..]$ in $O(1)$ time. Then we proceed the traversal from the position of T at which the mismatch occurred. The traversal will end at the root of the suffix tree.

All nodes visited during the traversal are labelled by substrings of T, and thus belong to $\mathcal{A}$. For each i the traversal visits the deepest node of $ST(\mathcal{S})$ labelled by a prefix of $T[i..]$. Hence, conditions (i) and (ii) of the lemma hold. We now estimate the running time. Obviously, the number of successful matches is no more than $|T|$. We estimate the number of operations made due to unsuccessful matches by amortised analysis. During the traversal we follow at most $|T|$ suffix links and each time the depth of the current node decreases by at most one [6]. Hence, the number of up-walks is also bounded by $|T|$. Each up-walk decreases the current node-depth by one as well. On the contrary, traversal of an edge at step 3) increases the current node-depth by one. Since the maximal depth of a node visited by the traversal is at most $|T|$, the total number of down-walks is $O(|T|)$. $\qquad\qquad\square$

2.2 Difference Cover Sparse Suffix Arrays

A *difference cover modulo* τ is a set of integers $DC_\tau \subseteq \{0, 1, \ldots, \tau - 1\}$ which for any i, j contains two elements i', j' such that $j - i \equiv j' - i' \pmod{\tau}$. For any τ a difference cover DC_τ of size at most $\sqrt{1.5\tau} + 6$ can be computed in $O(\sqrt{\tau})$ time [4]. Note that this size is optimal to within constant factors, since any difference cover modulo τ must contain at least $\sqrt{\tau}$ elements.

For a string T of length n and a fixed difference cover modulo τ, DC_τ, we define a *difference cover sample* $DC_\tau(T)$ as the subset of T's positions that are in the difference cover modulo τ, i.e.,

$$DC_\tau(T) = \{i \mid 1 \le i \le n \wedge i \bmod \tau \in DC_\tau\} \,.$$

The following lemma captures two important properties of difference cover samples that we will use throughout the paper. The proof follows immediately from the above definitions.

Lemma 2. *The size of $DC_\tau(T)$ is $O(n/\sqrt{\tau})$, and for any pair p_1, p_2 of positions in T there is an integer $0 \le i < \tau$ such that both $(p_1 + i)$ and $(p_2 + i)$ are in $DC_\tau(T)$.*

We will consider difference cover samples of the string $T = T_1 \$_1 T_2 \$_2 \cdots T_k \$_k$, i.e., the string obtained by concatenating and delimiting the input strings with unique characters $\$_1, \ldots, \$_k$. See Figure 1 for an example of a difference cover sample of two input strings.

$$SA_\tau = [\,14\,,21\,,17\,,26\,,\ 6\ ,\ 1\ ,16\,,22\,,11\,,12\,,19\,,24\,,\ 4\ ,27\,,\ 7\ ,\ 2\ ,\ 9\ ]$$
$$LCP_\tau = \quad[\ 0\ ,3\ ,1\ ,2\ ,2\ ,0\ ,1\ ,2\ ,1\ ,2\ ,3\ ,4\ ,0\ ,1\ ,1\ ,0\ ]$$

$$SA_\tau^R = [\,14\,,\ 1\ ,17\,,21\,,26\,,\ 6\ ,16\,,22\,,11\,,19\,,12\,,24\,,\ 4\ ,\ 2\,,27\,,\ 7\ ,\ 9\ ]$$
$$LCP_\tau^R = \quad[\ 0\ ,1\ ,1\ ,4\ ,3\ ,0\ ,2\ ,4\ ,1\ ,3\ ,2\ ,1\ ,0\ ,2\ ,4\ ,0\ ]$$

Fig. 1. The string $T = T_1 \$_1 T_2 \$_2 =$ aggctagctacct$\$_1$acacctaccctag$\$_2$ sampled with the difference cover $DC_\tau = \{1,2,4\}$ modulo 5. The resulting difference cover sample is $DC_\tau(T) = \{1,2,4,6,7,9,11,12,14,16,17,19,21,22,24,26,27\}$. Below the arrays SA_τ, LCP_τ, SA_τ^R and LCP_τ^R are shown. Sampled positions in T_1 and T_2 are marked by white and black dots, respectively.

The *difference cover sparse suffix array*, denoted SA_τ is the suffix array restricted to the positions of T sampled by the difference cover, i.e., it is an array of length $n/\sqrt{\tau}$ containing the positions of the sampled suffixes, sorted lexicographically. Similarly, we define the *difference cover sparse lcp array*, denoted LCP_τ, as the array storing the longest common prefix (lcp) values of neighbouring suffixes in SA_τ. Moreover, for a sampled position $p \in DC_\tau(T)$ we denote by $RB(p)$ the reversed substring of length τ ending in p, i.e., $RB(p) = T[p]T[p-1]\ldots T[p-\tau+1]$, and we refer to this string as the *reversed block* ending in p. As for the sampled suffixes, we define arrays SA_τ^R and LCP_τ^R for the reversed blocks. The first contains the sampled positions sorted according to the lexicographic ordering of the reversed blocks, and the latter stores the corresponding longest common prefix values. See Figure 1 for an example of the arrays SA_τ, LCP_τ, SA_τ^R and LCP_τ^R.

The four arrays can be constructed in $O(n\sqrt{\tau} + (n/\sqrt{\tau})\log(n/\sqrt{\tau}))$ time and $O(n/\sqrt{\tau})$ space [3,11]. To be able to compute the longest common prefix between pairs of sampled suffixes and pairs of reversed blocks in constant time, we use the well-known technique of constructing a linear space *range minimum query* data structure [5,1] for the arrays LCP_τ and LCP_τ^R.

3   Longest Common Substring of Two Strings

In this section we prove Theorem 1(i). We do so by providing two algorithms both using $O(n/\sqrt{\tau})$ space which are then combined to obtain the desired trade-off. The first one correctly computes the LCS if it has length at least τ, while the second one works if the length of the LCS is less than τ. In the second algorithm we must assume that $\tau \le n^{2/3}$, which translates into the $\varepsilon \le 1/3$ bound on the trade-off interval.

$$\overbrace{}^{I_1}\quad\overbrace{}^{I_2}\quad\overbrace{}^{I_3}\quad\overbrace{}^{I_4}$$

$$SA_\tau = [\,14\,,21\,,17\,,26\,,\ 6\ ,\ 1\ ,16\,,22\,,11\,,12\,,19\,,24\,,\ 4\ ,27\,,\ 7\ ,\ 2\ ,\ 9\ ]$$

$$LCP_\tau = [\ 0\ ,\ 3\ ,\ 1\ ,\ 2\ ,\ 2\ ,\ 0\ ,\ 1\ ,\ 2\ ,\ 1\ ,\ 2\ ,\ 3\ ,\ 4\ ,\ 0\ ,\ 1\ ,\ 1\ ,\ 0\ ]$$

Fig. 2. The intervals of SA_τ containing the pairs with lcp values at least $\ell = 2$ for the string shown in Figure 1. The pair maximising the lcp value of the corresponding reversed blocks is $p'_1 = 11, p'_2 = 22$, which happens to be the LCS of T_1 and T_2: ctacc.

3.1 A Solution for Long LCS

We first compute a difference cover sample with parameter τ for the string $T = T_1 \$_1 T_2 \$_2$, where $\$_1, \$_2$ are special characters that do not occur in T_1 or T_2. We then construct the arrays and the range minimum query data structures described in Section 2.2 for computing longest common prefixes between pairs of sampled suffixes or pairs of reversed blocks in constant time.

The LCS is the longest common prefix of suffixes $T[p_1..]$ and $T[p_2..]$ for some $p_1 \le |T_1|$ and $p_2 > |T_1| + 1$. If $|\mathrm{LCS}| \ge \tau$ then from the property of difference cover samples (Lemma 2) it follows that there is an integer $r < \tau$ such that $p'_1 = p_1 + r$ and $p'_2 = p_2 + r$ are both in $DC_\tau(T)$, and the length of the LCS is thus $r + \mathrm{lcp}(T[p'_1..], T[p'_2..]) - 1$. In particular, this implies that

$$|\mathrm{LCS}| \quad = \quad \max_{\substack{p'_1 \le |T_1| \\ p'_2 > |T_1| + 1}} \Big(\mathrm{lcp}\big(RB(p'_1), RB(p'_2)\big) + \mathrm{lcp}\big(T[p'_1..], T[p'_2..]\big) - 1 \Big).$$

The -1 is necessary since a sampled suffix overlaps with the reversed block in one position. We will find the LCS by computing a pair of sampled positions $p_1^* \le |T_1|$, $p_2^* > |T_1| + 1$ that maximises the above expression. Obviously, this can be done by performing two constant time range minimum queries for all $O((n/\sqrt{\tau})^2)$ pairs of sampled positions, but we want to do better.

The main idea of our algorithm is to exploit the observation that since $\mathrm{lcp}(RB(p_1^*), RB(p_2^*)) \le \tau$, it must hold that $\mathrm{lcp}(T[p_1^*..], T[p_2^*..])$ is in the interval $[\ell_{max} - \tau + 1; \ell_{max}]$, where ℓ_{max} is the longest common prefix of two sampled suffixes of T_1 and T_2. Thus, we can ignore a lot of pairs with small lcp values.

First, we compute ℓ_{max} in $O(n/\sqrt{\tau})$ time by one scan of LCP_τ. We then compute the pair p_1^*, p_2^* in τ rounds. In a round i, $0 \le i \le \tau - 1$, we only consider pairs $p'_1 \le |T_1|, p'_2 > |T_1| + 1$ such that the length of the longest common prefix of $T[p'_1..]$ and $T[p'_2..]$ is at least $\ell = \ell_{max} - i$. Among these pairs we select the one maximising $\mathrm{lcp}\big(RB(p'_1), RB(p'_2)\big)$.

The candidate pairs with a longest common prefix of length at least ℓ are located in disjoint intervals $I_1, I_2, \ldots, I_k$ of SA_τ. We compute these intervals by scanning LCP_τ to identify the maximal contiguous ranges with lcp values greater than or equal to ℓ. For each interval I_j we will find a pair $p'_1 \le |T_1|, p'_2 > |T_1| + 1$ in I_j that maximises $\mathrm{lcp}\big(RB(p'_1), RB(p'_2)\big)$. If $\mathrm{lcp}\big(RB(p'_1), RB(p'_2)\big) + \ell - 1$ is greater than the maximum value seen so far, we store this value as the new maximum. See Figure 2 for an example.

Instead of searching the k intervals one by one, we process all intervals simultaneously. To do so, we first allocate an array A of size $n/\sqrt{\tau}$ and if r is the rank of a reversed block $RB(p)$, $p \in I_j$, we set $A[r]$ to be equal to j. We then scan A once and compute the longest common prefixes of every two consecutive reversed blocks ending at positions $p'_1 \le |T_1|, p'_2 > |T_1| + 1$ from the same interval. We can do this if we for each interval I_j keep track of the rightmost r such that $A[r] = j$.

The intervals considered in each round are disjoint so each round takes $O(n/\sqrt{\tau})$ time and never uses more than $O(n/\sqrt{\tau})$ space. The total time is $O(n\sqrt{\tau})$ in addition to the $O(n\sqrt{\tau} + (n/\sqrt{\tau})\log(n/\sqrt{\tau}))$ time for the construction. Hence we have showed the following lemma:

Lemma 3. *Let* $1 \le \tau \le n$. *If the length of the longest common substring of* T_1 *and* T_2 *is at least* τ, *it can be computed in* $O(n/\sqrt{\tau})$ *space and* $O(n\sqrt{\tau} + (n/\sqrt{\tau})\log n)$ *time, where* n *is the total length of* T_1 *and* T_2.

3.2   A Solution for Short LCS

In the following we require that $\tau \le n^{2/3}$, or, equivalently, that $\tau \le n/\sqrt{\tau}$. Let us assume, for simplicity, that $n_1 = |T_1|$ is a multiple of τ. Note that if $|LCS| \le \tau$ then the LCS is a substring of one of the following strings: $T_1[1..2\tau], T_1[\tau + 1..3\tau], \ldots, T_1[n_1-2\tau+1..n_1]$. Therefore, we can reduce the problem of computing the LCS to the problem of computing the longest substring of T_2 which occurs in at least one of these strings.

We divide the set $\mathcal{S} = \{T_1[1..2\tau], T_1[\tau + 1..3\tau], \ldots, T_1[n_1 - 2\tau + 1..n_1]\}$ into disjoint subsets $\mathcal{S}_i$, $i = 1, \ldots, \sqrt{\tau}$, such that the total length of strings in $\mathcal{S}_i$ is no more $2n/\sqrt{\tau}$ (note that we can do this since $\tau \le n/\sqrt{\tau}$). For each $\mathcal{S}_i$ we compute the longest substring t_i^* of T_2 which occurs in one of the strings in $\mathcal{S}_i$, and take the one of the maximal length.

To compute t_i^* for S_i we build the generalised suffix tree $ST(\mathcal{S}_i)$ for the strings in $\mathcal{S}_i$. We traverse $ST(\mathcal{S}_i)$ with T_2 as described in Lemma 1. Any common substring of T_2 and one of the strings in S_i will be a prefix of the label of some visited node in $ST(\mathcal{S}_i)$. It follows that t_i^* is the label of the node of maximal string depth visited during the traversal.

We now analyse the time and space complexity of the algorithm. Since the total length of the strings in $\mathcal{S}_i$ is at most $2n/\sqrt{\tau}$, the suffix tree can be built in $O(n/\sqrt{\tau})$ space and time. The traversal takes $O(n)$ time (see Lemma 1). Consequently, t_i^* can be found in $O(n/\sqrt{\tau})$ space and $O(n)$ time. By repeating for all $i = 1, \ldots, \sqrt{\tau}$, we obtain the following lemma:

Lemma 4. *Let* $1 \le \tau \le n^{2/3}$. *If the length of longest common substring of* T_1 *and* T_2 *is at most* τ, *it can be computed in* $O(n/\sqrt{\tau})$ *space and* $O(n\sqrt{\tau})$ *time, where* n *is the total length of* T_1 *and* T_2.

Combining the Solutions. By combining Lemma 3 and Lemma 4, we see that the LCS can be computed in $O(n/\sqrt{\tau})$ space and $O(n\sqrt{\tau} + (n/\sqrt{\tau})\log n)$ time for $1 \leq \tau \leq n^{2/3}$. Substituting $\tau = n^{2\varepsilon}$ the space bound becomes $O(n^{1-\varepsilon})$ and the time $O(n^{1+\varepsilon} + n^{1-\varepsilon} \log n)$, which is $O(n^{1+\varepsilon})$ for $\varepsilon > 0$. This concludes the proof of Theorem 1(i).

4 Longest Common Substring of Multiple Strings

In this section we prove Theorem 1(ii). Similar to the case of two strings, the algorithm consists of two procedures that both use space $O(n/\sqrt{\tau})$. The first one correctly computes the LCS if its length is at least $\tau' = \frac{1}{11}\tau \log^2 n$, while the second works if the length of the LCS is at most τ'. We then combine the solutions to obtain the desired trade-off. The choice of the specific separation value τ' comes from the fact that we need $\tau' \leq n$, and since the general solution for long LCS requires a data structure with a superlinear space bound.

4.1 A General Solution for Long LCS

Note that we cannot use the same idea that we use in the case of two strings since the property of difference cover samples (Lemma 2) does not necessarily hold for d positions. Instead we propose a different approach described below.

If $d > n/\sqrt{\tau}$, the algorithm returns an empty string and stops. This can be justified by the following simple observation.

Lemma 5. *If $d > n/\sqrt{\tau}$ then $|LCS| < \tau$.*

Proof. From $d > n/\sqrt{\tau}$ it follows that among any d strings from $T_1, T_2, \ldots, T_m$ there is at least one string shorter than $\sqrt{\tau}$. Therefore, the length of LCS is smaller than $\sqrt{\tau} < \tau$.   □

This leaves us with the case where $d \leq n/\sqrt{\tau}$. We first construct the difference cover sample with parameter τ' for the string $T = T_1\$_1 T_2\$_2 \cdots T_m\$_m$, where $\$_i$, $1 \leq i \leq m$, are special characters that do not occur in $T_1, T_2, \ldots, T_m$. We also construct the arrays and the range minimum query data structures described in Section 2.2 for computing longest common prefixes between pairs of sampled suffixes or pairs of reversed blocks in constant time.

Suppose that the LCS is a prefix of $T_i[p_i..]$, for some $1 \leq i \leq m$, $1 \leq p_i \leq |T_i|$. Then to compute $|LCS|$ it is enough to find $(d-1)$ suffixes of distinct strings from $T_1, T_2, \ldots, T_m$ such that the lcp values for them and $T_i[p_i..]$ are maximal. The length of the LCS will be equal to the minimum of the lcp values. Below we show how to compute the minimum.

Let N_1 stand for zero, and N_i, $i \geq 2$, stand for the length of $T_1\$_1 \cdots T_{i-1}\$_{i-1}$. Consider the sampled positions $p_i^1, p_i^2, \ldots, p_i^z$ in an interval $[N_i + p_i, N_i + p_i + \tau']$ (see Figure 3).

From the property of the difference cover samples it follows that there is an integer $r < \tau'$ such that both $p_i' = (N_i + p_i) + r$ and $p_j' = (N_j + p_j) + r$ are in $DC_{\tau'}(T)$ — in particular, $p_i' = p_i^k$ for some k. Moreover, if $\operatorname{lcp}(T_i[p_i..], T_j[p_j..]) \geq$

Fig. 3. Sampled positions $p_i^1, p_i^2, \ldots, p_i^z$ of T in an interval $[N_i + p_i, N_i + p_i + \tau']$, and a reversed block $RB(p_i^k)$

τ', then the length of the longest common prefix of $RB(p_i^k)$ and $RB(p_j')$ is at least $r = (p_i^k - N_i) - p_i$.

Let lcp_j^k be the maximum length of the longest common prefix of $T_i[p_i^k - N_{i..}]$ and $T_j[p_j' - N_{j..}]$, taken over all possible choices of p_j', $N_j < p_j' \leq N_{j+1}$, such that $\text{lcp}\big(RB(p_i^k), RB(p_j')\big) \geq ((p_i^k - N_i) - p_i)$. For each k we define a list L^k to contain values $((p_i^k - N_i) - p_i) + \text{lcp}_j^k - 1$, $j \neq i$, in decreasing order. Note that since the number of the sampled positions in $[N_i + p_i, N_i + p_i + \tau']$ is at most $\sqrt{1.5\tau'} + 6$ (see Section 2.2), the number of the lists does not exceed $\sqrt{1.5\tau'} + 6$ as well.

We first explain how we use the lists to obtain the answer and then how their elements are retrieved. The lists L^k are merged into a sorted list L until it contains values corresponding to suffixes of $(d - 1)$ distinct strings from $T_1, T_2, \ldots, T_m$. The algorithm maintains a heap H_{val} on the values stored in the heads of the lists and a heap H_{id} on the distinct identifiers of strings already added to L. At each step it takes the maximum value in H_{val} and moves it from its list to L. Then it updates H_{val} and H_{id} and proceeds. The last value added to L will be equal to the length of the LCS.

We now explain how to retrieve values from L^k. Consider a set S of $|DC_{\tau'}(T)|$ coloured points in the plane, where a point corresponding to a position $p \in DC_{\tau'}(T)$ will have x-coordinate equal to the rank of $T[p..]$ in the lexicographic ordering of the sampled suffixes, y-coordinate equal to the rank of $RB(p)$ in the lexicographic ordering of the reversed blocks, and colour equal to the number of the string $T[p..]$ starts within.

We will show that after having retrieved the first $\ell - 1$ elements from L^k, the next element can be retrieved using $O(\log n)$ *coloured orthogonal range reporting* queries on the set S. For an integer ℓ and an axis-parallel rectangle $[a_1, b_1] \times [a_2, b_2]$, such a query reports ℓ points of distinct colours lying in the rectangle. We need only to consider the positions p such that $\text{lcp}\big(RB(p_i^k), RB(p)\big) \geq ((p_i^k - N_i) - p_i)$. These positions form an interval I^k of the reversed block array, SA_τ^R. For each L^k we maintain a rectangle $R = [x_1; x_2] \times I^k$ such that $x_1 \leq x \leq x_2$, where x is the x-coordinate of the point corresponding to the position p_i^k. After the first $(\ell - 1)$ elements of L^k have been retrieved, R contains points of $(\ell - 1)$ colours besides i and $L^k[\ell - 1] = ((p_i^k - N_i) - p_i) + \text{lcp}(x_1, x_2) - 1$, where $\text{lcp}(x_1, x_2)$ is the longest common prefix of suffixes of T with ranks x_1 and x_2 (see Figure 4). To retrieve the next element we extend R until it contains points of ℓ colours not equal to i. We do this by extending either its left or right side until it includes a point of a new colour. We keep the rectangle that maximises

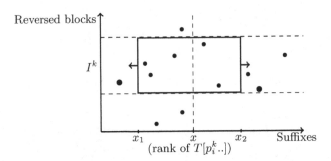

Fig. 4. Retrieving the ℓth element of L^k. A rectangle $R = [x_1; x_2] \times I^k$ contains points of $(\ell - 1)$ colours besides i. The two points of new colours shown in bold are the closest points of new colours from the left and from the right. We extend either the left or right side of the rectangle until it includes one of these points.

$\text{lcp}(x_1, x_2)$. Finding the two candidate rectangles can be done by performing two separate binary searches for the right and left sides using $O(\log n)$ coloured orthogonal range queries. Note that in each query at most ℓ points are to be reported.

The procedure described above is repeated for all $1 \leq i \leq m$ and $1 \leq p_i \leq |T_i|$. The maximum of the retrieved values will.be equal to the length of the LCS. We can compute the LCS itself, too, if we remember i and p_i on which the maximum is achieved.

Lemma 6. *Let $1 \leq \tau \leq 11n/\log^2 n$, and let LCS denote the longest substring that appears in at least d of the strings $T_1, T_2, \ldots, T_m$ of total length n. In the case where $|LCS| \geq \frac{1}{11}\tau \log^2 n$, the LCS can be found in $O(n/\sqrt{\tau})$ space and $O(nd\sqrt{\tau}\log^2 n(\log^2 n + d))$ time.*

Proof. If $d > n/\sqrt{\tau}$, the algorithm returns an empty string and thus is correct. Otherwise, $\tau' = \frac{1}{11}\tau \log^2 n \leq n$, and correctness of the algorithm follows from its description. The data structures for performing constant time lcp computations require $O(n/\sqrt{\tau})$ space and can be built in $O(n\sqrt{\tau}\log n)$ time.

Suppose that i and p_i are fixed. Each interval I^k can be found using $O(\log n)$ lcp computations. To perform coloured orthogonal range queries on the set S of size $|DC_{\tau'}(T)| = O(n/(\sqrt{\tau}\log n))$, we use the data structure [8] that can be constructed in $O(|S|\log^2 |S|) = O((n\log n)/\sqrt{\tau})$ time and $O(|S|\log |S|) = O(n/\sqrt{\tau})$ space and allows to report ℓ points of distinct colours in time $O(\log^2 |S| + \ell) = O(\log^2 n + \ell)$. Thus retrieving $L^k[\ell]$ takes time $O(\log n(\log^2 + \ell))$. The merge stops after retrieving at most d elements from each of the $O(\sqrt{\tau'})$ lists, which will take $O(d\sqrt{\tau'}\log n(\log^2 n + d)) = O(d\sqrt{\tau}\log^2 n(\log^2 n + d))$ time.

Merging the lists into L will take $O(\log \tau' + \log d)$ time per element, i.e., $O(d\sqrt{\tau'}(\log \tau' + \log d)) = O(d\sqrt{\tau}\log^{3/2} n)$ time in total, and $O(\sqrt{\tau'} + d) = O(n/\sqrt{\tau})$ space (remember that we are in the case $d \leq n/\sqrt{\tau}$). Therefore, computing the longest prefix of $T_i[p_i..]$ which occurs in at least $(d-1)$ other strings will take $O(d\sqrt{\tau}\log^2 n(\log^2 n + d))$ time. The lemma follows.   □

4.2   A General Solution for Short LCS

We start by proving the following lemma:

Lemma 7. *Given input strings $T_1, T_2, \ldots, T_m$ of total length n and a string S of length $|S|$. The longest substring t of S that appears in at least d of the input strings can be found in $O((|S| + n) \log |t|)$ time and $O(|S|)$ space.*

Proof. We prove that there is an algorithm that takes an integer i, and in $O(|S| + n)$ time and $O(|S|)$ space either finds an i-length substring of S that occurs in at least d input strings, or reports that no such substring exists. The lemma then follows, since by running the algorithm $O(\log |t|)$ times we can do an exponential search for the maximum value of i.

We construct the algorithm as follows. First we build the suffix tree $ST(S)$ for the string S, together with all suffix links. For every node of the suffix tree we store a pointer to its ancestor of string depth i (all such pointers can be computed in $O(|S|)$ time by post-processing the tree). Besides, for every node $v \in ST(S)$ of string depth i (explicit or implicit), we store a counter $c(v)$ and an integer $id(v)$, both initially set to zero. These nodes correspond exactly to the i-length substrings of S, and we will use $c(v)$ to count the number of distinct input strings that the label of v occurs in. To do this, we traverse $ST(S)$ with the input strings $T_1, T_2, \ldots, T_m$ one at a time as described in Lemma 1. When matching a character a of T_j, we always check if a node v of string depth i above our current location has $id(v) < j$. In that case, we increment the counter $c(v)$ and set $id(v) = j$ to ensure that the counter is only incremented once for T_j.

To prove the correctness note that for any i-length substring ℓ of T_j that also occurs in S there exists a node of $ST(T)$ labelled by it, and one of the descendants of this node will be visited during the matching process of T_j (see Lemma 1). The converse is also true, because any node $v' \in ST(T)$ visited during the traversal implies that all prefixes of the label of v' occur in T_j.

The suffix tree for S can be constructed in $O(|S|)$ time and space. The traversal with T_j can be implemented to take time $O(|T_j|)$, i.e., $O(n)$ time for all the input strings. In addition to the suffix tree, at most $|S|$ constant space counters are stored. Thus the algorithm requires $O(n + |S|)$ time and $O(|S|)$ space.   □

We now describe the algorithm for finding the LCS when $|\text{LCS}| \leq \tau' = \frac{1}{11}\tau \log^2 n$. Consider the partition of T into substrings of length $\delta n/\sqrt{\tau}$ overlapping in τ' positions, where δ is a suitable constant. Assuming that $\tau \leq n^{2/3-\gamma}$ for some constant $\gamma > 0$, implies that these strings will have length at least $2\tau'$, and thus the LCS will be a substring of one of them. We examine the strings one by one and apply Lemma 7 to find the longest substring that occurs in at least d input strings. It follows that we can check one string in $O(n/\sqrt{\tau})$ space and $O(n \log n)$ time, so by repeating for all $O(\sqrt{\tau})$ strings, we have:

Lemma 8. *Let $1 \leq \tau \leq n^{2/3-\gamma}$ for some constant $\gamma > 0$, and let LCS denote the longest substring that appear in at least d of the strings $T_1, T_2, \ldots, T_m$ of total length n. If $|\text{LCS}| \leq \frac{1}{11}\tau \log^2 n$, the LCS can be found in $O(n/\sqrt{\tau})$ space and $O(\sqrt{\tau} n \log n)$ time.*

Combining the Solutions. Our specific choice of separation value ensures that the assumption on τ of Lemma 8 implies the assumption of Lemma 6 (because $n^{2/3-\gamma} \leq 11n/\log^2 n$ for all n and $\gamma > 0$). Thus by combining the two solutions the LCS can be computed in $O(n/\sqrt{\tau})$ space and $O(d\sqrt{\tau}n\log^2 n(\log^2 n + d))$ time for $1 \leq \tau \leq n^{2/3-\gamma}$, $\gamma > 0$. Substituting $\tau = n^{2\varepsilon}$, we obtain the bound stated by Theorem 1(ii) with the requirement that $0 \leq \varepsilon < 1/3$.

5 Open Problems

We conclude with some open problems. Is it possible to extend the trade-off range of our solutions to ideally $0 \leq \varepsilon \leq 1/2$? Can the time bound for the general LCS problem be improved so it fully generalises the solution for two strings? The difference cover technique requires $\Omega(\sqrt{n})$ space, so the most interesting question is perhaps whether the LCS problem can be solved deterministically in $O(n^{1-\varepsilon})$ space and $O(n^{1+\varepsilon})$ time for any $0 \leq \varepsilon \leq 1$?

Acknowledgements. T. Starikovskaya has been partly supported by a grant 10-01-93109-CNRS-a of the Russian Foundation for Basic Research and by Dynasty foundation.

References

1. Bender, M.A., Farach-Colton, M.: The LCA Problem Revisited. In: Gonnet, G.H., Viola, A. (eds.) LATIN 2000. LNCS, vol. 1776, pp. 88–94. Springer, Heidelberg (2000)
2. Bille, P., Gørtz, I.L., Sach, B., Vildhøj, H.W.: Time-Space Trade-Offs for Longest Common Extensions. In: Kärkkäinen, J., Stoye, J. (eds.) CPM 2012. LNCS, vol. 7354, pp. 293–305. Springer, Heidelberg (2012)
3. Burkhardt, S., Kärkkäinen, J.: Fast Lightweight Suffix Array Construction and Checking. In: Baeza-Yates, R., Chávez, E., Crochemore, M. (eds.) CPM 2003. LNCS, vol. 2676, pp. 55–69. Springer, Heidelberg (2003)
4. Colbourn, C.J., Ling, A.C.H.: Quorums from difference covers. Inf. Process. Lett. 75(1-2), 9–12 (2000)
5. Gabow, H.N., Bentley, J.L., Tarjan, R.E.: Scaling and Related Techniques for Geometry Problems. In: Proc. 16th STOC, pp. 135–143 (1984)
6. Gusfield, D.: Algorithms on Strings, Trees and Sequences: Computer Science and Computational Biology. Cambridge University Press, New York (1997)
7. Ilie, L., Navarro, G., Tinta, L.: The longest common extension problem revisited and applications to approximate string searching. J. Discrete Algorithms 8(4), 418–428 (2010)
8. Janardan, R., Lopez, M.: Generalized Intersection Searching Problems. Int. J. Comput. Geom. Appl. 3(1), 39–69 (1993)
9. Karp, R.M., Rabin, M.O.: Efficient Randomized Pattern-Matching Algorithms. IBM J. Res. Dev. 31(2), 249–260 (1987)
10. Knuth, D.E., Morris, J.H., Pratt, V.R.: Fast Pattern Matching in Strings. SIAM J. Comput. 6(2), 323–350 (1977)
11. Puglisi, S.J., Turpin, A.: Space-Time Tradeoffs for Longest-Common-Prefix Array Computation. In: Hong, S.-H., Nagamochi, H., Fukunaga, T. (eds.) ISAAC 2008. LNCS, vol. 5369, pp. 124–135. Springer, Heidelberg (2008)

A Succinct Grammar Compression[*]

Yasuo Tabei[1], Yoshimasa Takabatake[2], and Hiroshi Sakamoto[2,3]

[1] ERATO Minato Project, JST, Japan
[2] Kyushu Institute of Technology, Japan
[3] PRESTO, JST, Japan
tabei.y.aa@m.titech.ac.jp, {takabatake,hiroshi}@donald.ai.kyutech.ac.jp

Abstract. We solve an open problem related to an optimal encoding of a *straight line program* (SLP), a canonical form of grammar compression deriving a single string deterministically. We show that an information-theoretic lower bound for representing an SLP with n symbols requires at least $2n + \log n! + o(n)$ bits. We then present a succinct representation of an SLP; this representation is asymptotically equivalent to the lower bound. The space is at most $2n \log \rho(1 + o(1))$ bits for $\rho \leq 2\sqrt{n}$, while supporting random access to any production rule of an SLP in $O(\log \log n)$ time. In addition, we present a novel dynamic data structure associating a digram with a unique symbol. Such a data structure is called a *naming function* and has been implemented using a hash table that has a space-time tradeoff. Thus, the memory space is mainly occupied by the hash table during the development of production rules. Alternatively, we build a dynamic data structure for the naming function by leveraging the idea behind the *wavelet tree*. The space is strictly bounded by $2n \log n(1 + o(1))$ bits, while supporting $O(\log n)$ query and update time.

1   Introduction

Grammar compression has been an active research area since at least the seventies. The problem consists of two phases: (i) building the smallest[1] context-free grammar (CFG) generating an input string uniquely and (ii) encoding an obtained CFG as compactly as possible.

The phase (i) is known as an NP-hard problem which can not be approximated within a constant factor [21]. Therefore, many researchers have made considerable efforts to design grammar compressions achieving better approximation results in the last decade. Charikar et al. [6] and Rytter [29] independently proposed the first $O(\log \frac{u}{g})$-approximation algorithms based on balanced grammar construction for the length u of a string and the size g of the smallest CFG. Later, Sakamoto [31] also developed an $O(\log \frac{u}{g})$-approximation algorithm based on an idea called pairwise comparison. In particular, Lehman [21] proved

---

[*] This study was supported by KAKENHI(23680016,20589824) and JST PRESTO program.`
[1] This is almost equal to minimizing the number of variables in G.

J. Fischer and P. Sanders (Eds.): CPM 2013, LNCS 7922, pp. 235–246, 2013.
© Springer-Verlag Berlin Heidelberg 2013

that LZ77 [35] achieved the best approximation of $O(\log n)$ under the condition of an unlimited window size. Since the minimum *addition chain* problem is a special case of the problem of finding the smallest CFG [20], modifying the approximation algorithms proposed so far is a difficult problem. Thus, the problem of grammar compression is pressing in the phase (ii).

A straight line program (SLP) is a canonical form of a CFG, and has been used in many grammar compression algorithms [36,19,35,1,23]. The production rules in SLPs are in Chomsky normal form where the right hand side of a production rule in CFGs is a *digram*: a pair of symbols. Thus, if n symbols are stored in an array called a *phrase dictionary* consisting of $2n$ fixed-length codes each of which is represented by $\log n$ bits, the memory of the dictionary is $2n \log n$ bits, resulting in the memory for storing an input string usually being exceeded. Although directly addressable codes achieving entropy bounds on strings whose memory consumption is the same as that of the fixed-length codes in the worst case have been presented [8,30,11], there are no codes that achieve an information-theoretic lower bound of storing an SLP in a phrase dictionary. Since a nontrivial information-theoretic lower bound of directly addressable codes for a phrase dictionary remains unknown, establishing the lower bound and developing novel codes for optimally representing an SLP are challenges.

We present an optimal and directly addressable SLP within a strictly bounded memory close to the amount of a plain representation of the phrase dictionary. We first give an information-theoretic lower bound on the problem of encoding an SLP, which has been unknown thus far. Let C be a class of objects. Representing an object $c \in C$ requires at least $\log |C|$ bits. A representation of c is succinct if it requires at most $\log |C|(1 + o(1))$ bits. Considering the facts and the characteristics of SLPs indicated in [23], one can predict that the lower bound for the class of SLPs with n symbols would be between $2n$ and $4n + \log n!$. By leveraging this prediction, we derive that a lower bound of bits to represent SLPs is $2n + \log n!$.

We then present an almost optimal encoding of SLPs based on *monotonic subsequence decomposition* of a sequence. Any permutation of $[1, n]$ is decomposable into at most $\rho \leq 2\sqrt{n}$ monotonic subsequences in $O(n^{1.5})$ time [33] and there is a 1.71-approximation[2] algorithm in $O(n^3)$ time [9]. While the previous encoding method for SLPs presented in [32] is also based on the decomposition, the size is not asymptotically equal to the lower bound when $\rho \simeq \sqrt{n}$. We improve the data structure by using the *wavelet tree* (WT) [12] and its improved results [3,10] such that our novel data structure achieves the smaller bound of $\min\{2n + n \log n + o(1), 2n \log \rho(1 + o(1))\}$ bits for any SLP with n symbols while supporting $O(\log \log \rho)$ access time. Our method is applicable to any types of algorithm generating SLPs including Re-Pair [19] and an online algorithm called LCA [24]. Barbay et al. [4] presented a succinct representation of a sequence using the monotonic subsequence decomposition. Their method uses the

---

[2] Minimizing ρ is NP-hard.

representation of an integer function built on a succinct representation of integer ranges. Its size is estimated to be the *degree entropy* of an ordered tree [14].

Another contribution of this paper is to present a dynamic data structure for checking whether or not a production rule in a CFG has been generated in execution. Such a data structure is called a *naming function*, and is also necessary for practical grammar compressions. When the set of symbols is static, we can construct a perfect hash as a naming function in linear time, which achieves an amount of space within around a factor of 2 from the information-theoretical minimum [5]. However, variables of SLPs are generated step by step in grammar compression. While the function can be dynamically computed by a randomization [17] or a deterministic solution [16] in $O(1)$ time and linear space, a hidden constant in the required space was not clear. We present a dynamic data structure to compute function values in $O(\log n)$ query time and update time. The space is strictly bounded by $2n \log n (1 + o(1))$ bits.

2   Preliminaries

2.1   Grammar Compression

For a finite set C, $|C|$ denotes its cardinality. *Alphabet* Σ is a finite set of letters and $\sigma = |\Sigma|$ is a constant. $\mathcal{X}$ is a recursively enumerable set of *variables* with $\Sigma \cap \mathcal{X} = \emptyset$. A sequence of symbols from $\Sigma \cup \mathcal{X}$ is called a string. The set of all possible strings from Σ is denoted by Σ^*. For a string S, the expressions $|S|$, $S[i]$, and $S[i,j]$ denote the length of S, the i-th symbol of S, and the substring of S from $S[i]$ to $S[j]$, respectively. Let $[S]$ be the set of symbols composing S. A string of length two is called a *digram*.

A CFG is represented by $\mathcal{G} = (\Sigma, V, P, X_s)$ where V is a finite subset of $\mathcal{X}$, P is a finite subset of $V \times (V \cup \mathcal{X})^*$, and $X_s \in V$. A member of P is called a production rule and X_s is called the start symbol. The set of strings in Σ^* derived from X_s by $\mathcal{G}$ is denoted by $L(\mathcal{G})$.

A CFG $\mathcal{G}$ is called *admissible* if exactly one $X \to \alpha \in P$ exists and $|L(\mathcal{G})| = 1$. An admissible $\mathcal{G}$ deriving S is called a grammar compression of S for any $X \in V$.

We consider only the case $|\alpha| = 2$ for any production rule $X \to \alpha$ because any grammar compression with n variables can be transformed into such a restricted CFG with at most $2n$ variables. Moreover, this restriction is useful for practical applications of compression algorithms, e.g., LZ78 [36], REPAIR [19], and LCA [24], and indices, e.g., SLP [7] and ESP [22].

The derivation tree of G is represented by a rooted ordered binary tree such that internal nodes are labeled by variables in V and the *yields*, i.e., the sequence of labels of leaves is equal to S. In this tree, any internal node $Z \in V$ has a left child labeled X and a right child labeled Y, which corresponds to the production rule $Z \to XY$.

If a CFG is obtained from any other CFG by a permutation $\pi : \Sigma \cup V \to \Sigma \cup V$, they are identical to each other because the string derived from one is transformed to that from the other by the renaming. For example, $P = \{Z \to XY, Y \to ab, X \to aa\}$ and $P' = \{X \to YZ, Z \to ab, Y \to aa\}$ are identical each other. On the other hand, they are clearly different from $P'' = \{Z \rightarrow aY, Y \rightarrow bX, X \rightarrow aa\}$ because their depths are different. Thus, we assume the following canonical form of CFG called *straight line program* (SLP).

Definition 1. *(Karpinsk-Rytter-Shinohara [18]) An SLP is a grammar compression over $\Sigma \cup V$ whose production rules are formed by either $X_i \rightarrow a$ or $X_k \rightarrow X_i X_j$, where $a \in \Sigma$ and $1 \leq i, j < k \leq |V|$.*

2.2   Phrase/Reverse Dictionary

For a set P of production rules, a *phrase dictionary* D is a data structure for directly accessing the phrase $X_i X_j$ for any $X_k \in V$ if $X_k \rightarrow X_i X_j \in P$. Regarding a triple (k, i, j) of positive integers as $X_k \rightarrow X_i X_j$, we can store the phrase dictionary consisting of n variables in an integer array $D[1, 2n]$, where $D[2k - 1] = D[2k] = 0$ if k belongs to an alphabet i.e., $1 \leq k \leq |\Sigma|$. X_i and X_j are accessible as $D[2k - 1]$ and $D[2k]$ by indices $2k - 1$ and $2k$ for X_k, respectively. A plain representation of D using fixed-length codes requires $2n \log n$ bits of space to store n production rules.

Reverse dictionary D^{-1} is a data structure for directly accessing the variable X_k given $X_i X_j$ for a production rule $X_k \rightarrow X_i X_j \in P$. Thus, $D^{-1}(X_i X_j)$ returns X_k if $X_k \rightarrow X_i X_j \in P$. A hash table is a representative data structure for D^{-1} enabling $O(1)$ time access and achieving $O(n \log n)$ bits of space.

2.3   Rank/Select Dictionary

We present a phrase dictionary based on the *rank/select dictionary*, a data structure for a bit string B [13] supporting the following queries: $\text{rank}_c(B, i)$ returns the number of occurrences of $c \in \{0, 1\}$ in $B[1, i]$ and $\text{select}_c(B, i)$ returns the position of the i-th occurrence of $c \in \{0, 1\}$ in B. For example, if $B = 10110100111$ is given, then $\text{rank}_1(S, 7) = 4$ because the number of 1s in $B[1, 7]$ is 4, and $\text{select}_1(S, 5) = 9$ because the position of the fifth 1 in B is 9. Although naive approaches require the $O(|B|)$ time to compute a rank, several data structures with only the $|B| + o(|B|)$ bit storage to achieve $O(1)$ time [26,27] have been presented. Most methods compute a select query by a binary search on a bit string B in $O(\log |B|)$ time. A data structure for computing the select query in $O(1)$ time has also been presented [28].

2.4   Wavelet Tree

A WT is a data structure for a string $S \in \Sigma^*$, and it can be used to compute the rank and select queries on a string S over an ordinal alphabet in $O(\log \sigma)$ time and $n \log \sigma(1 + o(1))$ bits [12]. Data structures supporting the rank and select queries in $O(\log \log \sigma)$ time with the same space have been proposed [10,3]. WT also supports access(S, i) which returns $S[i]$ in $O(\log \sigma)$ time. Recently, WT has been extended to support various operations on strings [25].

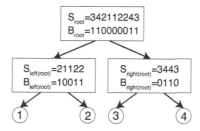

Fig. 1. Example of wavelet tree for a sequence $S = 342112243$ over an alphabet $\{1, 2, 3, 4\}$

A WT for a sequence S over $\Sigma = \{1, ..., \sigma\}$ is a binary tree that can be, recursively, presented over a sub-alphabet range $[a, b] \subseteq [1, \sigma]$. Let S_v be a sequence represented in a node v, and let $left(v)$ and $right(v)$ be left and right children of node v, respectively. The root v_{root} represents $S_{root} = S$ over the alphabet range $[1, \sigma]$. At each node v, S_v is split into two subsequences $S_{left(v)}$ consisting of the sub-alphabet range $[a, \lfloor \frac{(a+b)}{2} \rfloor]$ for $left(v)$ and $S_{right(v)}$ consisting of the sub-alphabet range $[\lfloor \frac{(a+b)}{2} \rfloor + 1, b]$ for $right(v)$ where $S_{left(v)}$ and $S_{right(v)}$ keep the order of elements in S_v. The splitting process repeats until $a = b$. Each node v in the binary tree contains a rank/select dictionary on a bit string B_v. Bit $B_v[k]$ indicates whether $S_v[k]$ should be moved to $left(v)$ or $right(v)$. If $B_v[k] = 0$, $S_{left(v)}$ contains $S_v[k]$. If $B_v[k] = 1$, $S_{right(v)}$ inherits $S_v[k]$. Formally, $B_v[k]$ with an alphabet range $[a, b]$ is defined as:

$$B_v[k] = \begin{cases} 1 \text{ if } S_v[k] > \lfloor (a+b)/2 \rfloor \\ 0 \text{ if } S_v[k] \leq \lfloor (a+b)/2 \rfloor \end{cases}.$$

An example of a WT is shown in Figure 1. In this example, since $S_{root}[2] = 4$ belongs to the higher half $[3, 4]$ of an alphabet range $[1, 4]$ represented in the root; therefore, it is the second element of S_{root} that must go to the right child of the root, $B_{root}[2] = 1$ and $S_{right(root)}[2] = S_{root}[2] = 4$.

3   Succinct SLP

3.1   Information-Theoretic Lower Bound

In this section, we present a tight lower bound to represent SLPs having a set of production rules P consisting of $n = |\Sigma \cup V|$ symbols. Each production rule $Z \rightarrow XY \in P$ is considered as two directed edges (Z, X) and (Z, Y), the SLP can be seen as a directed acyclic graph (DAG) with a single source and $|\Sigma|$ sinks. Here, we consider (Z, X) as the left edge and (Z, Y) as the right edge. In addition, P can be considered as a DAG with the single source and with a single sink by introducing a super-sink s and drawing directed left and right edges from any sink to s (Figure 2). Let $\mathcal{DAG}(n)$ be the set of all possible Gs

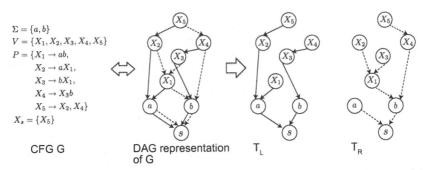

Fig. 2. Example of DAG representation of an SLP and its spanning tree decomposition. An SLP is represented by a DAG G. G is decomposed into the left tree T_L and right tree T_R.

with n nodes and $\mathcal{DAG} = \bigcup_{n\to\infty} \mathcal{DAG}(n)$. Since two SLPs are identical if an SLP can be converted to the other SLP by a permutation $\pi : \Sigma \cup V \to \Sigma \cup V$, the number of different SLPs is $|\mathcal{DAG}(n)|$. Any internal node of $G \in \mathcal{DAG}(n)$ has exactly two (left/right) edges. Thus, the following fact remarked in [22] is true.

Fact 1. *An in-branching spanning tree is an ordered tree such that the out-degree of any node except the root is exactly one. For any in-branching spanning tree of G, the graph consisting of the remaining edges and their adjacent vertices is also an in-branching spanning tree of G.*

The in-branching spanning tree consisting of the left edges (respectively the right edges) and their adjacent vertices is called the *left tree T_L* (respectively *right tree T_R*) of G. Note that the source in G is a leaf of both T_L and T_R, and the super-sink of G is the root of both T_L and T_R. We shall call the operation of decomposing a DAG G into two spanning trees T_L and T_R *spanning tree decomposition*. In Figure 2, the source x_5 in G is a leaf of both T_L and T_R, and the super-sink s in G is the root of both T_L and T_R.

Any ordered tree is an elements in $\mathcal{T} = \bigcup_{n\to\infty} \mathcal{T}_n$ where $\mathcal{T}_n$ is the set of all possible ordered trees with n nodes. As shown in [2,34], there exists an enumeration tree for $\mathcal{T}$ such that any $T \in \mathcal{T}$ appears exactly once. The enumeration tree is defined by the *rightmost expansion*, i.e., in this enumeration tree, a node $T' \in \mathcal{T}_{n+1}$, which is a child of $T \in \mathcal{T}_n$, is obtained by adding a rightmost node to T. In our problem, an ordered tree $T \in \mathcal{T}_{n+1}$ is identical to a left tree T_L with $n+1$ nodes for $n = |\Sigma \cup V|$ symbols.

Let $G \oplus (u,v)$ be the DAG obtained by adding the edge (u,v) to a DAG G. If necessary, we write $G \oplus (u,v)_L$ to indicate that (u,v) is added as a left edge. For a set E of edges, the DAG $G \oplus E$ is defined analogously. The DAG $G \oplus E$ is defined as adding all the edges $(u,v) \in E$ to G. The DAG $G \ominus E$ is also defined as deleting all the edges $(u,v) \in E$ from G.

Theorem 1. *The information-theoretic lower bound on the minimum number of bits needed to represent an SLP with n symbols is $2n + \log n! + o(n)$.*

Proof. Let $\mathcal{S}(n)$ be the set of all possible DAGs with n nodes and a single source/sink such that any internal node has exactly two children. This $\mathcal{S}(n)$ is a super set of $\mathcal{DAG}(n)$ because the in-degree of the sink of any DAG in $\mathcal{DAG}(n)$ must be exactly 2σ, whereas $\mathcal{S}(n)$ does not have such a restriction. By the definition, $|\mathcal{S}(n)|/n^{\sigma} \leq |\mathcal{DAG}(n)| \leq |\mathcal{S}(n)|$ holds.

Let $\mathcal{S}(n,T) = \{G \in \mathcal{S}(n) \mid G = T \oplus T_R,\ T_R \in \mathcal{T}_n\}$. We show $|\mathcal{S}(n,T)| = (n-1)!$ for each $T \in \mathcal{T}_n$ by induction on $n \geq 1$. Since the base case $n = 1$ is clear, we assume that the induction hypothesis is true for some $n \geq 1$.

Let T_L' be the rightmost expansion of T_L such that the rightmost node u is added as the rightmost child of node v in T_L, and let $G' \in \mathcal{S}(n+1, T_L')$ with a left tree T_L'. By the induction hypothesis, the number of $G \in \mathcal{S}(n, T_L)$ is $(n-1)!$ and T_L is embedded into G as the left tree. Then, G' is constructed by adding the left edge (u, v) and a right edge (u, x) for a node x in T_L.

Let s be the source of G. For $v = s$, each $G' = G \oplus (u,v)_L \oplus (u,x)_R \in \mathcal{S}(n+1, T_L')$ is admissible, and the number of them is clearly $n|\mathcal{S}(n, T_L)| = n!$. For $v \neq s$, if $x = s$, $G' = G \oplus (u,v)_L \oplus (u,x)_R \in \mathcal{S}(n+1, T_L')$ is admissible. Otherwise, there exists the lowest common ancestor y of s and x on T_R with $G = T_L \oplus T_R$. Then, $G' = G \oplus (u,v)_L \oplus (u,x)_R \ominus (y',y)_R \oplus (y',u)_R$ is an admissible DAG in $\mathcal{S}(n+1, T_L')$ where y' is the unique child of y in the path from y to s in T_R. In this case, the number of such G's is also $n!$ because no edge is changed in T_L and the pair (T_L', T_R') containing the edge $(u,x)_R$ is unique for any fixed T_L'. Thus, $|\mathcal{S}(n+1, T)| = n!$ is true for each $T \in \mathcal{T}_{n+1}$.

This result derives $|\mathcal{S}(n)| = C_n(n-1)!$ where $C_n = \frac{1}{n+1}\binom{2n}{n} \simeq 2^{2n}n^{-3/2}$ is the number of ordered trees with $n+1$ nodes. Combining this with $|\mathcal{S}(n)|/n^{\sigma} \leq |\mathcal{G}(n)| \leq |\mathcal{S}(n)|$ as well, we get the result that the information-theoretic minimum bits needed to represent $G \in \mathcal{DAG}(n)$ is at least $2n + \log n! + o(n)$. □

3.2   An Optimal SLP Representation

We present an optimal reresentation of an SLP as an improvement of the data structure recently presented in [32]. We apply the spanning tree decomposition to the DAG G of a given SLP, and obtain the DAG $T_L \oplus T_R(= G)$. We rename the variables in T_L by breadth-first order and also rename variables in T_R according to the T_L. Let G' be the resulting DAG from G. Then, for the array representation $D[1, 2n]$ of G', we obtain the condition $D[1] \leq D[3] \leq \ldots \leq D[2n-1]$. Since this monotonic sequence is encoded by $2n + o(n)$ bits, D is represented by $2n + n\log n + o(n)$ bits supporting $access(D, k)$ ($1 \leq k \leq 2n$) in $O(1)$ time. We focus on the remaining sequence of length n, i.e., $D[2], D[4], \ldots, D[2n]$. For simplicity, we write D instead of $[D[2], D[4], \ldots, D[2n]]$.

Let $\mathcal{S} = \{s_1, \ldots, s_\rho\}$ be a disjoint set of subsequences of $[1, n]$ such that any $i \in \{1, 2, ..., n\}$ is contained in some s_k and any s_i, s_j ($i \neq j$) are disjoint. Such an $\mathcal{S}$ is called a decomposition of D. A sequence $D[s_{k_1}], \ldots, D[s_{k_p}]$ is weakly monotonic if it is *increasing*, i.e., $D[s_{k_1}] \leq \ldots \leq D[s_{k_p}]$ or *decreasing*, i.e., $D[s_{k_1}] \geq \ldots \geq D[s_{k_p}]$. In addition, $\mathcal{S}$ is called *monotonic* if the sequence $D[s_{k_1}], \ldots, D[s_{k_p}]$ is weakly monotonic for any $s_k = [s_{k_1}, \ldots, s_{k_p}] \in \mathcal{S}$.

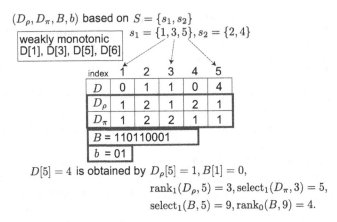

$D[5] = 4$ is obtained by $D_\rho[5] = 1, B[1] = 0$,

$$\text{rank}_1(D_\rho, 5) = 3, \text{select}_1(D_\pi, 3) = 5,$$
$$\text{select}_1(B, 5) = 9, \text{rank}_0(B, 9) = 4.$$

Fig. 3. **Encoded phrase dictionary:** D indicates the remaining sequence $D[2], D[4], \ldots, D[2n]$. D is encoded by $(D_\rho, D_\pi, \mathbf{B}, \mathbf{b})$ based on a monotonic decomposition $\mathcal{S}$ of D, i.e., each $s \in \mathcal{S}$ indicates a weakly monotonic subsequence in D; D_ρ is the sequence of i indicating the membership for some $s_i \in \mathcal{S}$, D_π is a permutation of D_ρ with respect to the corresponding value in D, $\mathbf{B}$ is a binary encoding of the sorted D in increasing order. We show only the case that $D[i]$ is a member of an increasing $s \in \mathcal{S}$, but the other case is similarly computed by $\mathbf{b}$.

Theorem 2. *Any SLP with n symbols can be represented using $2n \log \rho(1+o(1))$ bits for $\rho \leq 2\sqrt{n}$, while supporting $O(\log \log \rho)$ access time.*

Proof. It is sufficient to prove that any D of length n can be represented using $2n \log \rho + o(n)$ bits for some $\rho \leq 2\sqrt{n}$. By the result in [33], we can construct a monotonic decomposition $\mathcal{S}$ of D such that $\rho = |\mathcal{S}| \leq 2\sqrt{n}$.

We represent the sequence D as a four-tuple $(D_\rho, D_\pi, \mathbf{B}, \mathbf{b})$ using $\mathcal{S}$. For each $1 \leq p \leq n$, $D_\rho[p] = k$ iff p is a member of $s_k \in \mathcal{S}$ for some $1 \leq k \leq \rho$. Let $(D[1], D_\rho[1]), \ldots, (D[n], D_\rho[n])$ be the sequence of pairs $(D[p], D_\rho[p])$ $(1 \leq p \leq n)$. We sort these pairs with respect to the keys $D[p]$ $(1 \leq p \leq n)$ and obtain the sorted sequence $(D[\ell_1], D_\rho[\ell_1]), \ldots, (D[\ell_n], D_\rho[\ell_n])$. We define D_π as the permutation $D_\rho[\ell_1] \cdots D_\rho[\ell_n]$.

$\mathbf{B} \in \{0, 1\}^*$ is defined as the bit string

$$\mathbf{B} = 0^{D[\ell_1]} 10^{D[\ell_2] - D[\ell_1]} \ldots 10^{D[\ell_n] - D[\ell_{n-1}]} 1.$$

Finally, $\mathbf{b}[k] = 0$ if $s_k \in \mathcal{S}$ is increasing and $\mathbf{b}[k] = 1$ otherwise for $1 \leq k \leq \rho$. D and D_ρ are represented by WTs, respectively, and $\mathbf{B}$ is a rank/select dictionary.

We recover $D[p]$ using $(D_\rho, D_\pi, \mathbf{B}, \mathbf{b})$. When $D_\rho[p] = k$ and $\mathbf{b}[k] = 0$, i.e., $D[p]$ is included in the k-th monotonic subsequence $s_k \in \mathcal{S}$ that is increasing, we obtain

$$D[p] = \text{rank}_0(\mathbf{B}, \text{select}_1(\mathbf{B}, \ell))$$

by $\ell = \text{select}_k(D_\pi, \text{rank}_k(D_\rho, p))$. When $D_\rho[p] = k$ and $\mathbf{b}[k] = 1$, we can similarly obtain $D[p]$ replacing ℓ by $r = \text{select}_k(D_\pi, (\text{rank}_k(D_\rho, n) + 1 - \text{rank}_k(D_\rho, p)))$.

The total size of the data structure formed by $(D_\rho, D_\pi, \mathbf{B}, \mathbf{b})$ is at most $2n \log \rho(1 + o(1))$ bits. The rank/select/access operations of the WT for a static sequence over $\rho \leq 2\sqrt{n}$ symbols can be improved to achieve $O(\log \log \rho)$ time for each query [3,10]. □

In Figure 3, for the sequence $(0,1), (1,2), (1,1), (0,2), (4,1)$ of pairs $(D[p], D_\rho[p])$ $(1 \leq p \leq 5)$, the sorted sequence is $(0,1), (0,2), (1,2), (1,1), (4,1)$. Thus, D_π is 12211. $\mathbf{B} = 0^0 10^{(0-0)} 10^{(1-0)} 10^{(1-1)} 10^{(4-1)} 1 = 110110001$. $b[1] = 0$ because s_1 is increasing, and $b[2] = 1$ because s_2 is decreasing.

4   Data Structure for Reverse Dictionary

In this section, we present a data structure for simulating the naming function H defined as follows. For a phrase dictionary D with n symbols,

$$H(X_i X_j) = \begin{cases} D^{-1}(X_i X_j), & \text{if } D[k] = X_i X_j \text{ for some } 1 \leq k \leq n, \\ X_{n+1}, & \text{otherwise.} \end{cases}$$

For a sufficiently large V, we set a total order on $(\Sigma \cup V)^2 = \{XY \mid X, Y \in \Sigma \cup V\}$, i.e., the lexicographical order of the n^2 digrams. This order is represented by the range $[1, n^2]$. Then, we recursively define WT T_D for a phrase dictionary D partitioning $[1, n^2]$. On the root node, the initial range $[1, n^2]$ is partitioned into two parts: a left range $L[1, \lfloor (1+n^2) \rfloor /2]$ and a right range $R[\lfloor (1+n^2) \rfloor /2 + 1, n^2]$. The root is the bit string $\mathbf{B}$ such that $\mathbf{B}[i] = 0$ if $D[i] \in L$ and $\mathbf{B}[i] = 1$ if $D[i] \in R$. By this, the sequence of digrams, D, is decomposed into two subsequences D_L and D_R; they are projected on the roots of the left and right subtrees, respectively. Each sub-range is recursively partitioned and the subsequence of D on a node is further decomposed with respect to the partitioning on the node. This process is repeated until the length of any sub-range is one. Let $\mathbf{B}_i$ be the bit string assigned to the i-th node of T_D in the breadth-first traversal. In Figure 4, we show an example of such a data structure for a phrase dictionary D.

Theorem 3. *The naming function for phrase dictionary D over $n = |\Sigma \cup V|$ symbols can be computed by the proposed data structure D_T in $O(\log n)$ time for any digram. Moreover, when a digram does not exist in the current D, D_T can be updated in the same time and the space is at most $2n \log n(1 + o(1))$ bits.*

Proof. D_T is regarded as a WT for a string S of length n such that any symbol is represented in $2 \log n$ bits. Thus, $H(XY)$ is obtained by $\text{select}_{XY}(S, 1)$. The query time is bounded by the number of rank and select operations for bit strings performed until the operation flow returns to the root. Since the total range is $[1, n^2]$, i.e., the height of T_D is at most $2 \log n$, the query time and the size are derived. When XY does not exist in D, let $i_1, i_2, \ldots, i_k$ be the sequence of traversed nodes from the root i_1 to a leaf i_k and let $\mathbf{B}_{i_j}$ be the bit string on i_j. Given an access/rank/select dictionary for $\mathbf{B}_{i_j}$, we can update it for $\mathbf{B}_{i_j} b$ and $b \in \{0, 1\}$ in $O(1)$ time. Therefore, the update time of T_D for any digram is $O(k) = O(\log n)$. □

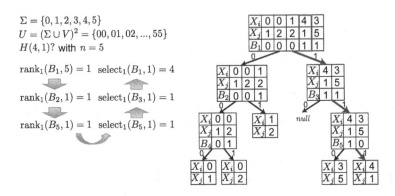

Fig. 4. WT for reverse dictionary: The bit string $\mathbf{B}_i$ is assigned to the i-th node in breadth-first order. For each internal node i, we can move to the left child by rank_0 and to the right child by rank_1 on $\mathbf{B}_i$. The upward traversal is simulated by select_0 and select_1 as shown. The leaf for an existing digram is represented by 1 and *null* is represented by 0, whereas these bits are omitted in this figure.

5   Discussion

We have investigated three problems related to the construction of an SLP: the information-theoretic lower bound for representing the phrase dictionary D, an optimal representation of a directly addressable D, and a dynamic data structure for D^{-1}. Here, we consider the results of this study from the viewpoint of open questions.

For the first problem, we approximately estimated the size of a set of SLPs with n symbols, which is almost equal to the exact set. This problem, however, has several variants, e.g., the set of SLPs with n symbols deriving the same string, which is quite difficult to estimate owing to the NP-hardness of the smallest CFG problem. There is another variant obtained by a restriction: Any two different variables do not derive the same digram, i.e., $Z \to XY$ and $Z' \to XY$ do not exist simultaneously for $Z \neq Z'$. Although such variables are not prohibited in the definition of SLP, they should be removed for space efficiency. On the other hand, even if we assume this restriction, the information-theoretic lower bound is never smaller than $\log n!$ bits because, given a directed chain of length n as T_L, we can easily construct $(n-1)!$ admissible DAGs.

For the second problem, we proposed almost optimal encoding of SLPs. From the standpoint of massive data compression, one drawback of the proposed encoding is that the whole phrase dictionary must be stored in memory beforehand. Since symbols must be sorted, we need a dynamic data structure to allow the insertion of symbols in an array, e.g., [15]. Such data structures, however, require $O(n \log n)$ bits of space.

For the last problem, the query time and update time of proposed data structure are both $O(\log n)$. This cost is considerable and it is difficult to improve it to $O(\log \log n)$ because D is not static. When focusing on the characteristics of SLPs, we can improve the query time probabilistically; since any symbol X appears in D at least once and $|D| = 2n$, the average of frequency of X is at most two. Thus, using an additional array of size $n \log n$ bits, we can check $H(XY)$ in $O(1)$ time with probability at least $1/2$. However, improving this probability is not easy. For this problem, achieving $O(1)$ amortized query time is also an interesting challenge.

References

1. Apostolico, A., Lonardi, S.: Off-line Compression by Greedy Textual Substitution. Proceedings of the IEEE 88, 1733–1744 (2000)
2. Asai, T., Abe, K., Kawasoe, S., Arimura, H., Sakamoto, H., Arikawa, S.: Efficient Substructure Discovery from Large Semi-structured Data. In: SDM, pp. 158–174 (2002)
3. Barbay, J., Gagie, T., Navarro, G., Nekrich, Y.: Alphabet Partitioning for Compressed Rank/Select and Applications. In: Cheong, O., Chwa, K.-Y., Park, K. (eds.) ISAAC 2010, Part II. LNCS, vol. 6507, pp. 315–326. Springer, Heidelberg (2010)
4. Barbay, J., Navarro, G.: Compressed Representations of Permutations, and Applications. In: STACS, pp. 111–122 (2009)
5. Botelho, F.C., Pagh, R., Ziviani, N.: Simple and Space-Efficient Minimal Perfect Hash Functions. In: Dehne, F., Sack, J.-R., Zeh, N. (eds.) WADS 2007. LNCS, vol. 4619, pp. 139–150. Springer, Heidelberg (2007)
6. Charikar, M., Lehman, E., Liu, D., Panigrahy, R., Prabhakaran, M., Sahai, A., Shelat, A.: The smallest grammar problem. IEEE Trans. Inform. Theory 51, 2554–2576 (2005)
7. Claude, F., Navarro, G.: Self-Indexed Grammar-Based Compression. Fundam. Inform. 111, 313–337 (2011)
8. Ferragina, P., Venturini, R.: A simple storage scheme for strings achieving entropy bounds. Theor. Comput. Sci. 372, 115–121 (2007)
9. Fomin, F.V., Kratsch, D., Novelli, J.-C.: Approximating minimum cocolorings. Inf. Process. Lett. 84, 285–290 (2002)
10. Golynski, A., Munro, J.I., Rao, S.S.: Rank/select operations on large alphabets: a tool for text indexing. In: SODA, pp. 368–373 (2006)
11. González, R., Navarro, G.: Statistical Encoding of Succinct Data Structures. In: Lewenstein, M., Valiente, G. (eds.) CPM 2006. LNCS, vol. 4009, pp. 294–305. Springer, Heidelberg (2006)
12. Grossi, R., Gupta, A., Vitter, J.S.: High-order entropy-compressed text indexes. In: SODA, pp. 841–850 (2003)
13. Jacobson, G.: Space-efficient Static Trees and Graphs. In: FOCS, pp. 549–554 (1989)
14. Jansson, J., Sadakane, K., Sung, W.-K.: Ultra-succinct representation of ordered trees with applications. J. Comput. Syst. Sci. 78, 619–631 (2012)
15. Jansson, J., Sadakane, K., Sung, W.-K.: CRAM: Compressed Random Access Memory. In: Czumaj, A., Mehlhorn, K., Pitts, A., Wattenhofer, R. (eds.) ICALP 2012, Part I. LNCS, vol. 7391, pp. 510–521. Springer, Heidelberg (2012)
16. Karp, R.M., Miller, R.E., Rosenberg, A.L.: Rapid Identification of Repeated Patterns in Strings, Trees and Arrays. In: STOC, pp. 125–136 (1972)

17. Karp, R.M., Rabin, M.O.: Efficient Randomized Pattern-Matching Algorithms. IBM Journal of Research and Development 31, 249–260 (1987)
18. Karpinski, M., Rytter, W., Shinohara, A.: An Efficient Pattern-Matching Algorithm for Strings with Short Descriptions. Nordic J. Comp. 4, 172–186 (1997)
19. Larsson, N.J., Moffat, A.: Offline Dictionary-Based Compression. In: DCC, pp. 296–305 (1999)
20. Lehman, E.: Approximation Algorithms for Grammar-Based Compression. PhD thesis, MIT (2002)
21. Lehman, E., Shelat, A.: Approximation algorithms for grammar-based compression. In: SODA, pp. 205–212 (2002)
22. Maruyama, S., Nakahara, M., Kishiue, N., Sakamoto, H.: ESP-Index: A Compressed Index Based on Edit-Sensitive Parsing. In: Grossi, R., Sebastiani, F., Silvestri, F. (eds.) SPIRE 2011. LNCS, vol. 7024, pp. 398–409. Springer, Heidelberg (2011)
23. Maruyama, S., Nakahara, M., Kishiue, N., Sakamoto, H.: ESP-Index: A Compressed Index Based on Edit-Sensitive Parsing. J. Discrete Algorithms 18, 100–112 (2013)
24. Maruyama, S., Sakamoto, H., Takeda, M.: An Online Algorithm for Lightweight Grammar-Based Compression. Algorithms 5, 213–235 (2012)
25. Navarro, G.: Wavelet Trees for All. In: Kärkkäinen, J., Stoye, J. (eds.) CPM 2012. LNCS, vol. 7354, pp. 2–26. Springer, Heidelberg (2012)
26. Navarro, G., Providel, E.: Fast, small, simple rank/Select on bitmaps. In: Klasing, R. (ed.) SEA 2012. LNCS, vol. 7276, pp. 295–306. Springer, Heidelberg (2012)
27. Okanohara, D., Sadakane, K.: Practical Entropy-Compressed Rank/Select Dictionary. In: ALENEX (2007)
28. Raman, R., Raman, V., Rao, S.S.: Succinct indexable dictionaries with applications to encoding k-ary trees and multisets. In: SODA, pp. 233–242 (2002)
29. Rytter, W.: Application of Lempel-Ziv factorization to the approximation of grammar-based compression. Theor. Comput. Sci. 302, 211–222 (2003)
30. Sadakane, K., Grossi, R.: Squeezing succinct data structures into entropy bounds. In: SODA, pp. 1230–1239 (2006)
31. Sakamoto, H.: A fully linear-time approximation algorithm for grammar-based compression. J. Discrete Algorithms 3, 416–430 (2005)
32. Takabatake, Y., Tabei, Y., Sakamoto, H.: Variable-Length Codes for Space-Efficient Grammar-Based Compression. In: Calderón-Benavides, L., González-Caro, C., Chávez, E., Ziviani, N. (eds.) SPIRE 2012. LNCS, vol. 7608, pp. 398–410. Springer, Heidelberg (2012)
33. Yehuda, R.B., Fogel, S.: Partitioning a Sequence into Few Monotone Subsequences. Acta. Inf. 35, 421–440 (1998)
34. Zaki, M.J.: Efficiently mining frequent trees in a forest. In: KDD, pp. 71–80 (2002)
35. Ziv, J., Lempel, A.: A Universal Algorithm for Sequential Data Compression. IEEE Trans. Inform. Theory 23, 337–343 (1977)
36. Ziv, J., Lempel, A.: Compression of individual sequences via variable-rate coding. IEEE Trans. Inform. Theory 24, 530–536 (1978)

Data Structure Lower Bounds on Random Access to Grammar-Compressed Strings*

Elad Verbin and Wei Yu

Aarhus University
34 Åbogade, 8200 Aarhus N, Denmark
{eladv, yuwei}@cs.au.dk

Abstract. In this paper we investigate the problem of building a static data structure that represents a string s using space close to its compressed size, and allows fast access to individual characters of s. This type of data structures was investigated by the recent paper of Bille et al. [3]. Let n be the size of a context-free grammar that derives a unique string s of length L. (Note that L might be exponential in n.) Bille et al. showed a data structure that uses space $O(n)$ and allows to query for the i-th character of s using running time $O(\log L)$. Their data structure works on a word RAM with a word size of $\log L$ bits.

Here we prove that for such data structures, if the space is $\mathrm{poly}(n)$, the query time must be at least $(\log L)^{1-\varepsilon}/\log \mathcal{S}$ where $\mathcal{S}$ is the space used, for any constant $\varepsilon > 0$. As a function of n, our lower bound is $\Omega(n^{1/2-\varepsilon})$. Our proof holds in the cell-probe model with a word size of $\log L$ bits, so in particular it holds in the word RAM model. We show that no lower bound significantly better than $n^{1/2-\varepsilon}$ can be achieved in the cell-probe model, since there is a data structure in the cell-probe model that uses $O(n)$ space and achieves $O(\sqrt{n \log n})$ query time. The "bad" setting of parameters occurs roughly when $L = 2^{\sqrt{n}}$. We also prove a lower bound for the case of not-as-compressible strings, where, say, $L = n^{1+\epsilon}$. For this case, we prove that if the space is $O(n \cdot \mathrm{polylog}(n))$, the query time must be at least $\Omega(\log n/\log \log n)$.

The proof works by reduction from communication complexity, namely to the LSD (Lopsided Set Disjointness) problem, recently employed by Pătraşcu and others. We prove lower bounds also for the case of LZ-compression. All of our lower bounds hold even when the strings are over an alphabet of size 2 and hold even for randomized data structures with 2-sided error.

---

* The authors acknowledge support from the Danish National Research Foundation and The National Science Foundation of China (under the grant 61061130540) for the Sino-Danish Center for the Theory of Interactive Computation, within which part of this work was performed. Part of the work was done while the authors were working in IIIS, Tsinghua University in China.

J. Fischer and P. Sanders (Eds.): CPM 2013, LNCS 7922, pp. 247–258, 2013.
© Springer-Verlag Berlin Heidelberg 2013

1   Introduction

In many modern databases, strings are stored in compressed form. Many compression schemes are grammar-based, in particular Lempel-Ziv [6,11,12] and its variants, as well as Run-Length Encoding.

A natural desire is to store a text using space close to its compressed size, but to still allow fast access to individual characters: can we do something faster than simply extracting the whole text each time we need to access a character? This question was recently answered in the affirmative by Bille et al. [3] and by Claude and Navarro [5]. These two works investigate the problem of storing a string that can be represented by a small CFG (context-free grammar) of size n, while allowing some basic stringology operations, in particular random access to a character in the text. The data structure of Bille et al. [3, Theorem 1] stores the text in space linear in n, while allowing access to an individual character in time $O(\log L)$, where L is the text's *uncompressed* size.[1] But is that the best upper bound possible?

In this paper we show a $(\log L)^{1-\varepsilon}$ lower bound on the query time when the space used by the data structure is $\text{poly}(n)$, showing that the result of Bille et al. is close to optimal. Our lower bounds are proved in the cell-probe model of Yao [10], with word size $\log L$, therefore they in particular hold for the model studied by Bille et al. [3], since the cell-probe model is strictly stronger than the RAM model. Our lower bound is proved by a reduction from Lopsided Set Disjointness (LSD), a problem for which Pătraşcu has recently proved an essentially-tight randomized lower bound [8]. The idea is to prove that grammars are rich enough to effectively "simulate" a disjointness query: our class of grammars, presented in Section 3.1, might be of independent interest as a class of "hard" grammars for other purposes as well.

In terms of n, our lower bound is $n^{1/2-\varepsilon}$. The results of Bille et al. imply an upper bound of $O(n)$ on the query time, since $\log L \leq n$, therefore in terms of n there is a curious quadratic gap between our lower bound and Bille et al.'s upper bound. We show that this gap can be closed by giving a better data structure: we show a data structure which takes space $O(n)$ and has query time $O(\sqrt{n \log n})$, showing that no significantly better lower bound is possible. This data structure, however, comes with a big caveat – it runs in the highly-unrealistic cell-probe model, thus serving more as an impossibility proof for lower bounds than as a reasonable upper bound. The question remains open of whether such a data structure exists in the more realistic word RAM model.

Our lower bound holds for a particular, "worst-case", dependence of L on n. Namely, L is roughly $2^{\sqrt{n}}$. It might also be interesting to explicitly limit the range of allowed parameters to other regimes, for example to non-highly-compressible text; in such a regime it might be that $L = n^{1+\epsilon}$. The above result does not imply any lower bound for this case. Furthermore, we show in another result that for any data structure in that regime, if the space is $n \cdot \text{polylog}\, n$,

---

[1] The result of Bille et. al. also allows other query operations such as pattern matching; we do not discuss those in this paper.

the query time must be $\Omega(\log n/\log\log n)$. This lower bound holds, again, in the cell probe model with words of size $\log n$ bits, and is proved by a reduction from two-dimensional range counting (which, once again, was lower bounded by a reduction from LSD [8]).

2   Preliminaries

In this paper we denote $[N] = \{1, \ldots, N\}$. All logarithms are in base 2 unless explicitly stated otherwise.

Our lower bounds are proved in Yao's cell-probe model [10]. In the cell-probe model, the memory is an array of cells, where each cell consists of w bits each. The query time is measured as the number of cells read, while all computations are free. This model is strictly stronger than the word RAM, since in the word RAM the operations allowed on words are restricted, while in the cell-probe model we only measure the number of cells accessed. The cell-probe model is widely used in proving data structure lower bounds, especially by reduction from communication complexity problems [7]. In this paper we prove our result by a reduction from the BLOCKED-LSD problem introduced by Pătrașcu [8].

An SLP (straight line program) is a collection of n derivation rules, defining the symbols $g_1, \ldots, g_n$. Each rule is either of the form $g_i \to$ 'σ', i.e. g_i is a terminal, which takes the value of a character σ from the underlying alphabet, or of the form $g_i \to g_j g_k$, where $j < i$ and $k < i$, i.e. g_j and g_k were already defined, and we define the nonterminal symbol g_i to be their concatenation. The symbol g_n is the *start symbol*. To derive the string we start from g_n and follow the derivation rules until we get a sequence of characters from the alphabet. The length of the derived string is at most 2^n. W.l.o.g. we assume it is at least n. As the same in Bille et al. [3], we also assume w.l.o.g. that the grammars are in fact SLPs and so on the righthand side of each grammar rule there are either exactly two variables or one terminal symbol. In this paper SLP, CFG and grammar all mean the same thing.

The grammar random access problem is the following problem.

Definition 1 (Grammar Random Access Problem). *For a CFG G of size n representing a binary string of length L, the problem is to build a data structure to support the following query: given $1 \leq i \leq L$, return the i-th character (bit) in the string.*

We study two other data structure problems, which are closely related to their communication complexity counterparts.

Definition 2 (Set Disjointness, SD_N). *For a set $Y \subseteq [N]$, the problem is to build a data structure to support the following query: given a set $X \subseteq [N]$, answer whether $X \cap Y = \emptyset$.*

Given a universe $[BN] = \{1, \ldots, BN\}$, a set X is called *blocked with cardinality N* if when we divide the universe $[BN]$ into N equal-sized consecutive blocks, X contains exactly one element from each of the blocks while Y could be arbitrary.

Definition 3 (Blocked Lopsided Set Disjointness, $BLSD_{B,N}$). *For a set $Y \subseteq [BN]$, the problem is to build a data structure to support the following query: given a blocked set $X \subseteq [BN]$ where $|X| = N$ containing 1 element from each size B block, answer whether $X \cap Y = \emptyset$.*

For proving lower bound for near-linear space data structures, we also need reductions from a variant of the range counting problem.

Definition 4 (Range Counting). *The range counting problem is a static data structure problem. We need to preprocess a set of n points on a $[n] \times [n^\epsilon]$ grid. A query (x, y) asks to count the number of points in a dominance rectangle $[1, x] \times [1, y]$ (a rectangle contains the lower left corner $(1, 1)$). Return the answer modulo 2.*

Note that the above problem is "easier" than the classical 2D range-counting problem, since it is a dominant query problem, it is a grid $n \times n^\epsilon$, and it is modulo 2. However, the (tight) lower bound that is known for the general problem, given by Pătraşcu [8], could be generalized for the problem we define.

3   Lower Bound for Grammar Random Access

In this section we prove the main lower bound for grammar random access. In Section 3.1 we show the main reduction from SD and BLSD. In Section 3.2 we prove lower bounds for SD and BLSD, based on reductions to communication complexity (these are implicit in the work of Pătraşcu [8]). Finally, in Section 3.3 we tie these together to get our lower bounds.

3.1   Reduction from SD and LSD

In this section we show how to reduce the grammar access problem from SD or BLSD, by considering a particular type of grammar. The reductions tie the parameters n and L to the parameters B and N of BLSD (or just to the parameter N of SD). In Section 3.3 we show how to choose the relation between the various parameters in order to get our lower bounds. We remark that the particular multiplicative constants in the lemmas below will not matter, but we give them nonetheless, for concreteness.

These reductions might be confusing for the reader, but they are in fact almost entirely tautological. They just follow from the fact that the communication matrix of SD is a tensor product of the 2 by 2 communication matrices for the coordinates, i.e., it is just a N-fold tensor product of the matrix $\begin{pmatrix} 1 & 1 \\ 1 & 0 \end{pmatrix}$. For BLSD, the communication matrix is the N-fold tensor product of the $(2^B) \times B$ communication matrix for each block (for example, for $B = 3$ this matrix is $\begin{pmatrix} 1 & 0 & 1 & 0 & 1 & 0 & 1 & 0 \\ 1 & 1 & 0 & 0 & 1 & 1 & 0 & 0 \\ 1 & 1 & 1 & 1 & 0 & 0 & 0 & 0 \end{pmatrix}$). We do not formulate our arguments in the language of communication matrices and tensor products, since this would hide what is really going on. To aid the reader, we give an example after each of the two constructions.

Lemma 1 (Reduction from SD_N). *For any set $Y \subseteq [N]$, there is a grammar G_Y of size $n = 2N + 1$ deriving a binary string s_Y of length $L = 2^N$ such that for any set $X \subseteq [N]$, it holds that $s_Y[X] = 1$ iff $X \cap Y = \emptyset$.*

Note that in this lemma we have indexed the string s by *sets*: there are 2^N possible sets X, and the length of the string s_Y is also 2^N – each set X serves as an index of a unique character. The indexing is done in lexicographic order: the set X is identified with its *characteristic vector*, i.e., the vector in $\{0, 1\}^N$ whose i-th coordinate is '1' if $i \in X$, and '0' otherwise, and the sets are ordered according to lexicographic order of their characteristic vectors. For example, here is the ordering for the case $N = 3$: $\emptyset, \{1\}, \{2\}, \{1, 2\}, \{3\}, \{1, 3\}, \{2, 3\}, \{1, 2, 3\}$.

Proof. We now show how to build the grammar G_Y. The grammar has N symbols for the strings $0, 0^2, 0^4, \ldots, 0^{2^{N-1}}$, i.e., all strings consisting solely of the character '0', of lengths which are all powers of 2 up to 2^{N-1}. Then, the grammar has $N + 1$ additional symbols $g_0, g_1, \ldots, g_N$. The terminal g_0 is equal to the character 1. For any $1 \le i \le N$, we set g_i to be equal to $g_{i-1}g_{i-1}$ if $i \notin Y$, and to be equal to $g_{i-1}0^{2^{i-1}}$ if $i \in Y$. The start symbol of the grammar is g_N.

We claim that the string derived by this grammar has the property that $s_Y[X] = 1$ iff $X \cap Y = \emptyset$. This is easy to prove by induction on i, where the induction claim is that for any i, g_i is the string that corresponds to the set $Y \cap \{1, \ldots, i\}$ over the universe $\{1, \ldots, i\}$. □

Example 1. Consider the universe $N = 4$. Let $Y = \{1, 3\}$. The string s_Y is 1010000010100000. The locations of the 1's correspond exactly to the sets that don't intersect Y, namely to the sets $\emptyset$, $\{2\}$, $\{4\}$ and $\{2, 4\}$, respectively.

We now show the reduction from blocked LSD. It follows along the same general idea, but the grammar is slightly more complicated.

Lemma 2 (Reduction from $BLSD_{B,N}$). *For any set $Y \subseteq [BN]$, there is a grammar G_Y of size $n = 2BN + 1$ deriving a binary string s_Y of length $L = B^N$ such that for any blocked set $X \subseteq [BN]$ of cardinality N, it holds that $s_Y[X] = 1$ iff $X \cap Y = \emptyset$.*

Recall that by "a blocked set $X \subseteq [BN]$ of cardinality N" we mean a set such that the universe $[BN]$ is divided into N equal-sized blocks, and X contains exactly one element from each of these blocks.

Note that in this lemma we have again indexed the string s by *sets*: there are B^N possible sets X and the length of the string is B^N. The indexing is done in *lexicographic order*, this time identifying a set X with a length-N vector whose i-th coordinate is chosen according to which element it contains in block i, and the sets are ordered according to lexicographic order of their characteristic vectors. For example, here is the ordering for the case $N = 2, B = 3$: $\{1, 4\}, \{2, 4\}, \{3, 4\}, \{1, 5\}, \{2, 5\}, \{3, 5\}, \{1, 6\}, \{2, 6\}, \{3, 6\}$.

The construction in this reduction is similar to that in the case of SD, but instead of working element by element, we work block by block.

Proof. We now show how to build the grammar G_Y. The grammar has N symbols for the strings $0, 0^B, 0^{B^2}, 0^{B^3}, \ldots, 0^{B^{N-1}}$, i.e., all strings consisting solely of the character '0', of lengths which are all powers of B up to B^{N-1}. We cannot simply obtain the symbols directly from each other: e.g., to obtain 0^{B^2} from 0^B, we need to concatenate 0^B with itself B times. Thus we use BN rules to derive all of these symbols. (In fact, $O(N \log B)$ rules can suffice but this does not matter).

Then, beyond these, the grammar has $N+1$ additional symbols $g_0, g_1, \ldots, g_N$, one for each block. The terminal g_0 is equal to the character 1. For any $1 \leq 1 \leq N$, g_i is constructed from g_{i-1} according to which elements of the i-th block are in Y: we set g_i to be a concatenation of B symbols, each of which is either g_{i-1} or $0^{B^{i-1}}$. In particular, g_i is the concatenation of $g_i^{(1)}, \ldots, g_i^{(B)}$, where g_i^j is equal to g_{i-1} if the j-th element of the i-th block is not in Y, and it is equal to $0^{B^{i-1}}$ if the j-th element of the i-th block is in Y. To construct these symbols we need at most BN rules, because we need $B-1$ concatenation operations to derive g_i from g_{i-1}. (Note that here we cannot get down to $O(N \log B)$ rules – $\Theta(BN)$ seem to be necessary.) The start symbol of the grammar is g_N.

We claim that the string produced by this grammar has the property that $s_Y[X] = 1$ iff $X \cap Y = \emptyset$. This is easy to prove by induction on i, where the induction claim is that for any i, g_i is the string that corresponds to the set $X \cap \{1 \ldots, iB\}$ over the universe $\{1, \ldots, iB\}$. □

Example 2. Consider the values $B = 3$ and $N = 3$. Let $Y = \{1, 3, 5, 9\}$. The string s_Y is "010000010 010000010 000000000"[2]. The locations of the 1's correspond exactly to the blocked sets that don't intersect Y, namely to the sets $\{2, 4, 7\}$, $\{2, 6, 7\}$, $\{2, 4, 8\}$ and $\{2, 6, 8\}$, respectively. A brief illustration for this example is in Figure 1.

$$g_2 = 010000010$$

$$g_3 = 010000010 \quad 010000010 \quad 000000000$$

$$\{7\} \cap Y = \emptyset \quad \{8\} \cap Y = \emptyset \quad \{9\} \cap Y = \{9\}$$

Fig. 1. An illustration of Example 2

3.2   Lower Bounds for SD and BLSD

In this subsection we show lower bounds for SD and BLSD that are implicit in the work of Pătrașcu [8]. Recall the notations from Section 2: in particular, in all of the bounds, w, S, and t denote the word size (measured in bits), the

---

[2] The spaces are just for easier presentation.

size of the data structure (measured in words) and the query time (measured in number of accesses to words), respectively.

Theorem 1. *For any 2-sided-error data structure for SD_N, $t \geq \Omega(N/(w + \log S))$.*

Note that this theorem does not give strong bounds when $w = O(\log L)$, but it is meaningful for bit-probe ($w = 1$) bound and a warm-up for the reader.

Theorem 2. *Let $\varepsilon > 0$ be any small constant. For any 2-sided-error data structure for $BLSD_{B,N}$,*

$$t \geq \Omega \left(\min \left(\frac{N \log B}{\log S}, \frac{B^{1-\varepsilon} N}{w} \right) \right) . \tag{1}$$

The proofs follow by standard reductions from data structure to communication complexity, using known lower bounds for SD and BLSD (the latter is one of the main results in [8]).

We now cite the corresponding communication complexity lower bounds:

Lemma 3 (See [2,9,1]). *Consider the communication problem where Alice and Bob each receive a subset of $[N]$, and they want to decide whether the sets are disjoint. Any randomized 2-sided-error protocol for this problem uses communication $\Omega(N)$.*

Lemma 4 (See [8], Lemma 3.1). *Let $\varepsilon > 0$ be any small constant. Consider the communication problem where Bob gets a subset of $[BN]$ and Alice gets a blocked subset of $[BN]$ of cardinality N, and they want to decide whether the sets are disjoint. In any randomized 2-sided-error protocol for this problem, either Alice sends $\Omega(N \log B)$ bits or Bob sends $B^{1-\varepsilon} N$ bits. (The Ω-notation hides a multiplicative constant that depends on ε.)*

The way to prove the data structure lower bounds from the communication lower bounds is by reductions to communication complexity: Alice and Bob execute a data structure query; Alice simulates the querier, and Bob simulates the data structure. Alice notifies Bob which cell she would like to access; Bob returns that cell, and they continue for t rounds, which correspond to the t probes. At the end of this process, Alice knows the answer to the query. Overall, Alice sends $t \log S$ bits and Bob sends tw bits. The rest is calculations, which we include here for completeness:

Proof (Lemma 3 $\Rightarrow$ Theorem 1). We know that the players must send a total of $\Omega(N)$ bits, but the data structure implies a protocol where $t \log S + tw$ bits are communicated. Therefore $t \log S + tw \geq \Omega(N)$ so $t \geq \Omega(N/(\log S + w))$.   □

Proof (Lemma 4 $\Rightarrow$ Theorem 2). We know that either Alice sends $\Omega(N \log B)$ bits or Bob sends $B^{1-\varepsilon} N$ bits. Therefore, either $t \log S \geq \Omega(N \log B)$ or $tw \geq B^{1-\varepsilon} N$. The conclusion follows easily.   □

3.3   Putting It Together

We now put the results of Section 3.1 and 3.2 together to get our lower bounds. Note that in all lower bounds below we freely set the relation of n and L in any way that gives the best lower bounds. Therefore, if one is interested in only a specific relation of n and L (say $L = n^{10}$) the lower bounds below are not guaranteed to hold. The typical "worst" dependence in our lower bounds (at least for the case where $w = \log L$ and $\mathcal{S} = \text{poly}(n)$) is roughly $L = 2^{\sqrt{n}}$.

   Theorem 1 together with Lemma 1 immediately give:

Theorem 3. *For any 2-sided-error data structure for the grammar random access problem, $t \geq \Omega(n/(w+\log \mathcal{S}))$. And in terms of L, $t \geq \Omega(\log L/(w+\log \mathcal{S}))$.*

   When setting $w = 1$ and $\mathcal{S} = \text{poly}(n)$ (polynomial space in the bit-probe model), we get that $t \geq \Omega(n/\log n)$. And in terms of L, $t \geq \Omega(\log L/\log\log L)$.

Proof. Trivial, since $n = \Theta(N)$ and $L = 2^{\Theta(N)}$. □

Theorem 2 together with Lemma 2 give:

Theorem 4. *Assume $w = \omega(\log \mathcal{S})$. Let $\varepsilon > 0$ be any arbitrarily small constant. For any 2-sided-error data structure for the grammar random access problem, $t \geq n/w^{\frac{1+\varepsilon}{1-\varepsilon}}$. And in terms of L, $t \geq \frac{\log L}{\log \mathcal{S} \cdot w^{\frac{\varepsilon}{1-\varepsilon}}}$.*

   When setting $w = \log L$ and $\mathcal{S} = \text{poly}(n)$ (polynomial space in the cell-probe model with cells of size $\log L$), there is another constant δ such that $t \geq n^{1/2-\delta}$. And in terms of L, $t \geq (\log L)^{1-\delta}$.

The condition $w = \omega(\log \mathcal{S})$ is a technical condition, which ensures that the value of B we choose in the proof is at least $\omega(1)$. For $w \leq \log \mathcal{S}$ one gets the best results just by reducing from SD, as in Theorem 3.

Proof. For the first part of the theorem, substitute $B = (w/\log \mathcal{S})^{1/(1-\varepsilon)} \log(w/\mathcal{S})$, $N = n/B$, $L = B^N$ into (1). For the second part of the theorem, substitute $N = \frac{B^{1-\varepsilon}\log n}{\log^2 B}$, $n = BN$ and $L = B^N$. And for the result, set $\delta = \frac{2\varepsilon}{1-\varepsilon}$. □

4   Lower Bound for Less-Compressible Strings

In the above reduction, the worst case came from strings that can be compressed superpolynomially. However, for many strings we expect to encounter in practice, superpolynomial compression is unrealistic. A more realistic range is polynomial compression or less. In this section we discuss the special case of strings of length $O(n^{1+\epsilon})$. We show that for this class of strings, the Bille et al. [3] result is also (almost) tight by proving an $\Omega(\log n/\log\log n)$ lower bound on the query time, when the space used is $O(n \cdot \text{polylog}\, n)$. This is done by reduction from the range counting problem on a 2D grid. We have the following lower bound for the range counting problem (see Definition 4 for details). Due to lack of space, we omit the proof. A similar proof for a problem with slightly different parameters could be found in [8, Section 2.1+Appendix A].

Lemma 5. *Any data structure for the 2D range counting problem for n points on a grid of size* $[n] \times [n^\epsilon]$ *using* $O(n \operatorname{polylog} n)$ *space requires* $\Omega(\log n / \log \log n)$ *query time in the cell probe model with cell size* $\log n$.

Recall that the version of range counting we consider is actually dominance counting modulo 2 on the $n \times n^\epsilon$ grid. The main idea behind our reduction is to consider the length-$n^{1+\epsilon}$ binary string consisting of the answers to all $n^{1+\epsilon}$ possible dominance range queries (in the natural order, i.e. row-by-row, and in each row from left to right); call this the *answer string* of the corresponding range counting instance. We prove that the answer string can be derived using a grammar of size $O(n \log n)$. The reduction follows obviously, since a dominance prange query can be answered by querying one bit of the answer string.

Lemma 6. *For any range counting problem in 2D, the answer string can be derived by a grammar of size* $O(n \log n)$.

The idea behind the proof of is to simulate a sweep of the point set from top to bottom by a dynamic one-dimensional range tree. The symbols of the grammar will correspond to the nodes of the tree. With each new point encountered, only $2 \log n$ new symbols will be introduced. Since there are n points, the grammar is of size $O(n \log n)$.

Proof. Assume w.l.o.g.p that n is a power of 2. It is easy to see that the answer string could be built by concatenating the answers in a row-wise order, just as illustrated in Figure 2.

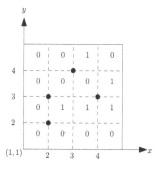

Fig. 2. The answer string for this instance is 0000 0111 0001 0010. The value in the grids are the query results for queries falling in the corresponding cell, including the bottom and left boundaries, excluding the right and top boundaries.

We are going to build the string row by row. Think of a binary tree representing the grammar built for the first row of the input. The root of the tree derives the first row of the answer string, whose two children respectively represent the answer string for the left and the right half of the row. In this way the tree is built recursively. The leaves of the tree are terminal symbols from $\{0, 1\}$. Thus there are $2n - 1$ symbols in total for the whole tree. At the same time we also maintain the *negations* of the symbols in the tree, i.e., making a new symbol g_i' for each symbol g_i in the tree, where $g_i' = 1 - g_i$ if g_i is a terminal symbol; or $g_i' = g_j' g_k'$ if $g_i = g_j g_k$.

The next row in the answer string will be derived by changing at most $2p \log n$ symbols in the grammar of the previous row, where p is the number of new points in the row. We process the new points one by one. For each point, the new symbols needed all lie in a path from a leaf to the root of the tree. Assuming the update introduced by the point is the path $h_1, h_2, \ldots, h_{\log n}$, the new tree will contain an update of $h_1, h_2, \ldots, h_{\log n}$. Also, all the right children of these nodes will be switched to their negations (this switching step does not actually require introducing any new symbols). An intuitive picture of the process is given in Figure 3. The first row has a grammar $g_7 = g_5 g_6$, $g_5 = g_1 g_2$, $g_6 = g_3 g_4$ and $g_1 = g_2 = g_3 = g_4 = 0$, as well as rules for g_i' when $1 \le i \le 7$. The second row has a grammar with new rules $h_3 = h_2 g_6'$, $h_2 = g_1 h_1$, $g_3' = g_4' = h_1 = 1$.

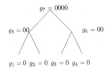
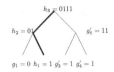

(a) The trees built for the first and second rows for the example in Figure 2.

(b) The general process illustrated by picture. The black parts stands for the negations of corresponding symbols.

Fig. 3. Examples for building answer strings

It is easy to see for each new point, $2 \log n$ additional rules are created. $\log n$ of them are the new symbols $(h_1, \ldots, h_{\log n})$, and another $\log n$ of them are their negations $(h_1', \ldots, h_{\log n}')$. After all, we use $2(2n - 1) + n \cdot 2 \log n = O(n \log n)$ symbols to derive the whole answer string. $\qquad \square$

By using the above lemma, we have the lower bound of the grammar random access problem when $L = n^{1+\epsilon}$.

Theorem 5. *Fix $\epsilon > 0$, any data structure using space $O(n \operatorname{polylog} n)$ for the grammar random access problem with n rules on strings of length $\Omega(n^{1+\epsilon})$ requires $\Omega(\log n / \log \log n)$ query time.*

Proof. For inputs of the range counting problem, we compress the answer string to a grammar of size $O(n \log n)$ according to Lemma 6. After that we build a data structure for the random access problem on this grammar using Lemma 8. For any query (x, y) of the range counting problem, we simply pass the query result on the index $(y - 1)n + x - 1$ on the answer string as an answer. Assuming there is a data structure using $O(n \log n)$ space and query time t, then it will also solve the range counting problem. According to Lemma 5 the lower bound for range counting is $\Omega(\log n / \log \log n)$ for $O(n \operatorname{polylog} n)$ space, thus $t = \Omega(\log n / \log \log n)$. $\qquad \square$

Note that natural attempt is to replace the 1D range tree that we used above by a 2D range tree and perform a similar sweep procedure, but this does not work for building higher dimensional answer strings.

5   LZ-Based Compression

In this section we discuss about what the lower bound means for LZ-based compression, which is a typical case for grammar-based compression by Lempel-Ziv [11,6]. First we look at LZ77. For LZ77 we have the following lemma.

Lemma 7 (Lemma 9 of [4]). *The length of the LZ77 sequence for a string is a lower bound on the size of the smallest grammar for that string.*

The basic idea of this lemma is to show that each rule in the grammar only contribute one entry for LZ77. Since LZ77 could compress any string with small grammar size into a smaller size, it can also compress the string s_Y in Lemma 1 and the answer string in Theorem 5 into a smaller size. Thus the both lower bounds for grammar random access problem also holds for LZ77.

The reader might also be curious about what will happen for the LZ78 [12] case. Unfortunately the lower bound does not hold for LZ78. This is because LZ78 is a "bad" compression scheme that even the input is 0^n of all 0's, LZ78 can only compress the string to length of $\sqrt{n}$. But a random access on an all 0 string is trivially constant with constant space. So we are not able to have any lower bounds for this case.

6   Optimality

In this section, we show that the upper bound in Bille et al. [3] is nearly optimal, for two reasons. First, it is clear that by Theorem 5, the upper bound in Lemma 8 is (almost) optimal, when the space used is $O(n \operatorname{polylog} n)$.

Lemma 8. *There is a data structure for the grammar random access problem with $O(n)$ space and $O(\log L)$ time. This data structure works in the word RAM with words of size $\log L$.*

Second, in the cell-probe model with words of size $\log L$ we also have the following lemma by Bille et al. [3].

Lemma 9. *There is a data structure for the grammar random access problem with $O(n)$ space and $O(n \log n / \log L)$ time.*

Proof. This is a trivial bound. The number of bits to encode the grammar is $O(n \log n)$ since each rule needs $O(\log n)$ bits. The cell size is $O(\log L)$, so in $O(n \log n / \log L)$ time the querier can just read all of the grammar. Since computation is free in the cell-probe model, the querier can get the answer immediately.   □

Thus, by using Lemma 8 when $n = \Omega(\log^2 L / \log \log L)$ and Lemma 9 in the case $n = O(\log^2 L / \log \log L)$, we have the following corollary. This corollary implies that our lower bound of $\Omega(n^{1/2-\varepsilon})$ is nearly the best one can hope for in the cell-probe model.

Corollary 1. *Assuming $w = \log L$, there is a data structure in the cell-probe model with space $O(n)$ and time $O(\sqrt{n \log n})$.*

Acknowledgement. We thank Travis Gagie and Pawel Gawrychowski for helpful discussions.

References

1. Babai, L., Frankl, P., Simon, J.: Complexity classes in communication complexity theory. In: FOCS, pp. 337–347 (1985)
2. Bar-Yossef, Z., Jayram, T.S., Kumar, R., Sivakumar, D.: An information statistics approach to data stream and communication complexity. Journal of Computer and System Sciences 68(4), 702–732 (2004)
3. Bille, P., Landau, G.M., Raman, R., Rao, S., Sadakane, K., Weimann, O.: Random access to grammar compressed strings. In: SODA (2011)
4. Charikar, M., Lehman, E., Liu, D., Panigrahy, R., Prabhakaran, M., Sahai, A., Shelat, A.: The smallest grammar problem. IEEE Transactions on Information Theory 51(7), 2554–2576 (2005)
5. Claude, F., Navarro, G.: Self-indexed text compression using straight-line programs. In: Královič, R., Niwiński, D. (eds.) MFCS 2009. LNCS, vol. 5734, pp. 235–246. Springer, Heidelberg (2009)
6. Lempel, A., Ziv, J.: On the complexity of finite sequences. IEEE Transactions on Information Theory 22(1), 75–81 (1976)
7. Miltersen, P.B., Nisan, N., Safra, S., Wigderson, A.: On data structures and asymmetric communication complexity. In: STOC, p. 111. ACM (1995)
8. Patrascu, M.: Unifying the Landscape of Cell-Probe Lower Bounds. SIAM Journal on Computing 40(3) (2011)
9. Razborov, A.A.: On the distributional complexity of disjointness. Theoretical Computer Science 106(2), 385–390 (1992)
10. Yao, A.C.C.: Should tables be sorted? Journal of the ACM 28(3), 615–628 (1981)
11. Ziv, J., Lempel, A.: A universal algorithm for sequential data compression. IEEE Transactions on information theory 23(3), 337–343 (1977)
12. Ziv, J., Lempel, A.: Compression of individual sequences via variable-rate coding. IEEE Transactions on Information Theory 24(5), 530–536 (1978)

Author Index